Understanding Environment

Understanding Environment

Arvind Kumar

Oxford Book Company
Jaipur ▪ New Delhi

Published By

Oxford Book Company

Regd. Off. "Bony Residency", Gate No. 2
Opp. Tilak Public School, Vishweswariya Nagar,
Gopalpura Road, Jaipur - 302018, India
Ph: 0141 - 2761280, Telefax: 2761381
email - oxfordbookjpr@gmail.com
website - oxfordbookcompany.com

Delhi Office: 102, 1st Floor, Satyam House,
4327/3, Ansari Road, Darya Ganj,
New Delhi - 110002, India
Ph: 011-45652440

Distributed by
KEYA INTERNATIONAL PRESS LLP
E-mail: keyainternationalpressllp@gmail.com

Published in India, 2025

ISBN : 9788189473853

Typeset by Shivangi Computers, Jaipur

Printed and bound in India by Replika Press Pvt. Ltd.

Preface

Environment is characterized by extreme spatial heterogeneity of conditions and organisms. Quite apart from the patterns caused by gradients, topography and ground cover interact over a wide range of spatial scales to create complex, nested mosaics of habitat conditions. These mosaics, in turn, often have considerable effects upon the spatial distributions of organisms as well as on their morphologies, growth, physiology, life cycles, behaviour patterns and the types and intensities of interactions between individuals. Such a high degree of specific heterogeneity and its dominating influences are major ways in which environment is to be considered unique.

Interactions between living beings in natural settings as well as brought about through induced changes as also through changes in physico-chemical environment have tremendous impact. The papers in the present book demonstrate clearly the role and impact of specific heterogeneity in creating the uniqueness of environment and the communities of living beings that inhabit them, whether they be plants, insects, microorganisms, small mammals or humans. The articles in the book also attempt at determining how organisms, from individuals to populations to communities are likely to respond to changes in natural well as physico-chemical environments.

Arvind Kumar

Contents

1

Effects of Ammonium Molybdate on Sodium Chloride Stress in *Vigna mungo* (L.) Hepper

A. Pragasam, M.I. Mohamada Begam and K. Nirmal Kumar

Department of Plant Science, K. M Centre for Post Granduate Stuides, Pondicherry-605 008, India.

Introduction

Salt affected soils were reported to occur extensively in the arid and semi arid regions in different agroclimatic zones of India. They occupy 9.38 million hectare and contributing to 2.9 per cent of the total geographical area of the country of which 5.50 million hectare soil is affected by salinity at varying degrees including coastal saline soil (Chaudhary and Bhragava, 2003). Saline soils are generally low in fertility and difficult conventional crop production. The plant growth in such soil is impaired mainly of osmotic effect of excess salts and nutritional imblalance. Higher salt concentration in soil causes low physiological availability of water to plants, indirect toxic effect of individual ions (Na^+, Cl^-, HCO_3^-, CO_3^-, etc.) and complex interaction among sodium, calcium and magnesium leading to disturbed equilibrium of these ions in the plants available to absorb water and nutrients in required amount. The reduced water availability at high salinity causes water deficit to plants in a manner similar to drought. The first reaction of plants to salinity is the poor germination, retarded growth, impaired physiological activities, delayed flowering and reduced yield.

A decrease in seed gemination per centage was reported by Kumar *et al.* (1989). Sharma and Gupta (1986) have reported a reduction in plant height, leaf number, leaf area and dry weight of root and shoot by salinity. Decrease in chlorophyll content (Sahoo and Sahu, 1994), carotenoid (Karadge and Chavan , 1981), carbohydrate (Shanta *et al.*, 1989), protein (Huber *et al.*, 1977) and glutamate dehydrogenase and nitrate reductase activity (Amrit Sagar and Vaidyanath, 1993) were reported due to salinity stress.

The first requisite for the crop production on saline is to lower the salt concentration in root zone and to maintain at optimum level that the crop does not suffer in future through reclamation. Removal of double salts can be done by various methods such as scraping, flushing and leaching. Scraping is the removal of salts by mechanical means. Flushing is the surface washing out of the salts with runoff water. Leaching is the method of solubilising and transporting the soluble salts by downward movement of water through soil. While leaching of salts is the ultimate answer to salinity, management of saline soils envisages selection of salt tolerant crops and their varieties.

Amelioration of stress due to sodium chloride by using various chemicals in crop plants is another successful method of overcoming the ill effects of soil salinity. The interactive effects of gibberellic acid and sodium chloride on the growth of ragi was revealed by Deepak Bhat and Homani (1991). Amrit Sagar and Vaidyanath (1993) have reported an amelioration effect of sodium choride stress in *Cajanus cajan* by 6-benzyl aminopurine. The studies of Sahoo and Sahu (1994) with Ca^{++} ions on NaCl stress in *Oryza sativa* have also revealed similar effects. Muthukumarasamy and Panneerselvam (1997) have suggested the use of triadimefon to overcome the stress in peanut seedlings. They have reproted an increase in root growth, dry weight, chlorophyll, carotenoid, protein and glycine betaine by treating the affected plants with 10 mg/l concentration of triadimefon. The activity of endogenous auxins and growth inhibitors, gibberellins and cytokinins was observed in the extracts of seedlings of *Pisum sativum* under sodium chloride stress by Mashad and Kamel (2001). They have reported an amelioration of stress caused by sodium chloride by treating the plants simultaneously with 10 ppm boron solution.

The study of literature revealed that there are several works on the amelioration of sodium chloride stress in various crop plants by using various chemicals. But there is no evidence on the amelioration of sodium chloride stress by ammonium molybdate. Molybdenum is a micronutrient available in the soil in the form of molybdate. It plays an important role in the fixation and assimilation of nitrogen. It is a cofactor of the enzyme nitrate reductase. It is essential for nodulation in leguminous plants. It regulates amino acid concentration in plants. So the present work has been taken up to study the effects of ammonium molybdate on sodium chloride stress in *Vigna mungo*. This study is more appropriate for the present condition of the coastal solid of Tamil Nadu and Pondicherry due to the impact of Tsunami.

Materials and Methods

The certified seeds of *Vigna mungo* (L.) Hepper cv. TAU-1 were used. The molar solution was prepared by dissolving one mole of sodium chloride (NaCl) in distilled water and the volume was made up to 1000 ml. From this, different concentrations (50mM, 75mM and 100mM) were prepared by diluting it with distilled water. 1000ppm concentration of ammonium molybdate $\{(NH_4)_2MoO_4\}$ was prepared by dissolving one gram of the salt in distilled water and the volume was made up to 1000 ml. From this 10ppm solution was prepared.

The pots were filled with equal amount of garden soil. In each pot 20 seeds of *Vigna mungo* were sown. The pots were treated daily with 10 ml each of distilled water (Control), 50mM sodium chloride solution, 10ppm ammonium molybdate solution and solutions prepared by combinations of different concentrations of sodium chloride and 10ppm ammonium molybdate (50mM + 10ppm, 75mM + 10ppm, and 100mM + 10ppm). After 10 days the seedlings were removed from the pots, washed thoroughly and used for estimation. The contents of pigments, sugars and proteins and the activity of the enzyme Nitrate reductase of the leaves were estimated following the procedures of Hiscox and Israelstam, 1979 (pigment extraction), Harborne, 1984 (pigment estimation), Snell and Snell, 1957 (reducing sugars), Nelson, 1944 (non-reducing sugars), Lowry *et al.*, 1951 (protein) and Joworski, 1971 (Nitrate reductase activity).

Result and Discussion

All the parameters studied were found to be more in the plants treated with the solution prepared by mixing 50mM sodium chloride and 10ppm ammonium molybdate than in the plants treated with distilled water and other solutions (50mM, 10ppm, 75mM + 10ppm and 100mM + 10ppm). An increase of 37.837 per cent of total chlorophyll, 39.436 per cent carotenoid, 42.845 per cent total sugars, 26.686 per cent protein and 62.980 per cent nitrate reductase activity over control were observed in the plants grown in 50mM + 10ppm treatment (Tables 1 and 2).

Table-1: Effects of ammonium molybdate on sodium chloride stress in *Vigna mungo* (Pigment contents of primary leaves–mg/g fresh weight)

Treatment	Chlorophyll-a	Chlorophyll-b	Total Chlorophyll	Carotenoid
Control	0.880±0.045	0.486±0.036	1.366±0.068	0.426±0.075
50mM	0.763±0.037	0.405±0.038	1.168±0.056	0.398±0.038
(NaCl)	(-13.295)	(-16.666)	(-14.494	(-6.572)
10ppm	0.936±0.054	0.510±0.053	1.446±0.075	0.476±0.042
{$(NH_4)_2$ MoO_4}	(+6.363)	(+4.938)	(+5.856)	(+11.737)
50mM + 10ppm	1.262±0.073	0.621±0.0666	1.883±0.081	0.594±0.071
	(+43.409)	(+27.777)	(+37.847)	(+39.236)
75mM + 10ppm	0.865±0.058	0.454±0.062	1.319±0.093	0.467±0.044
	(-1.704)	(-6.584)	(-3.440)	(+9.624)
100mM + 10ppm	0.773±0.062	0.408±0.045	1.188±0.015	0.421±0.053
	(-12.159)	(-16.049)	(-13.543)	(-11.737)

Note: Mean ± Standard Eerror

Change over control in parentheses

Table-2: Effect of ammonium molybdate on sodium chloride stress in *Vigna mungo* (sugars, protein–mg/g fresh weight; NRA–moles NO_2 formed/gram fresh weight/hour)

Treatment	Reducing sugars	Non-reducing sugars	Total sugars	Protein	NRA
Control	5.065±0.114	3.888±0.065	8.953±0.856	4.654±0.093	0.832±0.036
50mM	4.832±0.356	3.262±0.09	18.094±0.642	3.642±0.062	0.745±0.028
(NaCl)	(-4.600)	(-16.100)	(-9.594)	(-20.744)	(-10.456)
10ppm	5.896±0.356	4.965±0.108	10.861±0.954	5.460±0.075	1.215±0.033
{$(NH_4)_2MoO_4$}	(+16.406)	(+27.700)	(+21.311)	(+17.318)	(+46.033)
50mM + 10ppm	6.896±0.363	5.893±0.124	12.789±0.806	5.896±0.085	1.356±0.045
	(+36.150)	(+51.569)	(+42.845)	(+26.686)	(+62.980)
75mM + 10ppm	4.854±0.265	4.025±0.095	8.879±0.545	4.821±0.096	0.890±0.041
	(-4.165)	(+3.523)	(-0.826)	(+3.588)	(+6.971)
100mM+ 10ppm	4.259±0.108	3.256±0.036	7.515±0.623	4.540±0.045	0.785±0.028
	(-15.913)	(-16.255)	(-16.061)	(-2.449)	(-5.649)

Note: Mean± Standard Error

Change over control in parentheses

One of the important adverse effects of salt stress in the loss of chlorophyll in the leaves (Sahu and Mishra, 1987). According to Sudhir and Murthy (2004) the salt stress causes decrease in plant growth and productivity by disrupting physiological processes. When combined with 10ppm of ammonium molybdate the stress caused by sodium chloride might have been ameliorated, hence more pigment in the combined solution of 50mM + 10ppm. According to Dwivedi *et al.*, (1977) molybdenum has significant effect on chlorophyll biosynthesis.

Similar studies on the amelioration of sodium chloride by various chemicals have been reproted by many workers, Ca^{++} (Sahoo and Sahu, 1994), Boron (Mashad and Kamel, 2001), calcium nitrate and potassium nitrate (Cengiz Kaya, *et al.*, 2003). The enhanced pigment concentration in the plants treated with 50mM + 10ppm solution must also be due to the enhanced absorption of more essential elements (Mg and K) which are required for pigment synthesis.

Salinity in the soil affects the absorption of mineral elements and the ratio between K and Na is changed. It affects the bio-energetic process of photosynthesis. Due to decreased rate of photosynthesis, the photosynthates were also less in the plants affected by salt stress. Pulver and Ries (1973) have attributed this to the utilisation of sugars in metabolic activities in order to meet stress condition. This must also be due to the utilisation of sugars to compensate reduced rate of photosynthesis (Anabazhagan *et al.*, 1990). When the plants were treated with 50mM sodium chloride + 10ppm ammonium molybdate solution the stress effects of 50mM sodium chloride must have been antagonised by 10ppm ammonium molybdate solution and resulted in higher photosynthetic rate and accumulation of more photosynthates.

Increase in protein content in 50mM + 10ppm treatment must be due to the absorption of optimum quantities of potassium and nitrate ions. In the plants treated with higher concentrations of sodium chloride, there was a reduction in the absorption of essential elements leading to reduced physiological activities and finally resulted in reduced protein content (Swaminathan and Vaidheeswaran, 1991).

The Nitrate reductase is the first enzyme governing the nitrate assimilation and is generally susceptible to changes in environmental conditions (Campbell and Smarelli, 1986). More nitrate reductase activity indicates the stress tolerance mechanism prevailing in the plants. According to Sahulka (1978) the chloride ions were responsible for the reduced rate reductase activity. Pandey and Srivastava (1993) have reported that the inhibition in the nitrate reductase activity by salt stress was due to restricted transport of nitrate to the metabolic pool. This might be due to amelioration of sodium chloride stress by lower concentration of ammonium molybdate by mediating osmoregulation, absorption of more essential elements required for biosynthesis, increasing photosynthetic activities and reduced energy requirement. So the present work suggests the use of low concentration of ammonium molybdate (10ppm) to ameliorate the soil affected by sodium chloride stress.

References

Amrit Sagar and Vaidyanath, K. 1939. Effect of sodium chloride induced stress on GDH and NR enzymes in calli of certain salt tolerant and susceptiable varities of pigeon pea. *Ad Plant Sci.* **6(1) : 74-81.**

Anbazhagan, M,; Krishnamurthy, R. and Bhagwat, K.A. 1990. Changes in the carbohydrate levels of rice cultivars due to air pollution. *Indian J. Exp. Biol.* **28 : 395-396.**

Campbell, W.H. and Smarrelli, J. 1986. Nitrate reductase: Biochemistry and regulation. In: *Biochemical basis of plant breeding.* Vol. 2. C. A. Neyra (ed.). CRC Press Inc. Boca Ratan. Florida, pp. **1-39.**

Chaudhary, D.R. and Bhargava, G.P. 2003. Saline soil management for sustainable foodgrains production. *Agrbios.* **2(2) : 34-36.**

Cengiz Kaya; Bekir Erol, A.K. and David Higgs. 2003. Response of salt stresses strawberry plants to supplementary calcium nitrate and potassium nitrate. *J. Plant Nutrition.* **26(3) : 243-260.**

Deepak Bhat and Hosmoni, S.P. 1991. Interactive effects of gibberellic acid and sodium chloride on the growth or ragi. *Ad. Plant Sci.* **4(2) : 256-263.**

Dwivedi. S.K.; Meers Singh.; Nigram, P.K., Patel, R.S. and Agarwal. V.K. 1997. Effect of phosphorus and molybdenum application on leaf chlorophyll content in soybean. *Ad. Plant Sc.* **10(1) : 145-147.**

Harborne, J.B. 1984. Phytochemical methods. *A guide to modern technique of plant analysis* 2nd ed. (Chapman and Hole, London), pp. **215-127.**

Hiscox, J.D. and Israelstam, G.M. 1979. A method for the extraction of chlorophyll from leaf tissue without maceration. *Can J. Bot.* **57 : 1332-1334.**

Huber, W.; Kreutmerier, F. and Sankha, N. 1977 *Z. Pflanzenphsiol.* **81 : 234-247.**

Joworski, E.G. 1971. Nitrate reductase in intact plant tissue. *Biochem. Biophys. Res. Commun,* **43 : 1274-1279.**

Karadge, B.A. and Chavan, P.D. 1981. Salt tolerance studies in groundnut var. TMV. 10. *Biovigyanam.* **7 : 137-144.**

Kumar, A.; Bahadur, B. and Sharma, B.K. 1989. Effect of salt stress on germination and early seedling growth of *Secale cereale. Geobios.* **6 : 40-42.**

Lowry, O.H.; Rosenbrough, N.J.; Farr, A.L. and Randall, R.J. 1951. Protein measurement with the fohlin-phenol regent. *J. Biol. Chem.* **193 : 265-275.**

Mashad, A.A. and Kamel, E.A. 2001. Amelioration of sodium chloride stress in *Pisum sativum. Indian J. Exp. Biol.* **9(5) : 469-475.**

Muthukumarasamy. M. and Panneerselvam, R. 1997. Amelioration of sodium chloride stress by triadimefon in peanut seedlings. *Plant Growth Regulation.* **22(3) : 157-162.**

Nelson, N. 1944. A photometric adoption of the Somogyis method for the determination of reducing sugar. *Anal. Chem.* **31 : 426-428.**

Pandey, M. and Srivasthava, H.S. 1993. Inhibition of nitrate reductase activity and nitrate accumulation by mercury in maise leaf segments. *Indian J. Environ. Hlth.* **35 : 110-114.**

Pulver, E.L. and Ries, S.K. 1973. Action of simazine in increasing plant protein content. *Weed Sci. Soc. Amer.* **21 : 233-237.**

Sahoo, S.K. and Sahu, A.C. 1994. Ca^{++} does not antagonise the adverse effect of NaCl stress in excised rice leaves. *Ad. Plant Sci.* **7(1) : 100-104.**

Sahu, A.C. and Mishra, D. 1987. Changes in some enzymes activities during excised rice leaf senescence under NaCl stress. *Biochem.* Physiol. *Pflanzen.* **182 : 501-209.**

Sahulka, J. 1978. The effect of chloride on nitrate reductase level on anaerobic nitrate reduction and on nitrate content in excised *Pisum sativum* L. roots. *Biol. Plant.* **20 : 201-209.**

Shanta, S.; Salunkha and Karadge, B.A. 1989. Influence of salinity and water stress on growth and organic constituents of *Dodonaea viscosa* (L.). *Geobios,* **16 : 252-258.**

Sharma, S.K. and Gupta, I.C., 1986. *Saline environment and plant growth.* Agro Botanical Publishers India. pp. **170.**

Snell, F.D. and Snell, C.T. 1957. *Calorimetric method of analysis.* 3rd Ed. Van Nostrand, D., New York.

Sudhir, P. and Murhty, S.D.S. 2004. Effects of salt stress on basic processes of photosynthesis. *Photosynthetica.* **42(4) : 481-486.**

Swaminathan, K. and Vaidheeswaran, P. 1991. Effect of dyeing factory effluent on seed germination and seedling development of groundnut. *J. Environ. Biol.* **12 : 353-358.**

❑❑❑

2

Ultrastructural Changes in the Pituitary (Prolactin Cells) and Testis of *Glossogobius Giuris* (Ham), Induced by Organophosphorus Pesticode

Bela Zutshi
Department of Zoology, Jnana Bharathi,
Bangalore University, Bangalore—560 056,
e-mail: bela-zutshi@yahoo.com.in

Introduction

Due to the growth of intensive agriculture in and around fresh water bodies and tanks of Bangalore, pesticides used abundantly during the cultivation season find their way into water bodies. Further, fenthion, an organophosphorus pesticide, widely used for the eradication of mosquito larva in the fresh water tanks and ponds give rise to histophysiological alterations of various tissues of aquatic animals. In fishes, pesticides bring out disorders—including histopathological changes in the testis of *Puntius ticto* (Sahai 1989) and hyper-secretion in PRL cells initially and finally degeneration of the cells suggesting an acute stress response, since PRL is released under stress conditions. Similar observations have been made by, Jagdeesh and Sahai (1989) on pituitary of *H. fossilis* causing inhibition of gonadal growth.

Light microscopic stuides have been made on control and pesticide treated prolactin cells in pituitary gland (PRL cells) and testis of *G. giuris*, during breeding phase after exposing them to different sub-lethal concentration of fenthion (0.05 to 0.5ppm). Studies on ultrastructural cytotoxic changes in pituitary and testis have been carried out in *G. giuris* during breeding period. Although extensive light microscopic work has been done on pesticide effect on testis and pituitary of fish, very little is known about ultrastructure change in pituitary and testis in fishes especially during breeding phase, when PRL cells exhibit maximum histological changes following cytological events in the testis. The available investigations largely deal with mammalian testis and pituitary (Dutt and Kobayashi, 1980). Therefore, in the present investigation an attempt has been made to study the response of these cells to sub-lethal concentration of fenthion to correlate and compare the functional significance of the cells in relation to the testicular function.

Material and Method

The adult male fish, *G. giuris* (40-60g; length 90-120 mm) were collected from Sanathkumari tank, near Bangalore. They were maintained in the laboratory conditions in glass aquaria at 27±1°C and were fed with earthworms. Different concentrations of fenthion were prepared in alcohol medium (stock solution of fenthion — 0.5mg/l) and added separately to the test water to obtain the desired concentration by adopting serial dilution technique as outlined in APHA, AWWAH, W.P.C.F. (1971). The acclimated fishes were divided in 4 experimental groups of 10 each. Of these, three groups were placed separately in sub-lethal concentration of fenthion (0.05, 0.25 and 0.5ppm) while the fourth group served as control. They were straved for 24 hrs prior to the start of the experiment and were not fed during the course of the experiment (Dalela *et al.*, 1979). The fishes were vivisected at the interval of 24, 48, 72 and 96 hr from experimental as well as control groups.

Light Mircoscopic Studies

Pituitary intact with brain and pieces of testis from control and experimental groups were removed and fixed in Bouin's hollande and Bouin's fluid respectively. They were dehydrated in grades of alcohol and embedded in paraffin for histological examination. Sections (5um thick) were stained in CAHP, PAS and Ehrlich's haematoxylin and eosin stain respectively.

Electron Microscopic Studies

Since, cytotoxic effect was seen after 24 hours of treatment in higher concentration (0.5 ppm) of fenthion, electron micropscopic studies were conduced only for 24 hrs period during breeding period.

For these studies, pituitary and small pieces of testis, from control and treated fishes vivisected after 24 hr (0.5 ppm concentration) and fixed in 2.5 per cent glutaraldehyde (0.1M phosphate buffer, pH 7.4) for 48 hr. After post-fixation in 1 per cent osmium tetraoxide, the specimen was dehydrated in grades of alcohol at 4°C and embedded in epoxy resin (Sarsa Bharathi and Armugam S, 1979, 1982). The specimen block was kept in incubator for hardening Gross trimming and rough sections rewarmed. Thick sections (1μm) were stained with 1 per cent toulidine bule solution and examined under compound microscope. Ultrathin sections (700-800 μm) were placed on copper grid, stained with uranyl acetate lead citrate and were then observed under electron microscope.

Results

Light Microscopy

Control:

Morphology and Histology of Pituitary

Hypophysis of G. giruis is of laptobasel type and is composed of a nervous part – the neurohypophysis and glandular part – the adenohypophysis (fig. a).

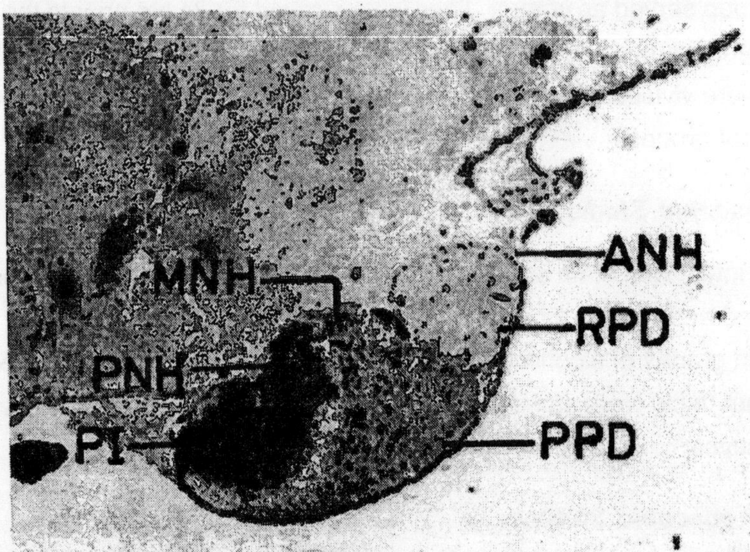

Fig. a: Median sagittal section of the hypothalamo-hypophyseal region of *G. giuris*. ANH-anterior neurophypophysis, MNH- middle neurohypophysis, PNH-Posterior neurohypophysis, RPD-rostal pars distalis, PPD-proximal pars distalis, PI- pars intermida. (CAHP) staining) X 600.

Histologically, adenohypophysis is divided into a pro–(rostral pars distalis-RPD), meso-(proximal pars distalis-PPD) and meta-(pars intermedia-PI) adenohypophysis. There are eight types of endocine secreting cells in the adenohypophysis, which differ from each other in size, staining reactions and state of sexual maturity. In the present work, an attempt has been made to study the effect of fenthion on prolactin secreting cells (PRL), and acidophil which is a major constituent of RPD. These cells exhibit striking changes during the sexual maturity and spawning cycle.

During breeding phase, prolactin cells contained varying quantities of granulation in the cytoplasm. Some cells were degranulated and chromophobic and others were with densely granulated cytoplasm. Few blood capillaries were also seen (fig. 1).

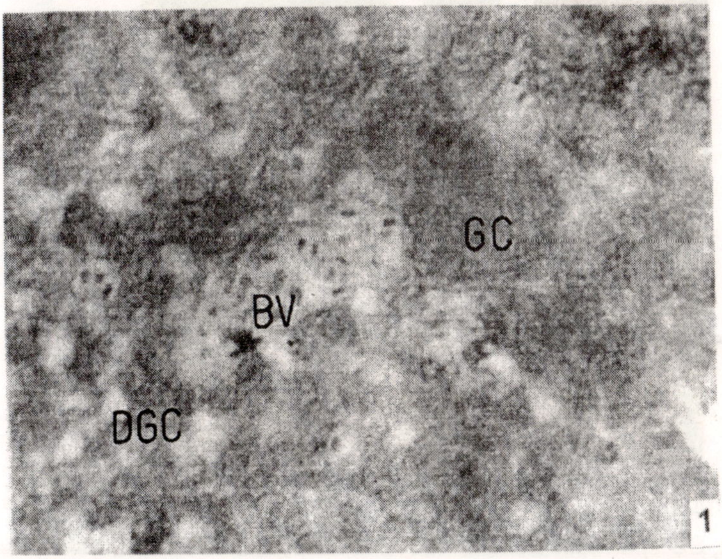

Fig. 1: Median sagittal section of the RPD region of *G. giuris* during spawning season, indicating degranulated prolactin cells (DGC) which are chromophobic. Note a few granulated cells (GC) and number of blood vessels (BV). (CAHP staining) X 600.

Morphology and Histology of Testis

The testis of *G. giuris* is an elongated organ, mainly composed of two tissues, the peripheral seminiferous tissue and central glandular tissue (fig. b). The seminiferous tissue is made up of single layer of seminiferous lobules whereas the glandular tissue consists of interstitial glandular cells.

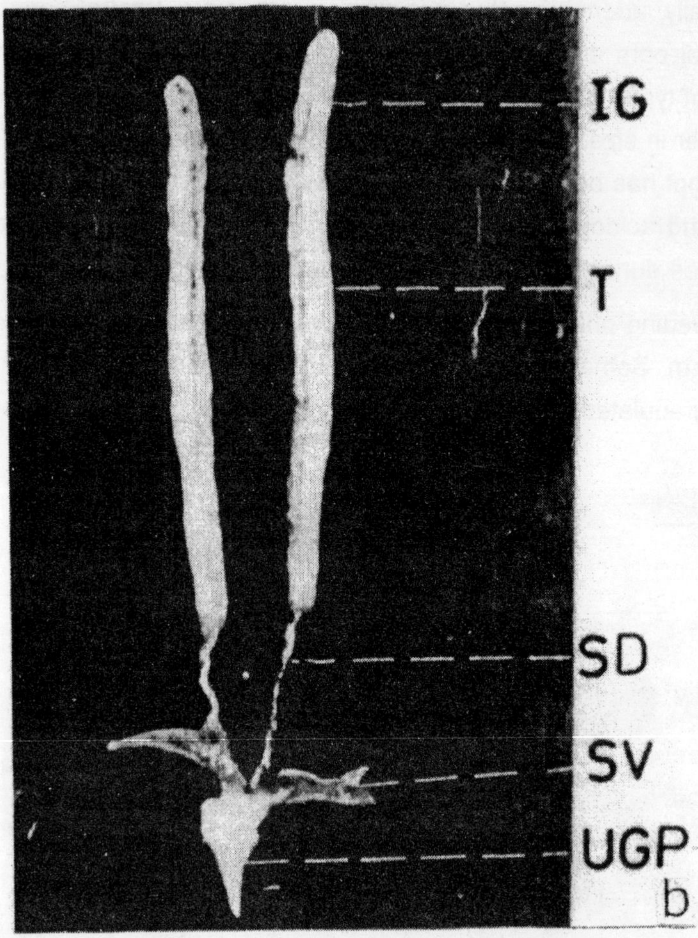

Fig. b: Male reproductive system of *G. giuris* during spawning season. T-testis, IG-interstitial gland, SD-spermatic duct, SV-seminal vesicle, UGP-urinogenital papillae.

During breeding phase, testis lobules has attained their maximum size and were uniformly packed with masses of mature spermatozoa. The lobules contained a few spermatocytes and spermatids. The tunica albugenia and interlobular connective tissue were thin (fig. 2).

Interstitial gland also increased in volume and contained granulated and degranulated cells. The gland cells were spherical with indistinct cell boundaries and contained irregular-shaped nuclei. Few vacuoles were also found in the inner region of the gland (fig. 2).

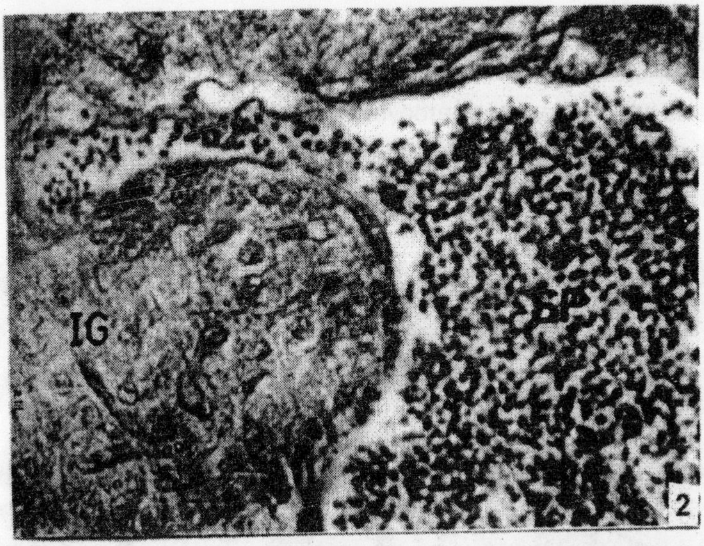

Fig. 2: Cross-section of testis of *G.giuris* showing seminiferous lobules packed with masses of spermatozoa. (SP). IG-Interestitial gland (H-E staining) X 600.

Fenthion Treated:

In prolactin cells of the pituitary of fishes, treated in 0.5 ppm concentration for 24 hrs, degranulative changes were more marked, than those of control ones. Vacuoles were observed in the depleted prolactin cells. blood sinusoids were also present (fig. 3).

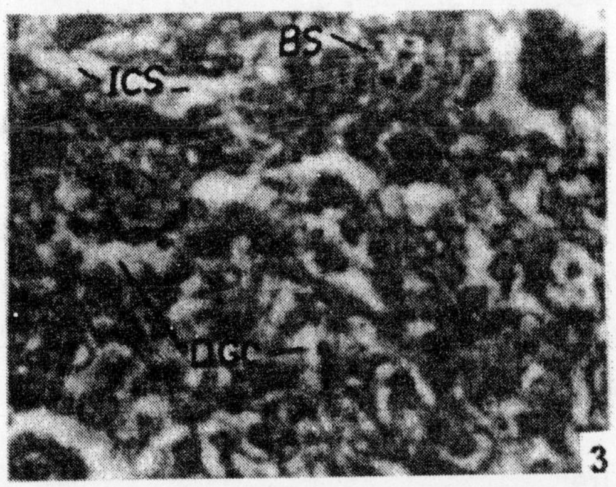

Fig. 3: Median sagittal section of the RPD region of fenthion treated fish showing large number of degranulated prolactin cells (DGC) and intercellular spaces (ICS). Note the blood sinusoids (BS) (CAHP staining) X 600.

Testis showed appreciable reduction in size, with most of the germ cells *viz.* spermatids and sperms in degenerating condition. Abundant spermatogonial cells and a few primary and secondary spermatocytes were present which corroborated reduction in the gonadosomatic index. Interstitial gland exhibited an increase in the degranulated cells and a few granulated cells (fig. 4).

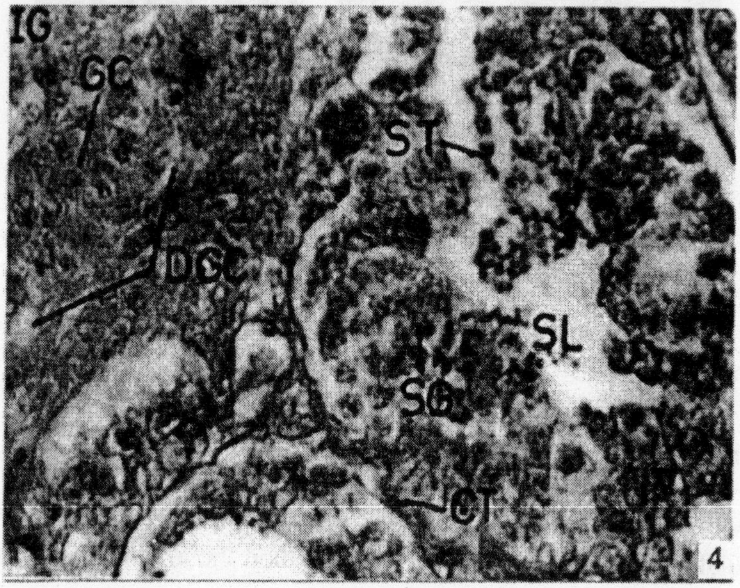

Fig. 4: Cross section of testis of fenthion treated fish showing degenerating sperms (SP) and spermatids (ST) with abundant spermatogonial (SG) cells. Note the number of degranulated (DGC) and sparsely granulated (GC) interstitial cells in the interstitial gland (IG). CT-Connective tissue, SL-Seminiferous lobule. (H-E staining) X 600.

Electron Microscopy

Light microscopic studies revealed that fenthion induced atrophy in the testis and pituitary. Ultrastructural studies confirmed the extensive cytotoxic effect in the tetis (sperm) and pituitary (PRL cells) in the fishes treated with 0.5 ppm concentration of fenthion for 24 hrs during breeding period.

Control: Ultrastructure:

Prolactin Secreting Cell

The prolactin cells of normal fish form a compact mass in the pituitary. Nucleus of prolatin cell contains uniformly distributed chromatin material, with one or two nucleoli in it. The cytoplasm is characterised by the presence of large number of membrane bound secretory granules. Golgi complex region contains few granules. Rough endoplasmic reticulum is present either in an array of layers or as scattered pieces of cisternae. Lysomes and mitochondria are also present in the cytoplasm (fig. 5).

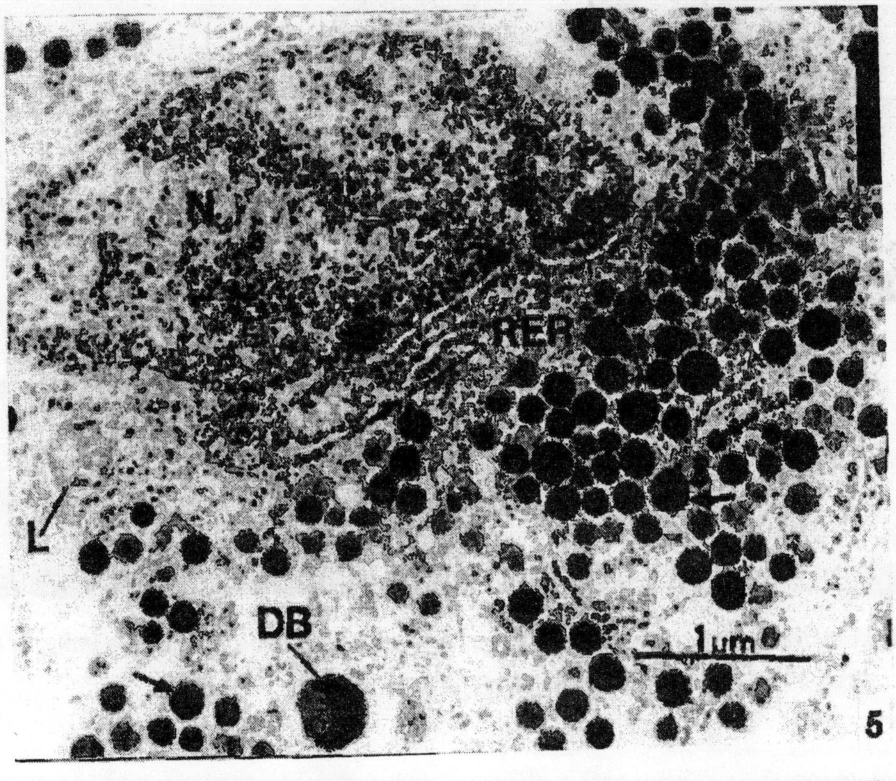

Fig. 5: Electron micrograph of prolactin cell cytoplasm of the control fish showing uniformally distributed chromatin material in the nucleus (N), large number of membrane bound granules (arrows), array of layers of scattered pieces of RER, a lysosome (L) and desne body (DB).

Seminiferous Lobule

The inner wall of each seminiferous lobule is covered with cysts in which germ cells develop. After metamorphosis from spermatids, spermatozoa are discharged from the cyst into the cavity of seminiferous lobule (fig. 6).

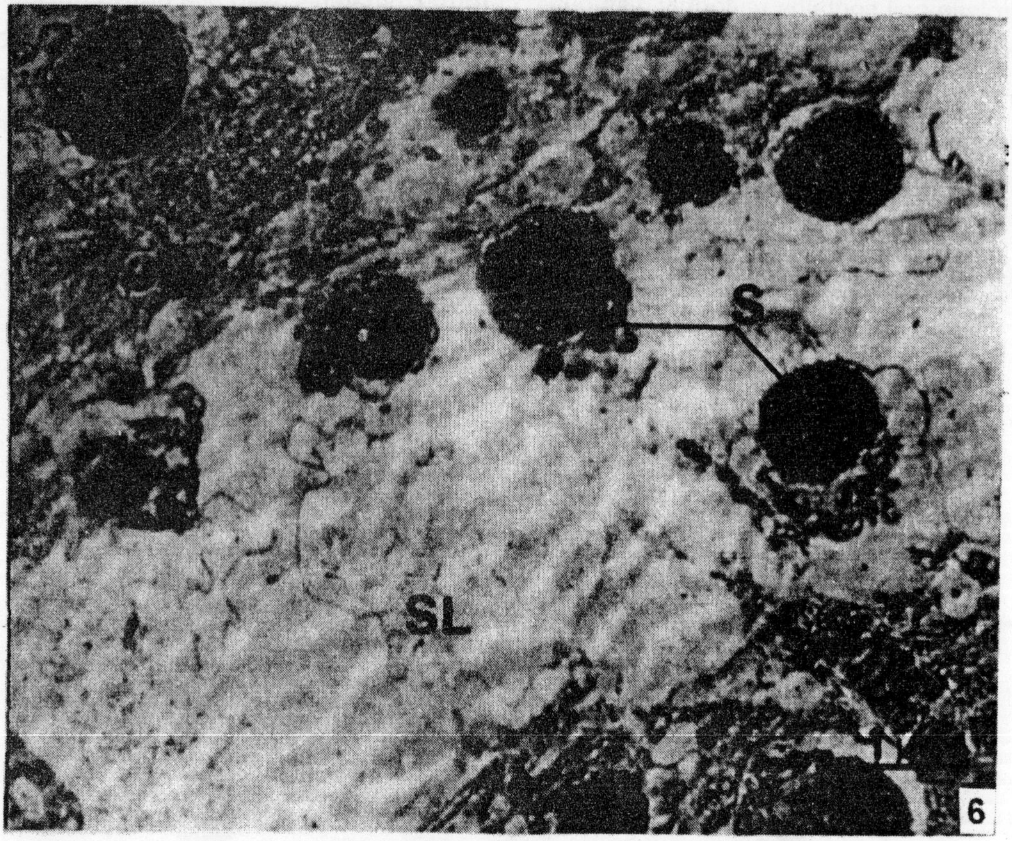

Fig. 6: Electron micrograph of a portion of seminiferous lobule (SL) of the testis of control fish. Note the sperms (S) in different stages of development.

Spermatozoa

Each spermatozoa consists of a prominent head, neck and tail consisting of axonemal complex. The nucleus of spermatozoa contains condensed chromatin material (fig. 7) and in higher magnification the residual cytoplasm around the nucleus show few rod shaped mitochondria with distinct vesicular cristae while that surrounding the middle piece contains large number of mitochondria. The axonemal complex of the tail consists of a pair of centrally placed tubular filament and nine pairs of circumferentially arranged peripheral axial filaments. A few dilated smooth endoplasmic reticulum are seen surrounding the complex.

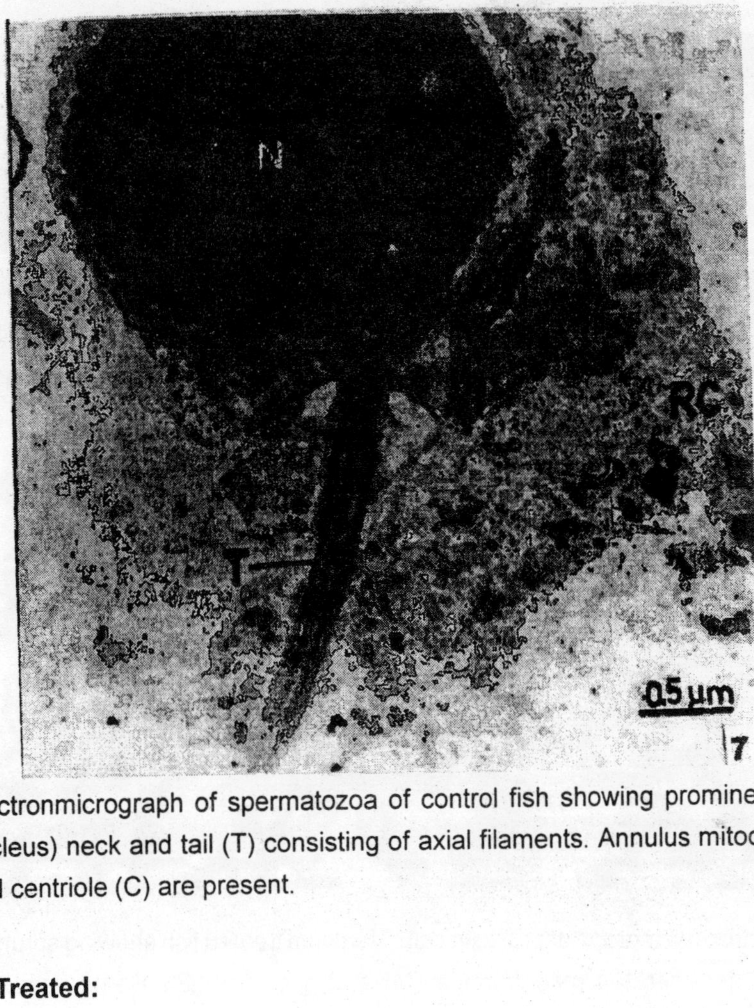

Fig. 7: Electronmicrograph of spermatozoa of control fish showing prominent head (N-nucleus) neck and tail (T) consisting of axial filaments. Annulus mitochondria (M) and centriole (C) are present.

Fenthion Treated:

Prolactin Secreting Cell

The prolactin cells of the fishes show conspicuous changes in their size and shape after 24 hrs of treatment with 0.5 ppm of fenthion. The nuclei of the prolactin cells appear shrunken. The cell membrane and the nuclear membrane are fragmented with reduced chromatin material. The nucleolus appears disinegrated. In the cytoplasm the degenerating secretory granules and mitochondria (having disintegrated cristae) show reduction in number and size. Endoplasmic reticulum and other organelles appear to have degenerated. An increase in intercellular spaces and vacuoles has also been observed (fig. 8).

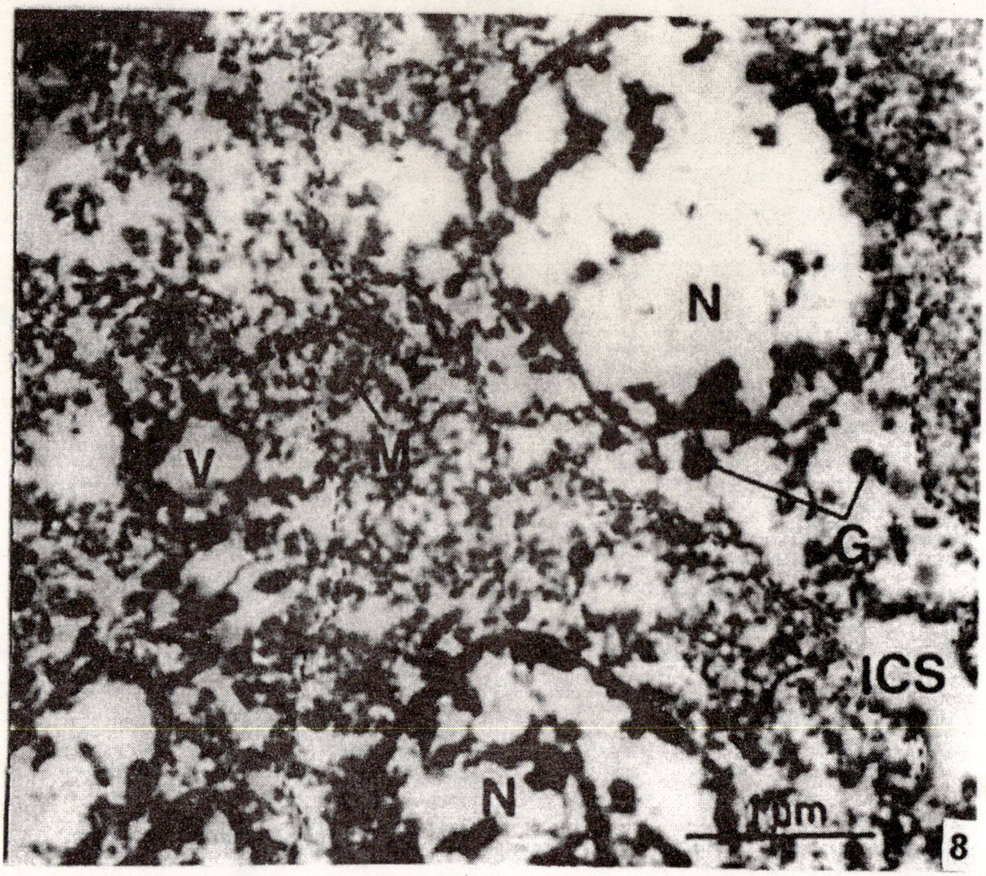

Fig. 8: Electron micrograph of prolactin cell of fenthion treated fish showing shrunken nucleus (N), degenerated mitochondria (M) and granules (G). Some vacuoles (V) and intercellular spaces (ICS) are also seen.

Spermatozoa

The spermatozoa of the fenthion treated fish showed atrophy in the head and tail region. The shape of the head became irregular with disintegrating nuclear membrane. The nucleoplasm consisted of granular material. The mitochondria showed degenerative changes mainly reduction in stainability and hypoplastic cristae. The tail was reduced to thread—like structure (fig. 9).

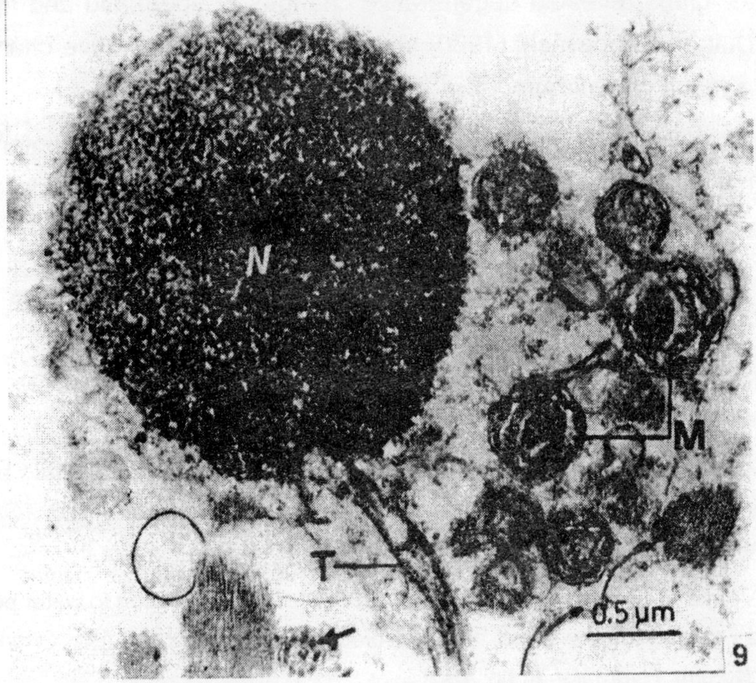

Fig. 9: Electronmicrograph of spermatozoa of fenthion treated fish showing less dense chromatin in nucleus (N), few empty mitochondria (M) and tail with discontinuous filaments (arrow).

Discussion

The results presented here clearly show that the fenthion exerts various toxic effects upon the pituitary-testicular-arix. Light microscopic studies on pituitary and testis of fenthion treated (for 24 hrs in 0.5 ppm) fish, *G. giuris* revealed necrisus of prolactin secreting cells, interstitial cells and sperms indicating, reduction in synthesis/secretion of androgen. This suggests that fenthion blocked the spermatogenic processes at the spermatid level. Numerous authors have reported cytological changes in the adenohypophyseal cells (prolactin cells) and tetis (including interstitial cells) in the pesticide exposed fishes (Ram and Sathyanesan, 1983).

The present ultrastructural studies on prolactin of fenthion (0.5 ppm for 24 hrs) exposed fish, *G. giuris* showed degeneration and fragmentation in rough endoplasmic reticulum and golgi apparatus. Oztan (1966b) reproted presence of rudimentary RER cisterna in 1/3 sea water exposed PRL cells of *Zoarces viviparous*. Testis if fenthion treated fish *G. giuris* revealed degenerative changes in spermatid and sperm-head and its tail. Dutt and Kobayashi (1980) showed various degenerative changes in the rat testis as a result of cadimum.

The abnormalities in the pituitary and testis affect the reproductive potential of fish, suggesting that spraying of any pesticide during breeding season of fish is lethal for its reproduction.

References

APHA, AWWAH, W.P.C.F. 1979: *Standard methods for the examination of water and waste water.* 13th edn. (Tarus) (A.P.H.A. Washington, D.C.).

Ashina, K.; Suzuki, K.; Aida, K.; Hibiya, T. and Tamaoki, B. 1985: Relationship between the structures and steroidogenic functions of the testis of the Urohaze – goby (*Glossogobius olivaceus*). *Gen. and Comp. Endrnol.,* **57,** pp. **281.**

Delela, R.C.; Verma, S.R. and Bhatangar, M.C. 1986: Biocides in relation to water pollution Part I, Bioassay studies on the effect of a few biocides on fresh water fish, *Channa gachua. Acta. Hydrochim. Hydrobiol,* **6,** pp. **15.**

Dutt, N.H.G. and Kan, Kobayashi 1980: Abnormalities in the fine structure of the spermatids of rats injected with cadmium, *J. Biosci.* Vol. **2 (4), 361-367.**

Jadgeesh, C.G. and Sahai, S. 1989: Effect of malathion on the pituitary gland of *Heteropneustes fossilis* (Bloch). *Environ. and Toxicolog.,* **283-287.**

Oztan, N. 1996 b: The fine structure of the adenohypophysis of *Zoarces viviparous L.Z. Zellforsch,* **69, 699-718.**

Ram, R.N. and Sathyanesan, A.G. 1983: Effect of mercuric chloride on the reproductive cycle of the teleostean fish, *Channa punctatus. Bull. Envirom. Contam. Toxicol.,* **30, 24-27.**

Sahai, S. 1989: *Accumulation of induced histopathological effects of pesticides in the testis of Puntius ticto* (Teleostei). Folia. Morph.

Sarsa Bharathi and Armugam S. 1979, 1982: *Practical techniques in Electron Microscopy,* Madras Medical College, Chennai.

❑❑❑

3

Rattans of Manipur : Ecological Prospective with Special Reference to Sustainable Use

S. Padmabati Devi and P. K. Singh
Ethnobotany and Plant Physiology Laboratory
Department of Life Sciences, Manipur University,
Canchipur, Imphal–795 003, Manipur

Introduction

Geographically, Manipur state is a part of Manipur, Nagaland orogenic belt, flanked in the east by the Central Myanmar basin. The Indo-Myanmar ranges came into existence due to collision of the Myanmar Plate with the Indian Plate; Manipur is located at the confluence of the two plates. The region has been the vivilovian centre of diversity and origin of many plant species. The regions incidentally also happen to be one of the two mega centres of endemism in India. The Western Ghat and Eastern Himalaya, with 26 micro endemic centres and Manipur occupies a unique number of places where indigenous cane species are available in the state. Rattan is one of the minor forest products of Manipur, which belongs to the family Palmeae. It comprises of 13 genera and about 600 species (Dransifies 1981). However, from Manipur state Deb (1961), reproted six cane species viz., *Calamus erectus* Roxb., *C. floribundus* Griff., *C. Letifolius* Roxb., *C. leptospadix* Griff., *C. tenius* Roxb., *Daemonorop jenkensianua* Mart.

Materials and Methods

Methods followed for the Phytosociological studies of the various cane species of Manipur state were investigated district-wisely by following technique of Cain 1939, Goodal (1952) as the Rattan Species are in the creeping and lens group, the Method of Singh (1969) was also adopted for further ecological study, the vegetation data were quantitatively analysed for frequency density and basal area, following the method of Curtis and McIntosh (1950), Oosting (1958). The importance value index (IVI) of all the Rattan species along with the non-rattan associated plant was determine as the sum of relative frequency, relative density and relative basal area by following the method of Phillips (1958), Curtis (1959) and Misra (1969).

In addition to these calculations of IVI of five dominant plants species among the 14 species were incorporated. But among these species the IVI value of *C. guruba* was found to be 74.8 as dominant one. Its population type is expending type. And the IVI value of the remaining four species was also calculated ith its associated species.

In Table (3) A, the calculation of IVI of *C tenius* with it associated species is presented. In this table the IVI value is found to be highest in the *Musa Peradesica* of about (193.1) and the population type is expanding type. The lowest population is found in *Ficus hispida* of about (17.84) and the type of population is accidental type.

In Table (3) B the IVI value is found to be highest in *Ficus hispida* of about (157.2) and the type of population is expanding type. Lowest population is found in *Mangifersa indica* of about (28.1) and the type of population is accidental type. In Table (3) C the highest population is found in *Bambusa tulda Roxb.* of about (94.43) the type of population is expanding type, and the lowest IVI value is found in *Lantana Camera* of about (23.63) and the type of population is accidental type. In Table (3) D the highest IVI value is found in *C. floribundus* of about (52.6) the type of population is expending type and the lowest IVI value is found in *Bambusa tulda* of about (10.33) and the type of population is accidental type.

In Table 3 (E) the IVI value is found to be highest in *Arundo donax* Linn. of about (67.22) the type of population is expanding type and lowest IVI value is found in *Citrus maxima* Linn. of about (20.46) type of population is accidental type. Table 3 (F) shows the calculation of IVI value of all the Rattan species in this table the highest IVI value is found in *Calamus guuba* Roxb of about (114.2) and the type of population is expanding type. And the lowest IVI value is found in *C. erectus* Roxb. of about (23.04) and the type of population found is Accidental type.

Ethnobotanical Aspect

Regarding ethnobotanical aspect, cane has lots of ethnobotanical aspect based on edible and non-edible Rattan sold in the market. The second aspect is medicinal value of cane and the third aspect is in the field of handicraft. Some species of Rattan are edible but species like *calamus* are mostly edible. The edible part is the fruits pulp. Species like *Daemonorop jenkensianus* are not edible. But in *calamus* also species like *C. tenius, C. erectus* are not edible Fruits are mostly sour in taste (Table 4).

Medicinal Value

Medicinal value of Rattan is not too high as compared to the other medicinal value. However, some species have high medicinal value. Some of the Rattan species having medicinal value among the 14 species collected ten species had medicinal value. They are *C. letifolius* Roxb., *C. flagellum* Griff., *C. guruba* Roxb., *C. tenius* Roxb. Some of the Rattan species having medicinal value are shown in the Table 5.

Handicraft (Rattan used as a raw material for handicraft)

Most of the Rattan species like *Calamus* are commonly used in handicraft in small cable industries. For making sofa, table, rag mat, *mora* and other decorative items so it helps in icreasin income to the people who are working in the small scale indicters. Now Rattan is one of the minor forest products of Manipur. It plays an important role in socio-economic activates. The types of Rattan which are mainly used in the handicraft industries are *C. flagellum* Grigg., *C. tenius* Robx., *C. leptospadix* Griff. Canes are commercially classified in the market as A-Class, B-Class and C-Class. In Manipur there are lots of cottage industries. The cottage industries are mainly confined in Imphal East dist. Churachandpur dist., Tamenglong, Bishnupur and Imphal West dist. The following data shown in the table.

From the above data we can observe that the distribution of Rattan species is more higher in forest area than the disturb area. The IVI value is found to be highest in *Calamus guruba* which are collected from forest area of Andro. Imphal East dist and the IVI value is found to be lowest in Rattan species like *C. erectus* Roxb (13.04) which are collected from Andro but from the locality disturb area. These show that the distribution of Rattan species is more in forest area then the disturb area.

Nowadays Rattan became one of the minor forest products of Manipur. It helps in increasing income to he people who depend their livelihood on the handicraft industry. It plays an important role in socio-economic activity. So, it is important to conserve these kinds of species.

Table-1: Showing Distribution of Rattan in 9 districts of Manipur

No. of Sp	Scientific Name	Local Name	Place of Collection	District
1	*Calamus erectus* Roxb	Leeyetabi	Andro, Jiribam	Imphal East Dist., Imphal East
2	*Calamus leptospadix* Griff	Tangchoubi	Thanga, Khonthoujam, Bammon kampu	Bishnupur, Imphal West, Imphal East
3	*Calamus flagellum* Griff	Hunphop	Kamaga, Thanga, Wangkhei	Imphal East Dist., Bishnupur, Imphal East
4	*Calamus letifolius* Griff	Leeren	Churachanpur, Khonthoujam, Phayeng, Leimaram	Churachanpur Dist., Imphal West, Bishnupur Dist.
5	*Daemonorop Jenkensainus* Mart	Lee Tangchoudabi	Baruni, Tamei	Imphal West, Tamenglong Dist.
6	*Calamus floribundus* Griff	Hunphop	Sekmai, Kongba, Phayeng, Thoubal	Imphal West, Imphal West Thoubal Dist.
7	*Calamus tenius* Roxb	Yairee	Kakching, Andro	Thoubal Dist. Imphal East
8	*Calamus guruba* Roxb	Lithet	Andro kharam leikai	Imphal East
9	*Calamus pseudotenius* Mart	Yeiree Manbi	Sekmai Yairipok	Imphal West, Thoubal Dist.
10	*Calamus palustris* Griff	Patlee	Motbung	Senapati District

11 *Calamus longisetus* Mart	Leeree	Tipaimukh	Churachandpur District
12 *Calamus acanthospathus* Griff	Lee-tingkhangakupi	Kamagya Jiri	Imphal East District
13 *Calamus vimilalis* Roxb	Leephop Macha	Thanlon	Churachandpur District
14 *Daemonorop kurzianus* Becc	Leeham	Baruni	Imphal East District

Table-2: Showing Ecological Parameters

	Scientific Name	Habitate	Height	Soil Type	Temp.	Altitude	No. of Plant Forming Clumps	Time of Harves-ting	Correspon-ding Plant
1	*Calamus leptosapdix* Griff	Climber	40-45 ft	Loamy Soil	20	782 ft	8	Oct-Nov	Mango, Ficus, hespida
2	*Calamus flagellum* Griff	Climber	30-35 ft	Clay Loamy	19	780 ft	6	Sept-Oct	Mango, Parkia, roxburghuii, Santalum album
3	*Calamus guruba* Roxb	Climber	26-30 ft	Loamy Soil	18	820 ft	11	Oct-Nov	Mango, Eryterina
4	*Daemono rop jenkensianus* Mart	Solitary not climbing type	10-15 ft	Swampy Place	17	1210 ft	1 or 2 sparsely distributed	Sept-Oct	Mango, Juglan, regia Ficus plant
5	*Calamus erectus* Roxb	Solitary not climbing type	20-25 ft	Clay Loamy soil	18	830 ft	Only 1 solitary	Nov-Dec	Parkia, Mango

6	*Calamus letifolius* Roxb	Climbing type	20-25 ft	Clay Loamy Soil	19	790 ft	9	Oct-Nov	Ficus, hespida
7	*Calamus tenius* Roxb	Not climbing type	15-20 ft	Loamy Soil	19	950 ft	5 solitary	Sept-Oct	Santalum, album, Lantena

Table-3 (A): Ecological Parameter showing IVI of *C. tenius* and Associated Plants of Kakching area (1 Transec = 1m^2)

	Name of the Sp.	Relative Density (%)	Relative Frequency (%)	Relative Basal (%)	Importance Value Index	Remark
1	*C.tenius* Roxb	42.4	1.856	0.014	43.87	Expanding Type
2	*Arundo* donax Linn	28.7	25.14	0.83	54.67	Expanding Type
3	*Ficus hespida* Linn	2.7	14.97	2.63	17.87	Accidental Type
4	*Musa peradisica* Linn	.181	10.77	0.950	193.18	Accidental Type
5	*Bambusa tulda* Roxb	14.28	11.97	3.341	29.59	Accidental Type

Table-3 (B): Ecological parameter showing IVI of *C. flagellum* and associated plants of Physyeng area (1 Transec = 1m^2)

Name of the Sp.	Relative Density (%)	Relative Frequency (%)	R.B.A. (%)	Importance Value Index	Remark
C. flagellum Griff	16.0	11.0	1.184	28.1	Accidental Type
Parkia roxburguhii G. (Don)	31.0	33.3	0.8226	65.17	Expanding Type
Mangifera indica Linn	12.0	11.1	2.31	25.41	Enleseted Type
Ficus hespida Linn	12.0	11.2	11.34	157.2	Expanding Type
Santalum album Linn	31.0	33.3	82.6	146.9	Expanding Type

Table-3 (C): Ecological Parameter showing IVI Sp. from Andro [*C. guruba*]

1 Transec = 1m²

Name of the Sp.	Relative Density (%)	Relative Frequency (%)	R. B. A. (%)	Importance Value Index	Remark
C. guruba Roxb	51.2	49.41	31.64	74.2	Expanding Type of Population
B. tulda Roxb	36.58	35.29	22.56	94.43	Expanding Type
Lantena camera Linn	3.6	10.04	10.03	23.63	Accidental Type
Mangifera indica Linn	2.43	2.35	40.12	44.90	Expanding Type
Citrus maxima Linn	5.88	6.09	13.65	25.62	Accidental Type

Table-3 (D): Ecological Parameter showing IVI Sp. from Sekmai [*C. floribundus*] 1 Transec = 1m²

Name of the Sp.	Relative Density (%)	Relative Frequency (%)	R. B. A. (%)	Importance Value Index	Remark
C. floribundus Griff	4.96	50	0.122	52.6	Expanding Type
Clerodendron arborium Linn	8.06	8.33	6.01	22.4	Incestral type
Lantena camera Linn	16.66	16.66	3.07	36.07	Accidental Type
Bambusa tulda Roxb.	16.73	16.6	70.76	10.33	Expanding Type
Cinamomum tamala Nus and Ebern	8.06	8.33	7.86	24.25	

Table-3 (E): Ecological Parameter showing IVI Sp. from Andro [*C. erectus*]

1 Transec = 1m^2

Name of the Sp.	Relative Density (%)	Relative Frequency (%)	R. B. A. (%)	Importance Value Index	Remark
C. erectus Roxb	5.55	15.78	1.84	23.04	Accidental
Citrus Maxima Linn	511.1	0.26	9.096	20.46	Accidental
Mangifera indica Linn	16.66	15.78	0.960	33.4	Accidental Type
Arundo donax Linn	5.5	52.63	9.096	67.22	Expanding Type
Parkia roxburghuii G. (Don)	11.11	10.52	0.568	22.19	Accidental Type

Table-3 (F): Ecological Parameter showing calculation of IVI for 5 Rattan species

Name of the Sp.	Relative Density (%)	Relative Frequency (%)	R. B. A. (%)	Importance Value Index	Remark
C.flagillum Griff.	16.0	11.0	1.184	28.1	Accidental Type
C. Floribundus Griff.	4.96	50	0.122	52.6	Expanding Type
C. guruba Roxb.	51.2	49.41	13.64	114.2	Expanding Type
C. erectus Roxb.	5.5	15.78	1.84	23.04	Accidental Type
C. tenius Roxb.	42.46	1.856	0.014	43.87	Expanding Type

Table-4: List of Commonly Available Rattan Species of Manipur

No. of Species	Edible	Non-edible
1	*C. floribundus* Griff	*C. tenius* Roxb
2	*C. letifolius* Roxb	*C. erectus* Roxb
3	*C. leptospadix* Griff	*Daemonorop Jenkensanus* Mart

Table-5: Showing List of Rattan Species having Medicinal Value

No. of Species	Scientific Name	Part Used	Medicinal Value
1	*C. letifolius* Roxb	Tender leaf decoction is used	Cured nose bleeding
2	*C. flagillum* Griff.	Tender leaf or root decoction	Cured chronic fever, Vomiting
3	*C. guruba* Roxb.	Aerial stem produce sweet juice which are alkaloid	Used as a tonic by local people
4	*C. tenius* Roxb	Tender leaf and and aerial shoot	Used to cured bowel disorder and urinary trouble.
5.	*C. palustris* Griff	Tender leaf decoction	Cure arthritis biddy irritation leprosy
6	*C. Viminalis* Willd	Tender leaf and root	Chromic fever, blood diseases and urinary tract infection
7	*C. pseudotenius* Becc	Tender leaf and shoot	Curve hameoptric and chronic fever
8	*C. flaribundus* Griff	Tender leaf	Cure dizziness
9	*C. leptospadix* Griff	Tender leaf	Apply in poisonous bite as first aid remedy
10	*C. erectus* Roxb.	Tender Shoot	Stomach, ulcer and muscular sprain

Table-6: Showing Distribution of Handicraft Industry and Income of the Worker

Sl. No.	Name of the Sp (S.N)	Local Name	Type of Class	Price (Rs)	Item made by Cane	Dist where the Cottage Industries are Established	No. of Small Scale Industry based on Handicraft Society of Manipur	Salary of the Worker Month
1	*C. flagilhum* Griff.	Hunphop	B-Class	55	Chair, Sofa	Imphal East Dist.	8	750
2	*C. tenius* Roxb	Yairee	B-Class	60	Sofa, Rag Decorative Item	Churachandpur Dist.	3	450-500
3	*C. floribunbus* Griff.	Leekhet	C-Class	53	Bascet Decorative Item	Tamenglong	2	500
4	*C. letifolius* Roxb.	Leeren	A-Class	50	Sofa chair Mora	Bishnupur	3	500-600
5	*C. lepto-spadix* Griff.	Tangchoubi	C-Class	45	Rag Chair	Imphal West dist.	6	600-700
6	*C. guruba* (Becc hem).	Leethit	B-Class	52	Mora Decorative Item	Imphal- East District	6	400-450
7	*C. palustris* Griff.	Patlee	B-Class	50	Sofa, Chair	Senapati District	7	500
8	*C.longisetus* Griff.	Leeree	C-Class	50	Rag, Chair	Churachandpur District	4	500-550
9	*C. aconthos-pathus* Griff.	Leetingkang akupi	B-Class	45	Mora, Chair	Jiri Imphal-East	3	500-550
10	*C. Viminalis* Griff	Leephop macha	B-Class	52	Chair, Sofa	Churachandpur District	2	400-450
11	*C. erectus* Roxb	Lee-yetable	B-Class	45	Chair, Rag	Imphal-East	2	500

12	*D. kurzianus* Hook	Lee-ham	C-Class	35-40	Container	Imphal-East	2	450
13	*C. pseudo-tenius* Becc	Yairee-manbi	B-Class	50	Mora, Sofa	Imphal-East	2	550
14	*D. jenkensi-anus* Mart	Lee Tang-chaudabi	B-Class	45	Sofa, Chair	Imphal-East	3	500

References

Deb, D.B. 1961: Monocotyledoneous plants of Manipur. *Territory, Bull, Bot. Surv. India*, **3(2) : 115-138.**

Mishra, R. 1968: *Ecology Workbook*. Oxford and I.B.H. Pub. Co., New Delhi.

Curtis, J.T. and R.P. McIntosh 1950: *The introduction of certain analytic and systhetic phytosociological characters ecology,* **31: 434-455.**

Oosting H.J. 1958: *The studies of plant communities*, W.H.F. Freeman and Co., San Francisco, U.S.A.

Phillips, E.A. 1958: *Methods of vegetation study*. Henry Holt and Co. Inc. New York, U.S.A.

Cain, S.A. 1938: The species area curve. *Am. Midl. Nat* **19 : 573-581.**

Dransfield, J. 1981: *A Manual of the Rattans of the Malaya. Peninsular.* Forest Department, Ministry of Primary Industries, Malaysia.

Singh 1969: Influence of biotic disturbance on the preponderance and interspecific association of the two common Forb in the grassland of Varanasi. *Trop. Ecol.* **10 : 59-71.**

Goodal, D.W. 1952: Quantitative aspects of plant distribution. *Biol. Rev.* **27 : 194-245.**

□□□

4

Diversity and Composition of Insecta in Rice Agro Ecosystem in Barak Valley of Assam (N.E. India)

D.C. Ray and Partha P. Bhattacharjee
Department of Ecology and Environmental Science,
Assam University, Silchar–788 011

Introduction

Insects are the dominant components of most ecosystems and agro-ecosystems are no exception. They play a key role in pollination and natural pest control, but it is a fact that most of the agricultural pests are insects (LaSalle, 1999). Thus, insect biodiversity form a vital component of a variety of ecosystem functioning.

According to the status report of India's faunal wealth, large proportion of Indian fauna constitute Insecta (more than 40% of the total fauna) (Anonymous, 1991). But this faunal diversity is under estimated for improper surveys and lack of systematic identifications, particularly insects whose only larval and pupal stages occur in aquatic habitats (Brij Gopal, 1997). A number of works worked on insect diversity on various ecosystem levels (Engles and Wood, 1999; Wood and Lenne, 1999; Chesalmah *et al.*, 2000; Ogbelbu, 2001).

Barak valley-southern part of Assam belongs to North-east hotspot area and Himalayan foothills where the study is being carried out. A great insect diversity was noticed from rice agro-ecosystem. Considering the importance of insects in agro-ecosystem an attempt has been to study the diversity and composition of insects during different seasons.

Description of the Study Sites

The present investigation was undertaken in Cachar district—southern part of Assam. The geographical location of study sites are situated at longitudes of 24°41'29.9" N and latitude of 92°45'25.9" E. Altitude of the study areas are about 36 meter above msl. The entire study areas are low plain land and wetland with small hillocks. Rainfall is moderate to high. Paddy is cultivated in medium extent and mostly depends upon climatic conditions. The region has a plenty of faunal biodiversity. Vegetation is evergreen, semi-evergreen and subtropical. The overlying soil cover is mostly sandy laterite. In most of the areas land has poor nutrient status except the riverine plains of Barak River. Most of the hillocks are covered by tea plantation. Paddy is cultivated in the low plains between the hillocks.

Materials and Methods

Five sites *viz.*, Irongmara, Machghat, Bariknagar, Dudpatil and Binnakandi were selected for the study purpose during 2003-04. A quadrat (1m. × 1m) was used to collect the insect sample as a standard device (Ghosh, 1985). The diversity of insects were determined by counting the numbers and kept in 70 per cent alcohol for preservation. The known insects were released to the field in living conditions and unknown specimens were kept in preservative for identification and further study. The sample was drawn at an interval of ten days from all the sites. From each site five quadrats were drawn randomly from 0.5 hectare cropland during the Boro season.

Diurnal studies were conducted in Irongmara paddy field during Amon, Aus and Boro seasons on Sataki, Mayamoti and Basful varieties, respectively. The period of study were confined to 7-17 hrs. at Amon season, from 6 to 18 hrs. at Aus and Boro seasons at an interval of one hour. Here also five quadrats of same size were randomly drawn from 0.5 hectare of cropland.

Shannon - diversity index (H) was employed with the following formula (Shannon and Weaver, 1949):

$$H = -\Sigma \ Pi. \ Log \ Pi$$

Sorensen Co-efficient (Similarity) (Cs) was employed for the determination of similarity and dissimilarity of insects between communities of different seasons by the following formula (Sorensen, 1948):

$$Cs = \frac{2a}{(2a + b + c)}$$

't' - test was employed to assess the significance of diurnal activities of insects at various peak hours.

Results and Discussion

Group-wise abundance of insects was incorporated in Tabie-1. As regards the abundance it was found that more dominant groups was Coleoptera (35-53%) followed by Hemiptera (9-37%) in all the sites out of seven available orders. Lowest group was considered as Odonata (4-14%) (Table-1). Shannon diversity index indicated moderate diversity of insects in all the study sites during Boro season only (Table-2). Whereas season-wise diversity index (H) indicated also moderate diversity of insects between the community (Table-3).

Table-1: Group-wise Diversity of Insects on Rice Ecosystem during Boro Season (2003)

	Sites				
Orders	**Irongmara**	**Machghat**	**Bariknagar**	**Dudpatil**	**Binnakandi**
Coleoptera	35.55%	52.14%	47.37%	53.40%	36.49%
Lepidoptera	19.05%	3.11%	1.75%	4.85%	0
Hymenoptera	3.81%	3.11%	3.51%	0	5.40%
Diptera	3.49%	19.07%	0	0	2.70%
Hemiptera	12.38%	9.34%	0	33.01%	37.84%
Orthoptera	21.59%	5.06%	33.33%	3.88%	12.16%
Odonata	4.13%	8.17%	14.03%	4.85%	5.40%

Note: '0' means nil.

Table-2: Showing the Shannon-diversity Index (H) in Five Sites of Barak Valley (2003)

	Boro Season	
Study Sites	**Diversity Index**	**Paddy Varieties Studied**
Irongmara	H = 1.72547	Basful
Machghat	H = 1.9050	Basful
Barik nagar	H = 1.5441	Tupa Sali.
Dudpatil	H = 1.6339	Asatti
Binnakandi	H = 1.6757	Jaya

Table-3: Shannon-diversity Index (H) Showing Insect Communities during Three Seasons (2003-04)

Diversity index	Seasons		
	Aus	Amon	Boro
Shannon diversity index (H)	1.8694	1.8765	1.8581

As regards the Sorensen Co-efficient (Cs) of similarity between the two seasons it was found more similarity (0.75-0.92) of insect population distribution (Table-4). Diurnal activities of insect population was studied during three seasons (Tables 5-7 and Fig. 1). During Aus season study was undertaken at hourly interval from 6-18 hrs. Maximum activities were recorded at 6-7 hrs. (67.0) and 17-18 hrs. (66.0). Most part of the day the activities if insects were moderate (Table-5). Diurnal activities were studied during Amon season, which indicated maximum activity (115.0) at 15-16 hrs. followed by 14-15 hrs. (100.0). Least activity 57.0 was noticed at 7-8 hrs. (Table-6). In Boro season the peak activity was recorded 159.0 at 17-18 hrs. followed by 16-17 hrs. where the insect population was recorded 132.0. The lowest activity (78.0) was recorded at 7-8 hrs. (Table-7). However, a 't' - test was employed between the peak hour activity of insects between two seasons. Although the peak hour activities varied in different seasons it did not significantly different from Aus vs. Amon vs. Boro and Amon vs. Boro ($P < 0.05$) (Table–5-7).

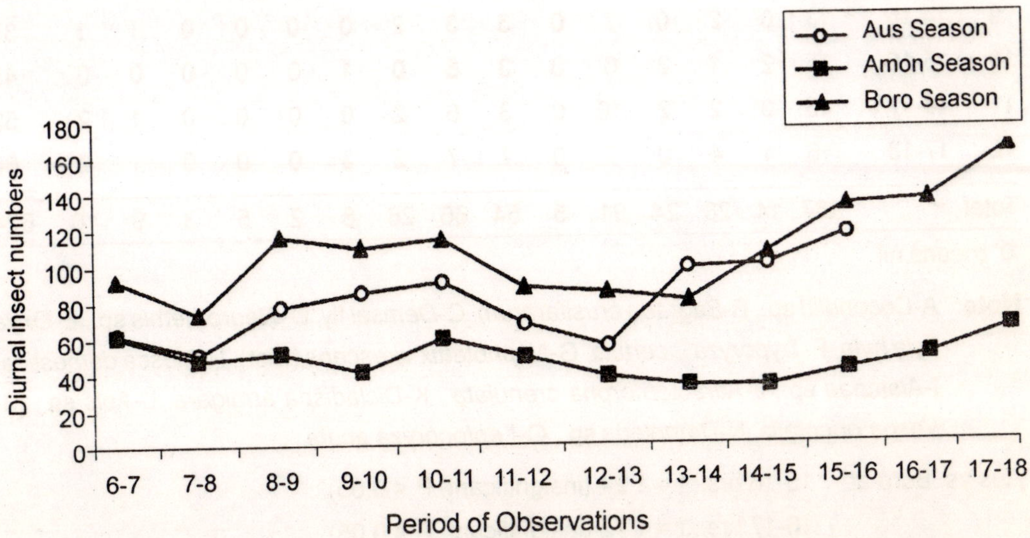

Fig. 1: Showing Diurnal Insect abundance during three seasons (2003-04)

Table-4: Sorensen Co-efficient of Insect Communities during Three Seasons (2003-04)

Soresen Co-efficient (Cs)	Aus/Amon	Amon/Boro	Boro/Aus
Similarity	0.758	0.846	0.928
Dissimilarity	0.242	0.154	0.072

Table-5: Diurnal Variation of Insects from Paddy field of Irongmara during Aus Season (2003)

No. of obs.	Time of obs.	A	B	C	D	E	F	G	H	I	J	K	L	M	N	O	Total
1	6-7	18	5	2	2	5	3	7	6	1	1	0	0	0	0	0	67
2	7-8	13	2	3	2	5	0	3	4	5	2	2	0	0	0	0	54
3	8-9	17	1	2	2	8	0	2	7	3	0	0	1	1	1	0	57
4	9-10	17	1	4	2	4	0	4	4	3	0	0	2	1	0	0	47
5	10-11	10	1	3	7	11	1	7	6	3	0	0	1	1	1	2	64
6	11-12	13	1	2	4	10	0	6	5	3	0	0	0	0	1	2	54
7	12-13	12	0	1	1	8	1	4	7	0	1	0	1	0	2	0	43
8	13-14	8	0	2	0	10	0	5	5	2	0	0	0	0	1	1	38
9	14-15	13	0	2	0	7	0	3	3	2	0	0	0	0	1	1	37
10	15-16	15	2	1	2	6	0	3	6	0	1	0	0	0	0	0	45
11	16-17	15	0	2	2	10	0	3	6	2	0	0	0	0	1	2	52
12	17-18	16	1	4	0	7	0	7	7	2	3	0	0	0	1	2	66
Total		167	14	28	24	91	5	54	66	26	8	2	5	3	9	10	624

'0' means nil.

Note: A-*Coconelid* sp., B-*Bagrada crusifararum*, C-*Demsel* fly, D -*Neurothemis* sp., E-*Oxya hyla hyla*, F-*Tryporyza incertula*, G-*Nephotettix virescens* (Dist), H-*Musca domestica*, I-*Alstonae* sp., J-*Attractomorpha crenulata*., K-*Dicladispa armigera*, L-*Apis* sp., M-*Vespa orientalis*, N-*Demoleus* sp., O-*Leptocoryza acuta*.

Aus vs. Boro at 15-16 hrs.: $t = 1.24$ (insignificant: $P < 0.05$)

16-17 hrs.: $t = 1.72$ (insignificant: $P < 0.05$)

17-18 hrs.: $t = 0.91$ (insignificant: $P < 0.05$)

Table-6: Diurnal Variation of Insects from Paddy Field of Irongmara during Amon Season (2003)

No. of obs.	Time of obs.	A	B	C	D	E	F	G	H	I	J	K	L	M	N	Total
1	6-7	12	22	6	0	2	4	0	8	2	10	2	0	0	0	68
2	7-8	5	21	7	0	0	0	4	2	0	13	5	0	0	0	57
3	8-9	16	39	0	0	0	4	2	0	5	13	2	0	0	0	81
4	9-10	20	44	2	1	0	3	4	0	0	10	3	0	0	1	88
5	10-11	13	51	2	1	0	0	0	7	3	12	4	0	0	0	93
6	11-12	12	31	1	1	0	0	5	2	5	10	3	0	1	0	71
7	12-13	10	23	3	2	0	3	2	8	0	4	3	1	0	0	59
8	13-14	14	39	3	4	0	0	3	1	8	22	5	0	0	0	99
9	14-15	20	39	1	14	1	3	0	3	3	16	0	0	0	0	100
10	15-16	14	51	8	19	1	6	1	5	3	7	0	0	0	0	115
Total		136	360	33	42	4	23	21	36	29	117	27	1	1	1	831

'o' meand nil

Note: A-*Oxya hyla hyla*, B-*Coccinelid* sp., C-*Neurothemis* sp., D-*Attractomorpha* sp., E-Demsel fly, F-*Leptocrozya acuta*, G-*Sitophilus oryzae*, H-*Musca domestica*, I-*Bagrada crusiferarum*, J-*Tryporzya incertula*, K-*Galerucella* sp., L-*Demoleus* sp., M-*Dicladispa armigera*, N-*Apis* sp.

Aus vs Amon at 15-16 hrs. : $t = 1.19$ (insignificant: $P < 0.05$)

Table-7: Diurnal Variation of Insects from Paddy Field of Irongmara during Boro Season (2003-04)

No. of obs.	Time of obs.	A	B	C	D	E	F	G	H	I	J	K	L	M	Total
1	6-7	49	11	4	1	3	9	12	2	2	0	0	3	0	96
2	7-8	43	9	3	2	5	4	7	3	0	1	0	1	0	78
3	8-9	60	8	6	4	5	11	14	2	3	0	0	3	1	117
4	9-10	55	9	3	5	12	6	9	4	3	1	2	2	0	111
5	10-11	63	5	2	3	12	4	17	1	1	1	2	4	0	115
6	11-12	47	5	2	9	6	4	8	1	2	1	3	2	0	90
7	12-13	42	8	5	3	8	3	10	1	5	0	0	1	1	87
8	13-14	41	8	6	3	7	3	6	2	2	1	2	1	0	82
9	14-15	51	8	5	7	6	6	10	0	4	3	0	5	1	106
10	15-16	59	8	5	10	13	9	9	3	5	1	2	6	0	130
11	17-18	61	8	9	14	12	11	6	7	0	0	0	4	0	132
12	18-19	88	7	11	7	11	4	16	2	2	1	4	6	0	159
Total		659	94	61	68	100	74	124	28	29	10	15	38	3	1303

'0' means nil.

Note: A-*Coccinelid* sp., B-*Leptocoryza actua,* C-*Neurothemis* sp., D-*Oxya hyla hyla,* E-*Tryporyza incertula,* F-*Musca domestica,* G-*Dicladispa armigera,* H-*Bagrada crusiferarum,* I-Demsel fly, J-*Attractomorpha crenulata,* K-*Nephotettix virescens* (Dist.), L-*Alstonae* sp., M-*Demoleus* sp.

Amon vs. Boro at 15-16 hrs: $t = 0.0043$ (insignificant: $P > 0.05$).

During three crop seasons the insect density was studied m^2 area basis (Table-8). Study indicated that maximum numbers were *Cocinelid* sp. in all the seasons. Peak population was recorded as $659/m^2$ during Boro season followed by $360/M^2$ during Amon season. Least numbers ($167/m^2$) were recorded in Aus season. The second dominant insect was *Tryporyza incertulus* was found in all the study seasons. Some insects *viz., Sitophilus oryzae, Gallerucella* sp. and *Vespa orientalis* were not found in all the seasons. Among the crops, rice cultivation is predominant in Barak Valley of Assam. Insect diversity and abundance are very rich in rice agro-ecosystem. It is perhaps tropical area, moderate to high humidity and rainfall prevails during the rainy season which helps in multiplication of insect population. Group-wise study reveals that out of seven orders Coleoptera occupies the dominant group. Ogbeibu (2001) studied group-wise insect on aquatic ecosystem in four sites in Nigeria. Site-wise study of Shannon diversity index (H) reveal moderate to high diversity in all the sites but in season-wise it also proves moderate to high diversity. Diversity index also reported by Ogbeibu (2001) and Nautiyal *et al.* (2000) on aquatic system and found higher diversity index.

Table-8: List of Insects with Density (m^{-2}) in Paddy field during Three Crop-seasons (2003-04)

Taxa Scientific Name	Common Name	Aus	Amon	Boro
Coccinelid sp.	Lady brid beetle	167	360	659
Alstonae sp.	Blue beetle	26	-	38
Dicladispa armogers O1	Rice hispa	2	1	124
Sitophilus orzyae Linn.	Rice weevil	-	21	-
Galerucella sp.	-	-	27	-
Tryporyza incertula	Yellow rice borer	5	117	100
Demoleus sp.	Butter fly	9	1	3
-	Demsel fly	28	4	29
Neurothemis sp.	Dragon fly	24	33	61
Oxya hyla hyla.	Grasshopper	91	136	68
Attractomorpha cremulata	Grasshopper	8	42	10
Bagrada crusiferarum	-	14	29	28
Leptocorzya acuta	Mahua	10	23	94
Nephotettix viresecens (Dist)	Green leaf hopper	54	-	15
Musca domestica	Fly	66	36	74
Apis sp.	Bee	5	1	-
Vespa orientalis	Wasp	3	-	-

Note: '–' means nil.

Soresen Co-efficient of similarity reveals high similarity of insect population among the three seasons. Ogbeibu (2001) reported significant dissimilarity among the four study sites. This may be because of aquatic ecosystem where the work is performed. As regards the diurnal study it reveals out of three seasons maximum activates (159.0) reported during Boro season (2003-04) at 17-18 hrs. followed by Amon season (115.0) at 15-16 hrs (Fig. 1). This may be because of less temperature preferred by insects which prevails at that time. 't' - test was employed to assess the activities of insect among the peak hour period between Aus vs. Boro, Aus vs. Amon and Amon vs. Boro seasons. It proves the difference of activities insignificant ($P < 0.05$) among all the seasons although the numbers of insect abundance is varying at different hour intervals. Density of insect taxa is found to be maximum in case of *Coccinelid* sp. (659/m^2) in Boro season followed by *Oxya hyla hyla* (136/m^2) in Amon season. It may be noted that in all the three seasons *Coccinelid* sp. is found to the dominating. Dominating numbers recorded may be because of species which can sustains in all types of environmental conditions. *Galerucella* sp., *Vespa orientalis* species may not available because of either seasonal insect or non-availability of food in off-season.

References

Anonymous, 1991. *Animal Resources of India: Protozoa to Mammalia*. State of the Art. Zoological Survey of India, Kolkata. pp. **694.**

Chesalmah, M.R.; Hassan, S.T.S. and Abu Hassan, A. 2000 Local movement and feeding pattern of adult *Neurothemis tullia* (Drury) (Odonata: Libellulidae) in a rainfed rice field. *Tropical Ecology,* **41(2): 233-241.**

Engles, J.M.M. and Wood, D. 1999. *Conservation of Agro biodiversity*. CAB International Agro biodiversity. **355-385.**

Ghosh, A.K. 1985. *Sampling techniques in rice research*. ICAR Publication. New Delhi, pp. **623.**

Gopal, B. 1997. Biodiversity in Inland Aquatic Ecosystems in India: An overview. *International Journal of Ecology and Environmental Sciences.* **23(4): 305-313.** International Scientific Publications, New Delhi.

LaSalle, J. 1999. *Insect Biodiversity in Agroecosystems: Function, value and optimisation*. CAB International Agrobiodiversity. **155-182.**

Nautiyal, C.; Nautiyal, P. and Singh, H.R. 2000. Species richness and diversity of epilithic diatom communities on different natural substrates in the cold water river Alaknanda. *Tropical Ecology,* **41(2): 255-258.**

Ogbeibu, A.E. 2001. Composition and diversity of Diptera in temporary pond in southern Nigeria. *Tropical Ecology*. **42(2): 259-268.**

Shannon, C.E. and Weaver, W. 1949. *The mathematical theory of communication*. Urbana, IL.: University of Illinoid Press.

Soresen, T. 1948. A method of establishing groups of equal amplitude in plant sociology based on similarity of species count and its application to analyses of the vegetation on Danish commons. *Biol. Skr.* (K. Danske Vidensk. Selsk. NS) **5, 1-34.**

Wood, D. and Lenne, J.M. 1999. *Agrobiodiversity and Natural biodiversity: Some Parallels*. CAB International Agrobiodiversity. **425-445.**

❑❑❑

5

Thermal and pH Stability of Dibutyl Phthalate, An Antimetabolite of Proline from *Streptomyces Albidoflavus* 321.2

R.N. Roy, S.K. Sen
School of Life Sciences, Department of Botany,
Visva-Bharati, Santiniketan–731 235

Introduction

Antimetabolites are the compounds, which are able to repress the mechanism for synthesis of cellular metabolites from simple nutrients; this repression can be reversed by concurrent administration of one or more biochemicals (Pruess and Scannell, 1974; Hossain *et al.*, 1987). Several antimetabolities are effective antimicrobials and showed their effectiveness in cancer treatment (Hossain *et al.*, 1987; Rosangkima and Prasad, 2004).

Antimicrobial (El-Naggar, 1997; Lee, 2000) efficacy and antimetabolic (Roy *et al.*, 2005) activity of dibutyl phthalate (DBP) have been studied from *Streptomyces*. This antimetabolic activity of DBP was found to overcome by application of proline. Being an important amino acid, proline plays a notable role in cell metabolism especially under stress *e.g.* water (Materinez *et al.*, 2004), heat (Rivero *et al.*, 2004) and salt (Tonon *et al.*, 2004). DBP is also effective as peroxisome proliferator (O'Brein *et al.*, 2001), controlling demodicidosis (Yuan *et al.*, 2001) and as drug channeling agent (Makhija and Vaiva, 2003). The compound shows no acute toxicity on the plant and animal system (Roy and Sen, 2004). Thus, DBP becomes an important compound for proper understanding of its role in the study of cell physiology.

This paper deals with the thermal and pH stability of DBP, an antimetabolite of proline from *Streptomyces albidoflavus* 321.2.

Materials and Methods

Microorganism—The producer, *Streptomyces albidoflavus* 321. 2 (Roy and Sen, 2002) a new soil isolate was maintained on Glucose asparagine agar (Krainsky, 1941); the test organism *Escherichia coli* ATCC. 25922 was grown on Nutrient agar medium (Hi Media). Organisms were maintained at 4°C.

Fermentation—Spores (1.75 x 10^7/ml) of the isolate 321.2 was inoculated into 100 ml of medium (pH 6.75) composed of 0.025% $NaNo_3$, 2.18% Glycerol, 0.05% KH_2PO_4, 0.15% NaCl, 0.03% $MgSO_4$ $7H_2O$, 0.001% Fe_2 $(SO_4)_3$ $6H_2O$, 0.0001% $CuSO_4$ $5H_2O$, 0.001% $MnSO_4$ H_2O, 0.0001% $ZnSO_4$ $7H_2O$ in 500 ml of Erlenmeyer flask, sterilised at 120±1°C and cultured in stationary condition at 32°C for 6 days.

Isolation and Purification of Active Compound—The active compound was isolated from fermented broth by solvent extraction method, using ethyl acetate (3:1). The crude antimicrobial principal was purified by ascending paper chromatography using a mixture of petroleum ether (60-80°C), ethyl acetate and methanol in the ratio of 10:0.5:1 as mobile phase. Using end guide strip, the active band was detected by bioassay. The semi purified compound, from the detected band, was eluted in ethanol. Then the compound was further purified by silica gel column of 10×2 cm and was eluted with methanol under a flow rate of 3ml/min. Fractions (of 1 ml each) were collected separately and the activity was detected by bioassay. The active fractions were pooled together and evaporated. One-dimensional TLC was performed using benzene: ethyl acetate (1:1) as mobile phase. The bands were detected by keeping the plate in iodine vapour chamber and were eluted with methanol: ethyl acetate (1:1). Confirming homogeneity of compound, the activity was detected by bioassay.

Determination of pH Stability—pH stability of DBP was determined as a measurement of retained activity after incubation at a given pH. The purified antibiotic principal (500 μg/ml) was treated in buffer (Sadasivam and Manickam, 2004) of pH between 2.0-10.0 and was incubated for 24 hr at 35°C. The loss of antibiotic activity was measured by bioassay.

Determination of Thermal Resistance—Thermal resistance was determined by exposing the purified DBP (500 μg/ml) at different temperatures (30-121°C) for 5 min and 10 min period. The loss of antimicrobial activity was detected by bioassay in each case.

Antimicrobial activity was determined by agar cup method (Higashide *et al.*, 1971), using glucose-asparagine agar along with 1 ml cell suspension (1.2×10^6 CFU) of the test organism. Each cup was filled with purified principal (0.1 ml) and incubated at 30°C for 24 hr.

Results and Discussion

Commercial feasibility of an active compound depends on maintenance of its self-life. The compound was quite stable in varied pH and temperature. The loss of activity of DBP was more pronounced in alkaline pH (Fig.-1) and 50% loss was recorded at pH 9.5. Augustine *et al.* (2005) reported about the effect of pH and temperature on the activity and stability of the antibiotices from the *Streptomyces albidoflavus* PU 23 where the active compound was incubated at pH in the range 5.7-8.0, maximum activity was observed at a pH range 6.8-8.0, the activity decreased at pH 6.3 to downward. Thermal loss of activity of DBP was only 32% at 100°C and retained its 100% activity up to 50°C (Table-1).

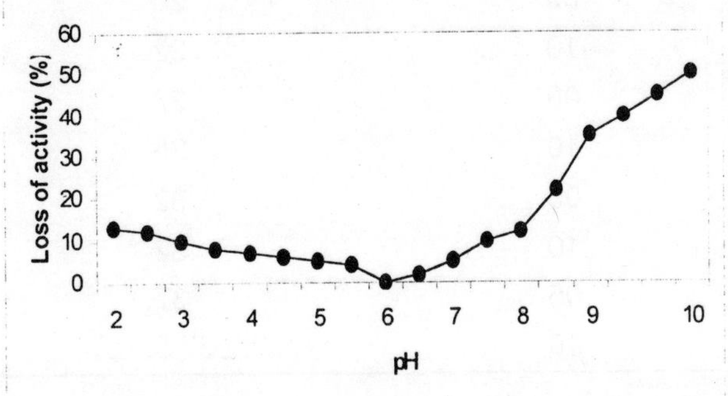

Fig. 1: Effect of pH on antimicrobila activity of DBP

The non-polyene antimicrobial principal from isolate PU 23 was stable at different temperature (30-80°C), however the principal compound lost its activity completely when treated in an autoclave at 121°C for 15 minutes (Augustine *et al.*, 2005) but DBP lost only 37.5% of its activity. So, DBP is less susceptible to degradation and/or modification in change in pH and temperature.

The active principal from the isolate 321.2, defined as dibutyl phthalate ester, was found stable probably due to its aromatic structure with no free reactive group. Further studies regarding the effect of various parameters on the activity of DBP are currently in progress.

Table-1: Effect of temperature on antimicrobial activity of DBP

Temperature (°C)	Treatment period (min)	Loss of activity (%)
30	05	0
	10	0
40	05	0
	10	0
50	05	0
	10	0
60	05	10
	10	15
70	05	15
	10	18
80	05	20
	10	22
90	05	22
	10	25
100	05	32
	10	32
121	05	37
	10	37.5

References

Augustine, S.K.; Bhavsar, S.P. and Kapadnis, B.P. 2005. A non-polyene antifungal antiontic from *Streptomyces slbidoflavus* Pu 23. *J. Biosci.* **30: 201-211.**

El-Naggar, M.Y.M. 1997. Dibutyl phthalate and the antitumor agent F5 AI, two metabolites produced by *Streptomyces nasri* submutant H35. *Biomed. Lett.* **55: 125-131.**

Higashide, E.; Fugono, T.; Hatano, K. and Shibata, M. 1971. Studies on T 2636 antibiotics. I Taxonomy of *Streptomyces rochei* var. *volubilis* var nov. and production of the antibiotics and an esterase. *J. Antibiot.* **24 : 1-12.**

Hossain, M.; Dastidar, S.G. and Chakrabarty, A.N. 1987. Antimetabolic activity of some commonly used drugs. *Indian J. Exp. Biol.* **25: 866-868.**

Krainsky, A. 1941. *Die Aktinomyceten und ihre Beduting in der Natur Center* Bakterio. *Parasitenk,* **41 : 649-688.**

Lee, D.S. 2000. Dibutyl Phthalate, an-Glucosidase Inhibitor from *Streptomyces melaniosprfaciens. J. Biosci. Bioeng.* **89 : 271-273.**

Makhija, S.N. and Vavia, P.R. 2003. Controlled porosity osmatic pump-based controlled released systems of pseudoephedrine. I. Cellulose acetate as a semi permeable membrane. *J. Control Release*. **89: 5-18.**

Martinez, J.P.; Lutts, S.; Schanck, A; Bajji, M. and Kient, J.M. 2004. Is osmotic adjustment required for water stress resistance in the Mediterranean shrub *Atriplex halimus* L? *J. Plant Physiol.* **161: 1041-1051.**

O'Brein, M.L.; Cunningham, M.L., Spear, B.T. and Glauert, H.P. 2001. Effects of peroxisome proliferators on glutathione and glutathion-related enzyme in rates and hamsters. *Toxicol. Appl. Pharmacol.* **171: 27-37.**

Pruess, D.L. and Scannell, J.P. 1974. Antimetabolites from microorganisms. *Adv. Appl. Microbiol.* **17 : 19-62.**

Rosangkima, G. and Prasad, S.B. 2004. Antitumour activity of some plants from Meghalaya and Mizoram against murine ascites Dalton's lymphoma. *Indian J. Exp. Biol.* **42: 981-988.**

Rivero, R.M.; Ruiz, J.M. and Romero, L.M. 2004. Importance of N source on heat stress tolerance due to the accumulation of proline and quaternary ammonium compounds in tomato plant. *Plant Biol. (Stutta),* **6: 702-707.**

Roy, R.N. and Sen, S.K. 2002. Survey of antimicrobial streptomycetes from soils of West Bengal: Characterisation and identification of potent broad spectrum antiotic producing *Streptomyces albiboflavus* 321.2, *Hind. Antibiot. Bull.* **44: 25-33.**

Roy, R.N. and. Sen, S.ZK. 2004. Evaluation of biological activity of dibutyl phthalate produced by *Streptomyces albidoflavus* 321.2. *J. Sci. Technol,* Sambalpur University, **16(A) : 125-132.**

Roy, R.N.; Laskar, S. and Sen, S.K. 2005. Dibutyl phthalate, the bioactive compound produced by *Streptomyces albidoflavus* 321.2. *Microbial. Res. Germany* (in press).

Tonon, G.; Kevers, C.; Faivre-Rampant, O.' Grazianil, M. and Gaspar, T. 2004. Effect of NaCl and mannitol iso-osmotic stresses on proline and free polyamine levels in embryogenic *Fraxinus angustifolia* callus. *J. Plant Physiol.* **161: 701-708.**

Sadasivam, S. and Manickam, A. 2004. *Biochemical Methods*. Second Edition. New Age International Publishers. pp. 246.

Yuan, F.S.; Guo, S.L.; Qin, Z.X.; Deng, S.H. and Huang. G.H. 2001. Effect of dibutyl phthalate on demodicidosis. *Zhongguo Ji Sheng Chong Xue Yu Ji Sheng Chong Bing Za Zhi*, **19 : 160-162.**

□□□

6

Effect of Foliar Application of Potassium Nitrate, Pyridoxine and Salicyclic Acid on Internal Leaf Water Relation and Seed Yield of Toria (*Brassica compestris* L.)

Milu Rani Das and K. Chandra
Department of Crop Physiology,
Assam Agricultural University
Jorhat- 785 013 (Assam)

Introduction

The oil seed crops play an important role in national economy of the country. Out of nine oil seed crops produced in India, rapeseed and mustard occupies second positions after groundnut. The productivity of rapeseed and mustard in Assam is far below the national average. The low productivity of the crop is accounted for various factors, in spite of an energy rich crop, rapeseed mustard are grown largely under energy starvation conditions. Moreover, the crop mostly cultivated in marginal and sub-marginal lands that to as rainfed crop, because of the non-availability of irrigation facilities during rabi season. Furthermore there is no suitable cultivar or genotype, which could be grown well under moisture stress conditions with poor fertility level. However in recent years variety of agro-chemicals in solution are directly sprayed on foliages of the crop plant, since they are readily absorbed and utilised more efficiently than soil application. Foliar fertilisations of chemicals involve a number of physiochemical changes, which lead to improve the interval water balance and ultimately increase the yield of the crops (Kumari and Bharti), 1988). Moreover information on the foliar application of chemicals, time of applications, optimum concentration and effect on metabolisms, growth development and yield of toria are scanty under the agro-ecological conditions of Jorhat (Assam). Keeping the above in mind our attempt has been made to sort out the most efficient and economical chemical and their concentration, which can improve the physiological efficiency and productivity potential of toria under rainfed ecosystem of Assam.

Materials and Methods

An experiment was conducted in field during *rabi* season in Instructional-cum-Research Farm, Department of Crop Physiology, Assam Agricultural University, Jorhat in 1999-2000 to investigate the effect of foliar fertilisation of a various chemicals on growth efficiency and productivity of rapeseed cv. M-27 under rainfed conditions. All together, there were 14 treatments comprised of KNO_3 5000, 10000, 15000, 20000 ppm, pyridoxine 0.05, 0.10, 0.15 and 0.20 ppm and salicylic acid 10, 100, 500, and 1000 ppm including two control with and without water treatment respectively. The foliar spraying was done twice, once at 15 days after sowing and another after 50% flowering. All treatments were replicated three times in a randomised Block Design. Seeds were sown in lines 30 cm apart at the rate of 10 kg/ha, plant to plant distance were maintained 10 cm by thinning. The soil of the field was sandy loam with pH 5.4, low in available nitrogen (294.24 g/ha), phosphorus (20.75 kg/ha) and potassium (52.40 kg/ha). The fertilisers doses of 40 kg N (as Urea), 35 kg P_2O_5 (as single super phosphate) and 15kg K_2O (as muriate of potash) per hectare were mixed together and applied to experimental area along with 10 kg borax per hectare. Ten plants having uniforms growth were selected randomly within one square meter area at the centre of each plot and marked for various observations. The water stress related parameters like RLWC, WSD and RSD were calculated following the formulas of Barrs and Wheatherly (1962). Free proline was estimated in leaves following the method of Bates *et al.* (1993), and *in vivo* nitrate reductase assay was done following the method of Jawarski (1971) and Kleeper *et al.* (1977). The influence of chemicals on internal water balance in conjunction with yield parameters are discussed here under agro-ecological conditions of Assam.

Results and Discussion

The relative leaf water content was significantly influenced by the applied chemicals at successive growth stages (Fig. 1). Moreover, a decreasing trend of RLWC with the advancement of crop age was observed. The highest 86.01%, RLWC was recorded with KNO_3, 10000 ppm concentration followed by KNO_3, 15000 ppm (85.77%), KNO_3, 20000 ppm (85.62%) and KNO_3, 5000 ppm (85.46%) at pre-anthesis stage, while at post anthesis stage KNO_3, 15000 ppm treated plants maintained highest 81.82% RLWC followed by KNO_3, 10000 ppm (81.51%). Among the next best treatments tested pyridoxine, 0.15 ppm, salicylic acid, 100 ppm were at par with each other and significantly superior over control at post anthesis stage.

Fig. 1: Effect of KNO_3, Pyridoxine and Salicylic acid on relative leaf water content (RLWC) at pre- and postanthesis stages

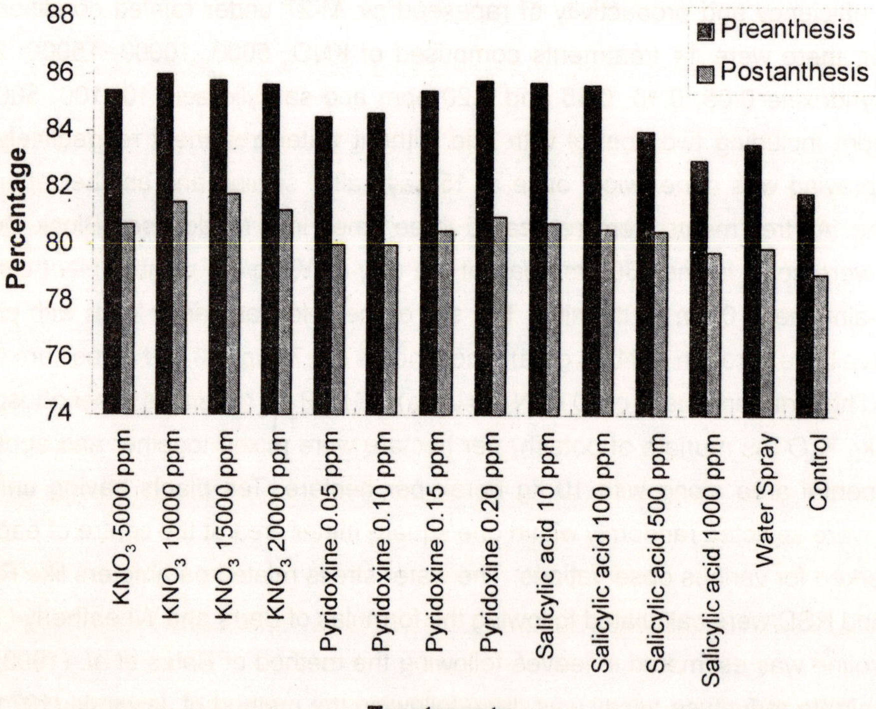

An increasing trend in WSD and RSD was observed with advancement in crop age (Fig. 2). The lowest 13.49 per cent and 13.99 per cent RSD and WSD was recorded with KNO_3, 10000 ppm treated plants at preanthesis stage respectively, whereas at post anthesis stage, it was lowest 14.88 per cent and 18.17 per cent RSD and WSD

Fig. 2: Effect of KNO_3, Pyridoxine and Salicylic acid on relative saturation defect (RSD) at pre- and postanthesis stages

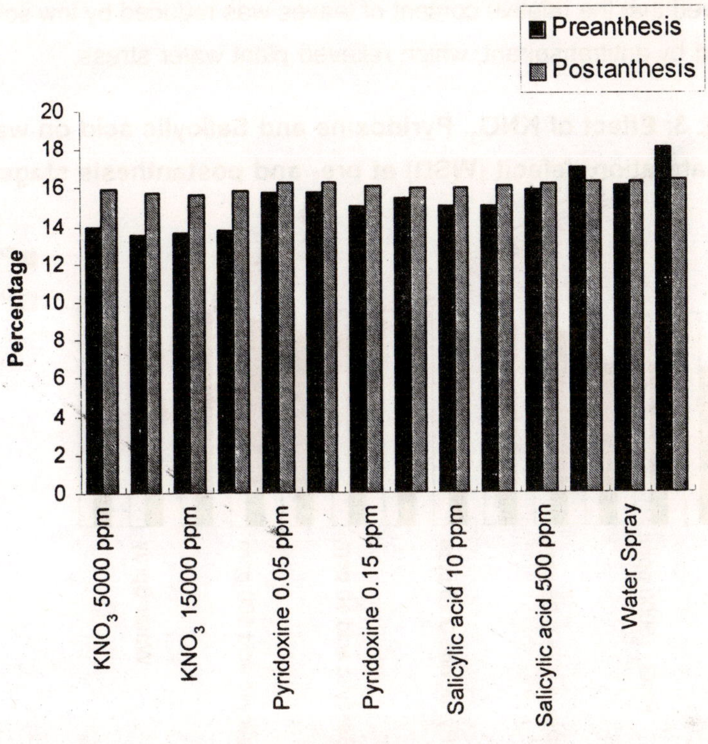

with KNO_3, 15000 ppm respectively. Furthermore, pyridoxine, 0.15 ppm and salicylic acid, 10 ppm treatment was at par with each other and differed significantly over control at preanthesis stage (Fig. 2). Similarly pyridoxine, 0.05 ppm and 0.10 ppm was at par with each other and showed superiority over control in respect of RSD and WSD at postanthesis stage. The RLWC, RSD and WSD in leaves appeared to better indicator for water status of plant as because, RLWC through its relation to cell volume more closely reflects the balance between water supply to the leaves and transpiration rate (Similair and Ludlow, 1985). Thus from the present investigation, it is evident that due to foliar application of chemicals like KNO_3 and pyridoxine not only improve the internal water status of plants, but also growth vigour which is essentially important for maintenance of turgidity and plant growth. These results are in close confirmation with the findings of Tripathi *et al.* (1995) in mustard. Plants well positive with potassium lost less water since potassium increased osmotic potential and had a positive influence on stomatal closure as well as (Gardener *et al.*, 1985). The application of aspirin spray reduces the consumptive use of water and increased the water efficiency in groundnut (Dhawse *et al.,* 1993). Patil (1993) observed that the relative content of leaves was reduced by low soil moisture but was increased by antitranspirant, which relieved plant water stress.

Fig. 3: Effect of KNO_3, Pyridoxine and Salicylic acid on water saturation defect (WSD) at pre- and postanthesis stages

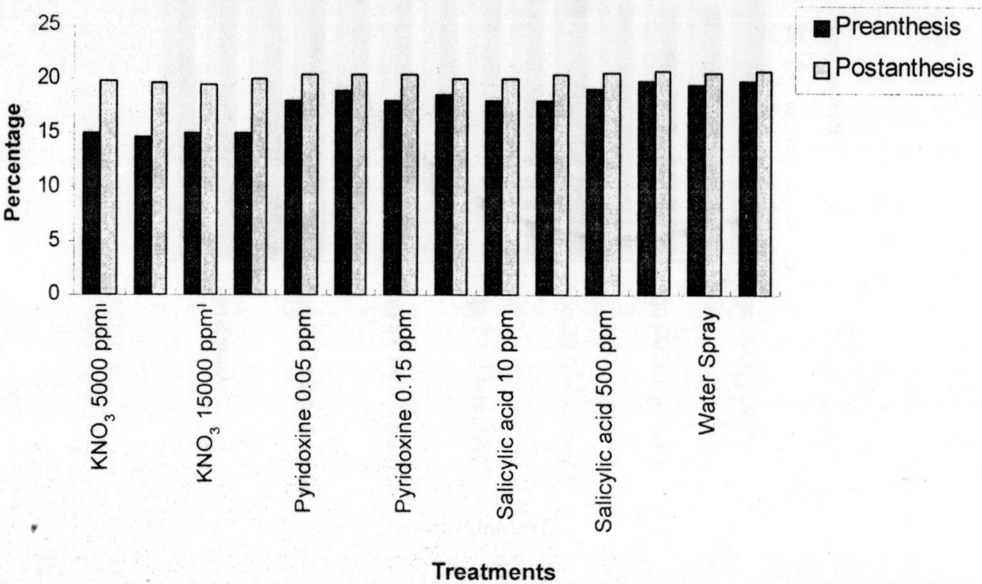

Data portrayed in Fig 4, showed that all the applied treatments had significant effect on NR activity both at pre- and postanthesis stage. The highest 73.98μ mol $NO_2 g^{-1}$ F.W. hr^{-1} NR activity was obtained with KNO_3, 15000 ppm treatment followed by KNO_3, 10000 ppm (73.91μ mol NO_2 g^{-1} F.W. hr^{-1}), while the same was found lowest 60.46μ mol NO_2 g^{-1} F.W. hr^{1} with salicylic acid, 1000 ppm concentration at preanthesis stage of crop growth. At postanthesis stage, highest 64.95 μ mol NO_2 g^{-1} F.W. hr^{-1} NR activity was recorded with KNO_3, 15000 ppm treated plants followed by KNO_3, 10000 ppm (64.82μ mol NO_2 g^{-1} F.W. hr^{-1}), KNO_3, 2000 ppm (64.81μ mol $N0_2$ g^{-1} F.W. hr^{-1}), and pyridoxine, 0.2 ppm (64.18μ mol NO_2 g^{-1} F.W. hr^{1}), while the lowest was recorded 52.12μ mol NO_2 g^{-1} F.W. hr^{1} in control plants. However, salicylic acid, 10 ppm and 100 ppm significantly increased the NR activity over control which has 63.92μ mol NO_2 g^{-1} F.W. hr^{1} and 63.06μ mol $NO_2 g^{-1}$ F.W. hr^{-1} NR activity at postanthesis stage respectively.

The highest activity of the enzyme appeared to be due to be a large amount of enzyme being present in the extract or due to a greater proportion of the enzyme being in an active form or due to both in the present investigation. In fact, NR activity in the leaf is controlled by demand for the products of nitrate assimilation in different metabolic pathway, which is controlled by age of leaves, duration and intensity of light. NR activity was also enhanced by foliar application of leaves, urea, DAP and K_2SO_4 (Kalarani and Teya Kumar, 1998). The increased NR activity by foliar application might be because of enhancement of N uptake and photosynthetic activity due to higher chlorophyll content, the essential one for increased enzyme activity. At lower concentration of salicylic acid increased the *in vivo* activity of nitrate reductose in maize seedling (Jain and Srivastava, 1981). Hence from the above, it is clear that salicylic acid increased nitrate reductose activity indirectly, by protecting the enzyme from inactivation. However, at higher concentration it inhibit the activity, it may disturb membrane potential, affect ion influx, block the induced synthesis of enzyme or directly modulate the physiochemical nature of the enzyme (Glass and Dunlop, 1974).

Fig. 4: Effect of KNO_3, Pyridoxine and Salicylic acid on nitrate reductase activity of leaves at pre- and postanthesis stages

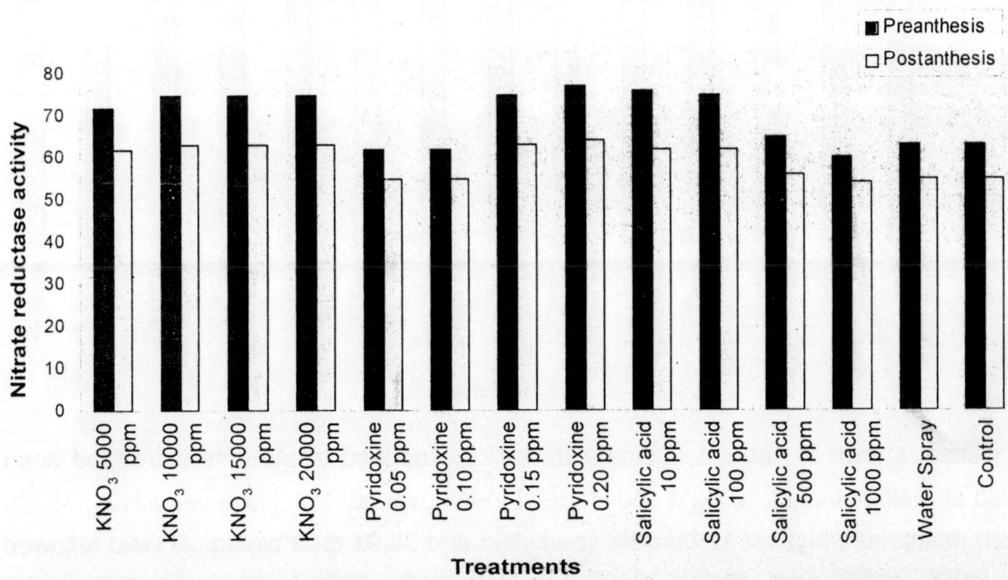

Similarly, proline accumulation in leaves tissues, differed significantly among the applied treatments in comparison to control as well as water treatment both at pre- and postanthesis stages (Fig. 5). The accumulation of proline content was higher during postanthesis stage than preanthesis stages in general. The highest 1.58μ mol NO_2 g^{-1} F.W. and 3.34μ mol NO_2 g^{-1} F.W. Proline content was observed with KNO_3, 15000 ppm treated plants which was at par with KNO_3, 20000 ppm and pyridoxine, 0.2 ppm treatments both at pre- and postanthesis stage respectively. The lowest 1.14μ mol NO_2 g^{-1} tissue and 2.68μ mol NO_2 g^{-1} tissue proline content was observed with salicylic acid, 1000 ppm at pre- and postanthesis stage, respectively. This is in agreement with the findings of Amaregauda *et al.* (1994). According to them, $CaCl_2$ (2%) treated seeds of wheat showed the maximum free proline content in the shoot tissue and observed a positive relationship between relative water content and proline accumulation in treated plants. The accumulated proline acts as solute and reduce the dehydration of the cells, thereby maintaining high water contents in leaves (Janagaudar *et al.*, 1983).

Fig. 5: Effect of KNO_3, Pyridoxine and Salicylic acid on proline content in leaf tissues at pre- and postanthesis stages

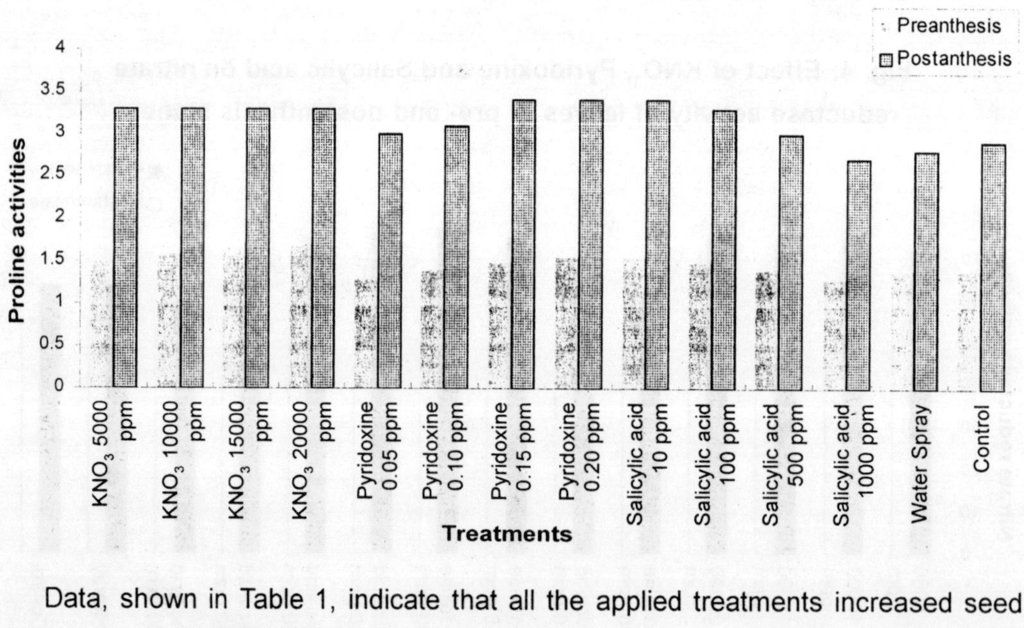

Data, shown in Table 1, indicate that all the applied treatments increased seed yield as well as biological yield and Harvest Index over control. However, KNO_3, 15000 ppm maintained highest 12.98 q/ha seed yield and 38.94 q/ha biological yield followed by KNO_3, 10000 ppm, KNO_3, 20000 ppm and KNO_3, 5000 ppm and pyridoxine, 0.2 ppm. The lowest seed yield (10.45 q/ha) was obtained with salicylic acid, 1000 ppm concentration, while the rest of the concentration of salicylic acid significantly increased

Table-1: Effect of foliar application of Potassium nitrate, Pyridoxine and Salicylic acid on economic yield, biological yield, harvest index and percentage yield increase over control

Treatments/Concentrations	Economic Yield (q/ha)	Biological yield (q/ha)	Harvest Index (%)	Percentage increase of yield over control
KNO, 5000 ppm	12.65	38.24	33.09	21.05
KNO, 5000 ppm	12.93	38.87	33.26	23.73
KNO$_1$ 5000 ppm	12.98	38.94	33.33	24.21
KNO$_1$ 5000 ppm	12.75	38.57	33.06	22.01
Pyridixine 0.05 ppm	11.60	37.21	31.17	11.00
Pyridoxine 0.10 ppm	11.75	37.58	31.27	12.44
Pyridoxine 0.15 ppm	11.98	37.75	31.74	14.04
Pyridoxine 0.20 ppm	12.38	37.92	32.65	18.47
Salicylic acid 10 ppm	12.12	37.88	31.99	15.98
Salicylic acid 100 ppm	11.88	37.63	31.57	13.68
Salicylic acid 500 ppm	11.21	37.42	29.65	7.27
Salicylic acid 1000 ppm	10.21	36.58	27.91	-2.29
Water spray (No chemical treatment)	11.15	37.15	30.01	6.69
Control (No treatment)	10.45	37.07	28.19	
S. Ed±	0.0308	0.0484	0.044	
CD for treatment (P-o.0.5)	0.095	0.0996	0.0914	
CD for treatment (P-0.01)	0.126	0.1347	0.123	
CV (%)	0.1510	0.1572	0.173	

the seed yield and biological yield. The increase in seed yield appeared to be the fact that all the applied chemicals at critical stages of crop growth are effectively absorbed by the crop and more efficiently translocate the photosynthates to the developing pods for proper filling of grains. This is well documented and reflected in higher values obtained for number of pods per plant, number of seeds per pod and the size of pod. Further due to foliar treatments number of factors such as increase relative leaf water content (Kumari and Bharti, 1988), enhanced carboxylation (Phulekar *et al.*, 1988) and thereby increase in carbon compounds and their utilisation leadi,g to higher siliqua number per plant has been also observed in the present study. The increase site of photosynthetic apparatus in terms of foliage weight and increased assimilation rate contributed for formation of better biological yield was observed in the present investigation.

The harvest index increased significantly with all the applied treatments except salicylic acid, 1000 ppm concentration (Table 1). The highest 33.33 per cent HI was obtained with KNO_3, 15000 ppm concentration followed by KNO_3, 10000 ppm (33.06%) and KNO_3, 5000 ppm (33.09%). The increase in harvest index was also reported by number of workers like Chandrababu *et al.* (1995) in grounnut and Emam and Moacid (1998) in wheat. The increase in harvest index is mainly due to the increase in grain yield and yielding components by the foliar application of chemicals was observed in the present study.

Results obtained from the above, it may be concluded that KNO_3, 15000 ppm concentration was found to be most effective and promising in improving the physiological efficiency and productivity potential of toria cv m27 under rainfed condition in terms of RLWC, RSD, WSD, BRA and proline content during successive stage of crop growth coupled with high gain yield. Further, it may also be advised that toria, (*Brassica compesris* L) can be grown successfully through foliar application of these applied nutrients where nutritional imbalances and water availability sparse.

References

Amaregauda, C.; Chetti, M.B. and Manjunath, S. 1994. Physiological basis of yield variation due to application of different chemicals in wheat. *Ann. Plant Physiol.* **8(1): 24-28.**

Bano. A.; Abdul Fattah, Q. and Hussain, Z. 1980. Relationship between nitrate reductase, nitrogen content, grain yield and protein content of potassium. Naphthenate treated maize plant, *Indian J. Plant Physiol.* **23(3) : 238-243.**

Chandrababu, R.; Manian, K; Nagarajan, M. and Ramachandra, T.K. 1995. Effect of mepiquetchloride on growth and yield of groundnut. *Madras Agric. J.* **82(3): 229-230.**

Dhacose, M.R.; Bhalerao, P.D. and Thorne, P.V. 1993. Response of groundnut to moisture stress, nitrogen application and antitranspirant in summer. *P.K.V., Res. J.* **17(2): 212-214.**

Emam, Y. and Moacid, G.R. 1998. The effects of foliar spray of urea at grain filling period on grain yield and grain protein content of winter wheat, *Triticum aestivum* L. *Crop Res.* **15(2 and 3) : 135-139.**

Gardner, F.P.; Pearce, R.B. and Mitchell, R.L. 1985. *Physiology of Crop Plants*. Sci, Publ, Jodhpur, pp **98-131.**

Glass, A.D.M. and Dunlop, K.J. 1994. Influence of phenolic on ion uptake. IV. Depolarisation of membrane potentials *Plant Physiol.* **53: 855-858.**

Jain, A. and Srivastava, H.S. 1981. Effect of salicylic acid and nitrate reductose activity in maise seedlings. *Physiol. Plant* **51: 339-342.**

Janagaudar, B.S.; Venkatsubbaiah, K.; Janardhan, K.V. and Pamechal. Y.C. 1983. Effect of short-term stress on free proline accumulation, relative water content and potassium content in different plant parts of three cotton genotypes. *Indian J. Plant Physiol.* **26: 82-87.**

Jawarski, E.J. 1971. Nitrate reductase assay in intact plant tissue. Biochkem. Biophys. *Res. Commun.* **43: 1247-1279.**

Kalaram, M.K. and Jeyakumar, P. 1988. Effect of nutrient and NAA spray on physiological changes in soybean (Glycine mark L. merill). *Indian J. Plant Physiol.* **3(3): 226-228.**

Kleeper, L.; Fleshu, D. and Hageman, R.H. 1977. Generation of reduced nicotinamide for nitrate reduction in green leaves. *Plant Physiol.* **48: 580-590.**

Kumari. S. and Bharti, S. 1988. Effect of CCC and FAP on water status and yield of sunflower (Helianthus annuus L.) under stimulated draught condition. *Indian J. Plant Physiol.* **31(4): 381-387.**

Patil, S.M. 1939. Effect of aspirin on summer groundnut. *P.K.V. Res. J.* **17(1): 117-118.**

Phulekar, C.S.; Chetti, M.B.; Nalini, A.S. and Patil, A.B. 1998. Relationship of leaf characters with total dry matter and yield in groundnut as influenced by growth reatrdent. *Indian J. Agric. Res.* **32(3): 195-200.**

Sinelair, T.R. and Ludlow, M.M. 1985. Who thought plant thermodynamics? *Aust. J. Plant. Physiol.* **12: 213-217.**

Tripathi, M.Z.; Kuruvanshi, S.M. and Namdeo, K.N. 1995. Effect of starch polymer and N-triacontanol on Indian mustard (Bransica Juncea). *Indian J. Agron.* **40(3) : 525-527.**

❑❑❑

7

Methane Efflux Studies from Different Paddy Cultivars Grown in Southern India

V. Rama Mohana Reddy[1], P.V. Krishna Rao[1], S.V. Subbaiah[2], and Venkataratnam[1]

[1]National Remote Sensing Agency, Hyderabad-500 037

[2]Directorate of Rice Research (ICAR), Hyderabad-500 030

Introduction

Public concern about global warming mostly concerned with carbon dioxide and trace gases such as Methane (CH_4), Nitrous oxide and Chlorofluoro carbons (CFC), which are considered as green house gases. Methane concentration in the atmosphere has more than doubled during the last 200 years. Its current atmospheric concentration of 1.7 ppm volume up from 0.7 ppm volume in preindustrial times, which is lower than the 345 ppm of carbon dioxide. But one molecule of metahne traps approximately 30 times as much heat as does Carbon dioxide. The heating effect of the atmospheric methane increase in approximately half as does Carbon dioxide increase (Dickinson and Cicerone 1986, Ramanathan *et al.*, 1985).

In recent past wetland paddy fields have been identified as a major source of atmospheric methane and the first comprehensive measurement of Methane flux from Paddy fields were reported only in early 1980s (Cicerone and Shetter, 1981, Cicerone *et al.*, 1983). In case of submerged soils, flooding of paddy fields cuts off the oxygen supply from the atmosphere to the soil, which results in anaerobic fermentation of soil organic matter. Methane is a major product of anaerobic fermentation and transported from submerged soils into the atmosphere by diffusion and ebullition and also through paddy plants itself. Methanogenic bacteria, which can metabolise only in the absence of free oxygen, produce methane and at redox potentials of less than -150mv and the methanogens are neutrophilic in nature with an optical pH of 6-8.

In wetland Paddy soils, Methane is largely produced by transmethylation of acetic acid and also by the reduction of carbon-dioxide (Takai, 1970). The formation of methane is preceded by the production of volatile acids on flooding and decreasing carbon dioxide concentration causes the methane formation (Neue and Scharpenseel, 1984). Methane production is negatively correlated with soil redox potential and positively correlated with soil temperature, soil carbon content and paddy growth stages. The rate of organic matter addition and decomposition determines the rate of methane formation in flooded paddy soils. Methane production generally increases during the cropping season, although the population density of methanogens remains stable (Schutz *et al.*, 1989).

Generally Paddy fields are prepared by flooding before Paddy is transplanted. If bare mud id flooded, most methane is trapped in the soil, and as long as the soil is not heavily amended with organic matter and remains undisturbed, only small amount of methane is released through ebullition. Under field conditions, large portions of the soil trapped methane escapes into the atmosphere through ebullition caused by such culture practices as wet tillage, pudding, harrowing, transplanting, fertilisation, weeding, pest management and harvest. In the wetland paddy fields, which are not disturbed by culture practices during the growing season, up to 70% of the methane released to the atmosphere through the aerenchyma cells of the Paddy plant. During the reproductive phase more than 90% of the emission attributed to the Paddy plant (Circerone and Shetter, 1981, Schutz *et al.*, 1989b). Methane emission rates are highly variable, emission rates mediated by paddy plants are highest in the early afternoon and lowest late at night, following the daily pattern of air and topsoil temperature and pH as well as oxygen concentrations in the floodwater and soil-water interface. Fertiliser containing farm yard manure may enhance growth and increase methane production.

However, the study on the influence of paddy cultivars on methane efflux was not taken up so far in the country. Hence, the present study was carried out at research farm of Directorate of Rice Research, ICAR, Rajendranagar, Hyderabad to characterise the methane efflux pattern from different paddy cultivars, which are being cultivated in various paddy-growing regions of the country. In addition the influence of various parameters like soil organic carbon content and plant biomass were also studied in the present study.

Materials and Methods

Experimental site

The field experiment was conducted during the kharif season (July-November) of 2002 in a RBD design with 3 replications. The farm is geographical situated at 17°25'N latitude and 78°26'E longitude at an altitude of 545 meters above MSL. The soil of the experiment field is clay loamy soils and the mechanical and chemical properties of the soils are given in Table 1. The different cultivars, which were Tulasi, Prasanna,

Table-1 Physical and Chemical properties of test site

Soil characteristics	Content
A. Texture composition (%)	
Clay	48
Sand	21
Silt	31
Class	Clay
B. Chemical properties	
Available N (kg/ha)	225
Available P (kg/ha)	103
Available K (kg/ha)	412
pH	8.4
EC (d sm^{-1})	0.65
Organic carbon (%)	0.57

Rasi, Swarna, Sambamasuri, Vibhava, Nagarjuna, Swarandhan, Tellahamsa, Krishnahamsa, Suraksha and IET 15359, whereas the high yielding and hybrid cultivars such as Ajaya and DRRHI were transplanted after puddling the fields and bunding. Water management and chemical fertiliser application was uniform for all plots in this experiment along with other package of practices for paddy cultivation. The puddle field was applied with basal dose of fertilisers consisting of 105 kg N ha^{-1}, 70 kg P ha^{-1}, 63 kg K ha^{-1} before tranplantation. The hybrid cultivars were applied with 10 ton of

FYM ha^{-1} before transplatation in order to study the influence of cultivars and FYM on the methane emission trends. Split application of remaining nitrogen fertilisers was done at the rate of 25 kg N ha^{-1} at 30 and 45 days after transplantation (DAT). A floodwater level of 5 to 7 cms was maintained during the entire experimentation. Paddy cultivars are broadly classified into early maturing, medium duration and late maturing depending on the growth cycle.

Sampling and Measurements

Air samples were collected at 18-21 days interval at 08.00-11.00 hours and 14.00-17.00 hours using perspex chamber of 100 liters capacity and with a surface area of about 1700 sq.cm to cover the plants. Whenever, the air samples were collected, the chamber was placed on a metal frame one day prior to collection of air samples in each plot and the grease free electric fan was used to ensure uniform air mixing inside the chamber. The air inside the chamber was withdrawn with a 50 ml syringe at 0.15, 30 minutes interval. Measurements are made at different times of the day as well as at different stages of the crop cycle. When air samples were collected, the other parameters such as plant biomass, plant height, soil Eh and soil temperature were also collected. The collected air samples were analysed for methane concentration on the gas chromatography (GC) attached with a flame ionisation detector (FID) available in Agriculture and Soils group of NRSA. Methane flux was calculated based on the equation gives by Lal *et al.*, 1993.

Results and Discussions

The results of methane emission tends from different cultivars during kharif season of 2002 for early maturing, medium duration, late maturing and hybrids were presented in Figures 1, 2, 3 and 4, respectively. In general two of three peaks of plant-mediated emission is observed in irrigated paddy fields during the cropping season. Methane emission increases during the vegetative phase, known as early intermediate peak during tillering stage; it decrease after panicle initiation and increases again after flowering, before declining at the maturity.

The present studies were conduced to study the influence of different cultivars on methane flux emissions and are broadly categorised into Early maturing. Mesium duration, Late maturing and High yielding-Hybrid cultivars. An analysis of Fig.1 revealed that all the early maturing cultivars has exhibited peak emission of methane at flowering stage of paddy crop. The methane peak emission was observed in Aditya cultivar followed by Tulasi, Rasi and Prasanna cultivars. However, maximum SIF was observed from Aditya and Tulasi cultivars, which are significantly differs in SIF from Prasanna and Rasi cultivars. All intra-seasonal methane emissions showed a bell shaped curve thus showing the influence of growth stage on the methane efflux.

A perusal of Fig. 2 revealted that high SIF of methane was observed in Krishnahamsa, followed by Tellahamsa, IET-15359 and Suraksha cultivars. Tellahamsa reported high Methane flux in flowering stage, while in case of IET-15359 cultivar the peak emission was observed during tillering stage, thus indicating the influence cultivar on the methane emission pattern. In the medium duration all the cultivars exhibited a bimodal peak at tillering and flowering stages. Thus the initial peak at tillering is related to biomass applied to field or previous crop residue, while the second peak during flowering is related to the decomposition of root exudes as well as leaf stubbles.

The late maturing varieties (Fig. 3) such as Swana, Sambhamasuri, Vibhava, Nagarjuna and Swarandhan cultivars and high SIF of methane was observed in Sambhamasuri variety as compared to other varieties. Low efflux of methane was observed during establishment and maturity stages in all varieties. Since, all the cultivars are of long duration and reported two peaks of methane efflux, the first peak at tillering stage and second peak at flowering stage. A perusal of Fig. 1 to Fig. 4 revealed that generally short duration and early maturing cultivars emitted more methane than medium duration and late maturing cultivars during paddy growing season.

In order to study influence of methane efflux from the application of farm yard manure, the hybrids and high yielding varieties of Ajaya and DRRHI was selected and applied with 10 tones of FYM. A perusal of Fig. 4 revealed that application of FYM reported high methane efflux associated with high biomass thus indicating the influence of plant mediated transport of methane from soil to atmosphere through plant aerenchyma cells. There are significant differences in methane efflux was observed at peak growth stage and comparison of FYM applied plots with control plots revealed that 4 times more SIF of methane was observed in case of FYM applied plots (Fig. 6).

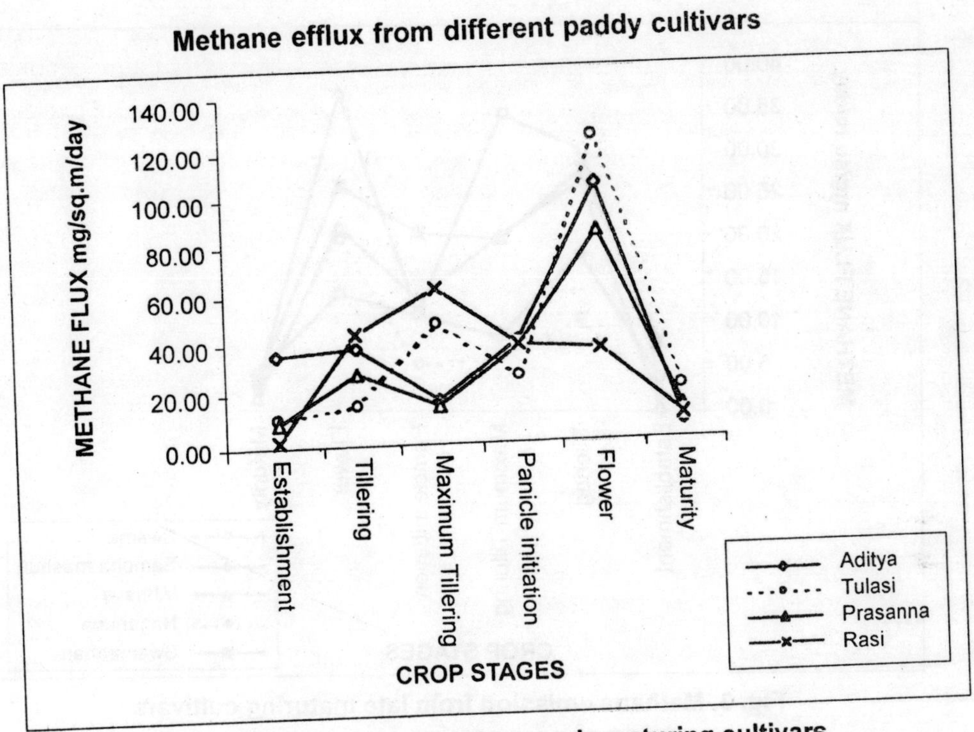

Fig. 1: Methane emission from early maturing cultivars

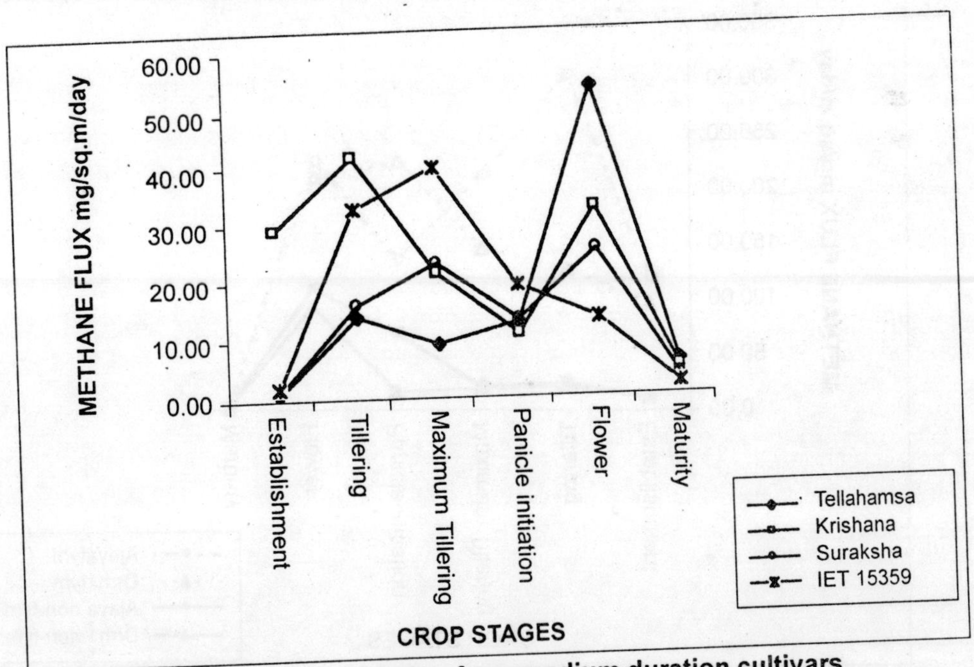

Fig. 2: Methane emission from medium duration cultivars

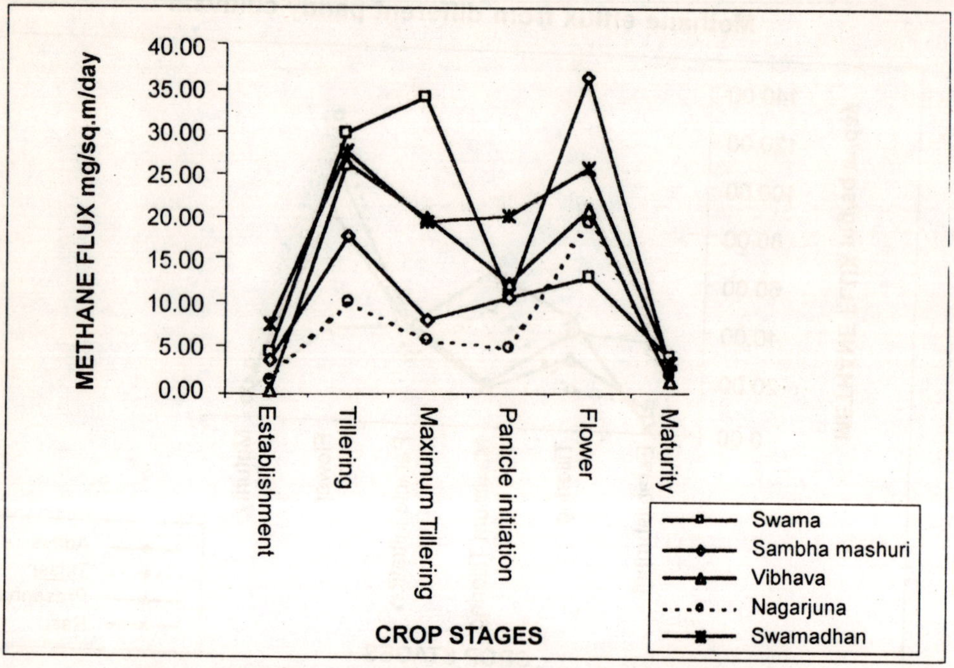

Fig. 3. Methane emission from late maturing cultivars

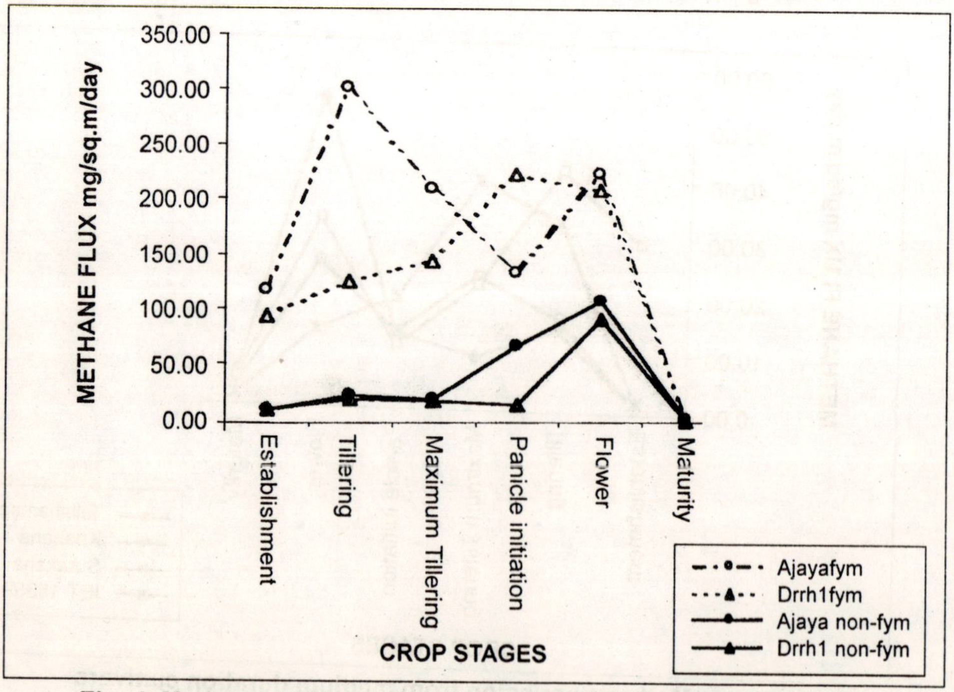

Fig. 4. Methane emission from Hybrid and high yielding cultivars

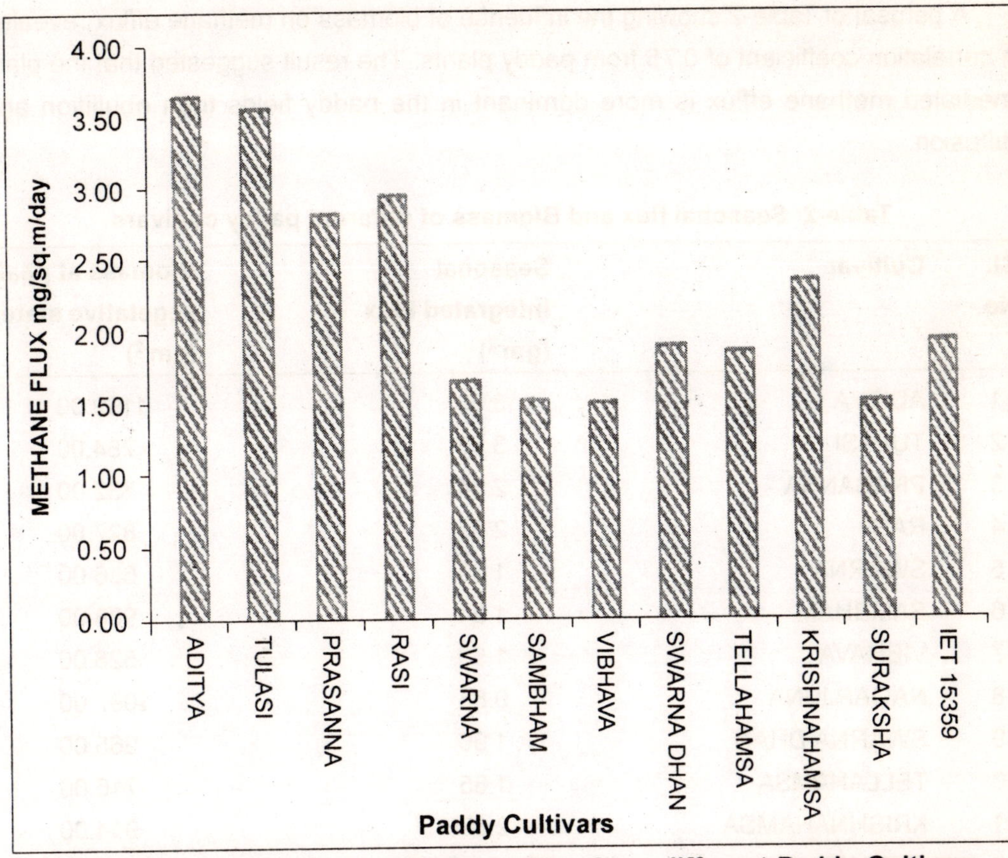

Fig. 5. Seasonal Integrated Methane Flux from different Paddy Cultivars

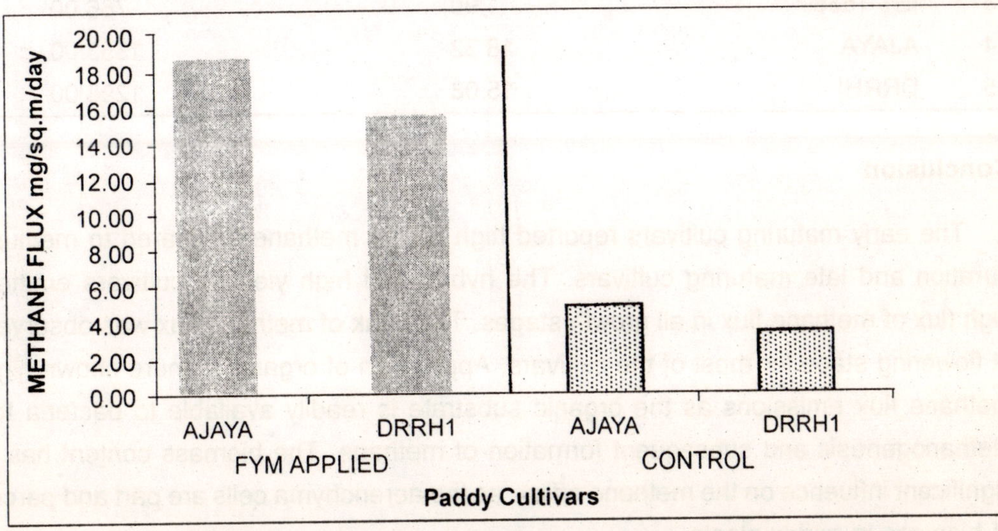

Fig. 6. Methane Efflux from Hybrid and High Yielding Cultivars

A perusal of Table-2 showing the influence of biomass on methane efflux revealed a correlation coefficient of 0.75 from paddy plants. The result suggested that the plant mediated methane efflux is more dominant in the paddy fields than ebullition and diffusion.

Table-2: Seasonal flux and Biomass of different paddy cultivars

Sl. No.	Cultivar	Seasonal Integrated Flux (gm^{-2})	Biomass at peak vegetative state (gm^{-2})
1	ADITYA	3.66	1169.00
2	TULASI	3.56	784.00
3	PRASANNA	2.84	732.00
4	RASI	2.95	827.00
5	SWARNA	1.69	656.00
6	SAMBHAM	1.54	908.00
7	VIBHAVA	1.53	528.00
8	NAGARJUNA	0.85	1097.00
9	SWARNA DHAN	1.90	965.00
10	TELLAHAMSA	1.85	716.00
11	KRISHNAHAMSA	2.32	914.00
12	SURAKSHA	1.49	1075.00
13	IET 15359	1.90	766.00
14	AJAYA	18.32	1353.00
15	DRRHI	15.08	1298.00

Conclusion

The early maturing cultivars reported high SIF of methane compared to medium duration and late maturing cultivars. The hybrid and high yielding cultivars emitted high flux of methane flux in all growth stages. The peak of methane flux was observed at flowering stage for most of the cultivars. Application of organic manure shown high methane flux emissions as the organic substrate is readily available to bacteria for Methanogenesis and subsequent formation of methane. The biomass content has a significant influence on the methane efflux, as the aerenchyma cells are part and parcel of biomass in paddy plants.

References

Bartlett, K.B., P.M. Crill, D.I. Sebacher, R.C. Harriss, J.O. Wilson and J. M. Melack, 1988. Methane fluxes from the central Amazonian floodplain. *Journal of Geophysical Research* **93: 1571-1582.**

Cicerone, R.J., J.D. Shetter and C.C. Delwiche, 1983. Seasonal variation of methane flux from a Calofonia rice paddy. *Journal of Geophysical Research* **88: 11022-11024.**

Cicerone, R.J., and J.D. Shetter, 1981. Sources of atomospheric methane: Measurements in rice paddies and a discussion, *Journal of Geophysical Research* **86: 7203-7209.**

Dickinson, R.E. and R.J. Cicerone, 1986. Future global warming from atmospheric trace gases. *Nature* **319: 109-114.**

Lal, S., Venkataramani, S. and Subbaraya, B.H., 1993. Methane flux measurements from paddy fields in the tropical Indian region. *Atoms. Environ*, **27: 1691-1694.**

Neue, H.U. and W.W. Scharpenseel, 1984. Gaseous products of decomposition of organic matter in submerged soils. In: *Organic Matter and Rice*. International Rice Research Institute, Los Banos. pp. **311-328.**

Ramanathan, V., R.J. Cicerone. H.B. Singh and J.T. Kiehl, 1985. Trace gas trends and their potential role in climate change, *Journal of Geophysical Research* **90: 5547-5566.**

Schutz, H., Holzapfel-Pschorrn, A., Conrad, R. Rennenberg and Seiler, W., 1989. A 3-year continuous record on the influence of daytime, season and fertiliser treatment on methane emission rates from an Italian rice paddy. *Journal of Geophysical Research* **94: 16405-16416.**

Schutz, H., W. Seiler and R. Conrad, 1989b. Processes involved in formation and emission of methane in rice paddies. *Biogeochemistry* **7:33-53.**

Selier, W., A. Holzapgel-Pschorn, E. Conrad and D. Scharffe, 1984. Methane emission from rice paddies, *Journal of Atmosphere Chemistry* **1: 241-268.**

Takai, Y., 1970. The mechanism for methane fermentation in flooded paddy soil. *Soil Sci. Plant Nutr.* **16: 238-244.**

Takai, Y. and T. Koyama, 1956. Composition of gases and organic acids contained is soil of paddy fields. *Journal of Japanese Soil and Ferti.*, **26: 509-512.**

Yagi, K. and K. Minami, 1990. Effect of organic matter application on CH_4 emission from Japanese paddy fields. In: *Soils and the Greenhouse Effect* (ed. A. F. Bouman). John Wiley and Sons, New York. pp. **467-473.**

❑❑❑

8

Effect of Paper and Pulp Mill Effluent on the Germination, Growth and Yield of *Cicer arietinum* L.

Anial K Rania, Mosmi Raina and Sunita Rani
Departmemt of Eniveronmental Sciences,
University of Jammu, Jammu–180 006, India

Introduction

Highly developed and well-spread network of industries are indices of national prosperity but industrial revolution has hastened the consumption of scare material on one end and thrown up a lot of unwanted wastes on the other end. Industrialisation along with some other factors *viz.* urbanisation; population explosion etc. has greatly accelerated the process of environmental degradation. The effluents generated through the manufacturing of wide variety of products by industries are being pumped either into the aquatic ecosystem or spilling over the agricultural field directly or through irrigation canals or even entering into the ground water leading to the deterioration of the land, agricultural crops and the soil system.

The use of industrial wastewater in irrigation of crops is a recent phenomenon with scientific attention towards this being focused only after 1940 when the problem of freshwater pollution due to wastewater disposal become particularly acute. A large number of waste *viz.* tannery waste (Mohlman, 1926), pulp and paper mill waste (Billings, 1959 and NEERI, 1979), dairy waste (Scott, 1962), food processing waste (Fisk, 1964), meat processing waste (Cohen, 1982) etc. have been successfully used for irrigation of crops with or without any treatment.

The present study has been carried out to determine the effect of the effluent coming out from a paper mill—M/s Jammu Paper (P) Ltd., Bari Brahamna, Jammu, on the germination, growth and yield of *Cicer arietimum* L.

Material and Methods

The analysis of various physico-chemical characteristics of effluents *viz*. colour, temperature, pH, electrical conductivity, alkalinity, total hardness, total solids, total suspended solids and total dissolved solids was done bringing untreated effluent at regular intervals from paper mill, Bari Brahmna.

The certified seeds of *Cicer arietinum* L. were procured from Directorate of Agriculture, Jammu. Seven sets of perforated polythene bags filled with prepared soil were made. Out of these one set was kept as control, two sets of seed treatment and four sets were kept for soil treatment. Ten replicate for each treatment were maintained. In each bag, 10 seeds were sown. Before sowing seeds for control and soil treatment, seeds were soaked in tap water. For seed treatment, some seeds were soaked in 100% concentration of effluent for 24 hrs and some seeds for 48 hrs. For the control and seed treatment sets *i.e.* set I, II, and III, normal tap water used for irrigation. For soil treatment, all four sets *i.e.* set IV, V, VI and VII were irrigated with the different concentrations of effluent, respectively. Irrigation of all the set with water and effluent was done regularly on weekly basis.

The number of seed that geminated were counted after 27 days and thus percentage gemination was calculated. The number of leaves and branches were recorded after every 10 days, beginning from 20th day upto 90th day. Numbers of flowers were counted after every 10 days from the initiation of flowering after 90th day upto 120th day.

One uprooting from each treatment set was done from 20th day onwards after every 10th day up to 120th days from recording the parameters like length of root and shoot and numbers of nodules and secondary roots. After the crop matured, total number of pods, their length and width, number of seeds/pod, weight of dried pods, dried seeds was recorded to work out their averages.

Result and Discussion

The results of physico-chemical analysis of paper mill effluent are presented in Table 1. The effects of different concentrations of effluent on germination, growth and yield of *Cicer arietinum* L. is presented in Table 2.

Table-1: Characteristics of Effluent

S. No.	Parameters	Characteristics
1	Colour	Milky White
2	Temperature	29°C
3	pH	7.6
4	Conductivity	0.76 mhos
5	Alkalinity (as $CaCO_3$)	68 mg/l
6	Total Hardness	362 mg/l
7	Total Solids	1642 mg/l
8	Total Suspended Solids	1034 mg/l
9	Total Dissolved Solids	608 mg/l

Table-2: Seed Germination (%), Morphological and Yield Parameters of C. arietinum

S. No.	Parameter	Experimental Sets						
		Control	Seed Treatment Sets			Soil Treatment Sets		
		I	II	III	IV	V	VI	VII
1	Seed Germination (%)	99	95	94	98	98	94	91
2	No. of Leaflets	564	590	560	600	644	600	580
3	No. of Branches/plant	4	4	4	4	6	4	3
4	No. of Secondary Roots/plant	42	40	41	42	48	38	30
5	No. of Nodules/plant	13	10	8	12	15	11	9
6	Shoot/Root Length/plant (cm)	41/39	40/39	38/35.8	36/37.4	43/44	39/41	34/35
7	Shoot/root ratio	1.05	1.02	1.06	0.96	0.97	0.95	0.97
8	Total No. of flowers/set (120th day)	161	237	148	184	321	277	270
9	Total No. of pods/set (120th day)	30	36	33	36	91	61	90
10	Length of pod (cm)	1.15	1.45	1.11	1.37	1.44	1.31	1.13
11	Width of pod (cm)	0.65	0.8	0.4	0.76	0.46	0.75	0.64
12	Dry weight of pods (gm)	71	61	69	116	111	116	93
13	Average no. of seeds/set	32	33	28	54	65	39	37
14	Range of sees/pod	0-2	0-2	0-1	0-2	0-2	0-2	0-2
15	Weight of dry seeds (mg)	42	39	34	35	75	53	68

The perusal of the table revealed that both the seed treatments *i.e.* for 24 and 48 hrs have retarding effect on the seed germination. Irrigation of the crop with different concentrations of effluent has also inhibited the seed germination, which increases with the increase in the concentration of the effluent being maximum in the 100% concentration of untreated effluent. Similar observations have also been recorded by Rajannan and Oblisami (1979), Somashekhar *et al.* (1984), Bishnoi and Gautam (1991) and Balashouri and Prameela Devi (1994). Inhibition of seed gemination at higher concentrations of effluent may be due to high levels of dissolved solids, which enrich the salinity and conductivity of absorbed solute by seed before germination. All growth parameters except root and shoot length, number of secondary roots and nodules as well as yield parameters except dry weight of pods and weight of dried seeds have enhancing effect when seed treatment was done for 24 hrs and retarding effect when seed treatment was for 48 hrs. as compared to control set.

The growth parameter such as number of secondary roots, shoot/root length and nodule/plant registered either equal or lower values in all the treatments, except for set V, *i.e.* the set irrigated with 50% concentration of effluent, were higher values have been recorded as compared to control set. Number of branches does not show much difference in treatment sets as compared to control set except in set V where there is an increase in number. Number of leaflets/plant registered an increase in all the treatment set with highest number at 50% concentration.

The yield parameters such as number of flowers/set and pods/set, length and width of pod, weight of pod, number of seeds/set, range of seeds/pod and weight of dried seed showed enhancing effect in almost all the treatment. The highest values for all these parameters were recorded in treatment sets originated with 50% concentrations of effluent.

Thus, on the basis of present study, it is concluded that the effluent from the paper and pulp mill either be treated or diluted to 50% concentration before its discharge into the crop.

References

Balashouri and Pramella Devi (1994). Effect of tanney effluent on germination of selected pulse and cereal crop plants. *J. Ecotoxicol. Environ. Monit.*, **4(2): 115-120.**

Billings, R.M. (1959). Stream improvement through spray disposal of sulphite liquor at the Kimberly —Clerk Corporation, Niagara, Wis. Mills. *Proc. 13 Indus. Waste Conf. Pnrdue. Univ. Ext. Ser.* **96, 71.**

Bishoni, S. and Gautam, D.D. (1991). Effect of dairy effluent on seed germination and seedling growth of some crop plants. *Int. J. Ecol. Environ. Sci.*, **17: 67-71.**

Cohen, C. (1982). Disposal of packing house waste by irrigation at Lared. *Texcas. Pub. Works*, **59: 377.**

Fisk, W.W. (1964). Food processing waste disposal. *Water and Sewage Work*, **417-420.**

Mohlman, F.W. (1926) Treatment of packing, tannery and corn wastes. *Industr. Engg. Chem.*, **18: 1076.**

NEERI (1979). Utilisation of pulp and paper mill wastewater for agriculture-*Report by NEERI*, Nagpur, India.

Rajannan, G. and G. Oblisami (1979). Effect of paper factory effluent on soil and crop plants. *Indian J. Environ. Hlth.*, **21: 120-130.**

Scott, R.H. (1962). Disposal of high organic content waste on land. *JWPCF*, **932-50.**

Somashekar, R.K.; M.T.G. Gowda; S.C.N. Shettigar and K.P. Srinath (1984). Effect of industrial effluent on crop plants. *Indian J. Environ. Hlth.*, **26: 130-146.**

❑❑❑

9

Rapid *in vitro* Propagation of Henna Plant—*Lawsonia inermis* L. using Nodal Explants

M. Malakar and P.J. Handique
Department of Biotechonology, Gawahati University,
Guwahati—781 014, Assam, India

Introduction

Lawsonia inermis Linn. Syn. *Lawsonia alba* Lam. (Family *Lyhraceae*) commonly known as "Henna" (in Assamese "Jetuka") is an important medicinal shrub or small tree and a native of North Africa and South West Asia. It grows on light loam to clay loam soil but prefers heavy soils which are retentive of moisture and can tolerate a little alkalinity in the soil. This glabrous shrub is much branched with grayish brown barks and the lateral branches often ends in a spinous point. The leaves are 1.3-3.3 cm by 0.6-1.6 cm seised, elliptic or broadly lanceolate, acute or obtuse, of the base tapering. Flowers numerous, less than 1.3 cm, fragment, white or rose coloured in large terminal pyramidal panicled cymes (Prain, 1963). Henna has long been used in India and Middle East countries for colouring palms of hands, soles of feet and eyes brows for personal adornment. Leaves are useful in headache, hemicranias, lumbago, ulcers, stomatis, ophtalima, syphilitic sores, scabies, disease of the spleen. The paste or decoction of Henna leaves are used externally against boils, burns, bruises and skin inflammations. The bark is widely used in the treatment of jaundice, leprosy, disease of spleen and skin. The flowers have a strong aroma and on steam distillation, it yields 0.01-0.02% of an essential oil with a strong fragrance and dark brown colour. The roots were considered diuretic and are given in gonorrhoea, bronchitis. The principal colouring matter present in Henna is "lowsone" which is present in dried leaves in a concentration of 1.0-1.4%.

Besides lewsone, other constituents reported to present in henna are tannin, glucose, gallic acid, mannitol, fat. resin, mucilage and traces of an alkaloid which have the high medicinal an commercial value (Anonymous, 1962; Kirtikar and Basu, 1981; Handique. 1994).

Propagation of *L. inermis* can be done by vegetative methods or cuttings and by seeds, but seed is unsuitable because of short viability and varous disease and pest infection. On the other hand, the species is becoming rare in natural habitat condition due to habitat destruction and over exploitation. Considering the utilisation of the species and its rarity, an attempt has been made to standardise a micropropagation protocol for *L. inermis* using nodal explants.

Materials and Methods

Fresh, healthy and uinfected shoots of *L. inermis* were collected from a two-year-old plant maintained in the departmental experimental garden. Shoot pieces consisted of a single node containing a small portion of internode on either side was used as explants. Explants were washed several times with tap water and soaked in Tween-20 (20%) for 20 minutes with occasional stirring followed by several washing with distilled water. Then the explants were soaked in 0.5% bavistin (fungicide) for 25 minutes and then washed several times with sterile double distilled water (SDDW). Therefore, under a laminar air flow cabinet, the explants were surface sterilised by dipping them in 70% ethanol for 30 seconds followed by 0.1% (w/v) $HgCl_2$ solution for 13 minutes. The explants were then thoroughly rinsed with SDDW for 5-6 times and were inoculated on MS medium (Murashige and Skoog, 1962) with 3% (w/v) sucrose, 0.8% agar, 100 mg/l myo-inositol, supplemented with different concentrations of Benzylaminopurine (BAP) and Kinetin (KN) in 25 × 150 mm culture tubes. The pH of the medium was adjusted to 5.8 before atuoclaving at 15 lb/inch 2 and 121°C for 15 minutes. The cultures were maintained at 25±2°C under fluorescent light of about 3000 Lux intensity with 16-hour photoperiod. For each treatment, 7 replicas were used and each treatment was repeated twice. The culture were maintained by sub-culturing at an interval of 30 days. For root induction, *in vitro* grown shoots (2-3 cm long) were cultured on rooting medium supplemented with Indole 3-butyric acid (IBA). For induction of callus, leaf and nodal explants were cultured in MS medium containing different growth regulators in different concentrations. After 30 days of callus culture, the leaf and nodal callus were sub-cultured in MS medium with the same hormone concentration and observed up to next 30 days. Then leaf and nodal callus were transferred to different MS medium containing various growth regulators for getting organogenic response.

After 35 days of rooting, the rooted shoots were transferred to hormone free full strength MS liquid medium and after a week, transferred to ½ strength MS liquid medium without hormone. After 12 days, they were transferred to SDDW for overnight and then transferred to pots containing mixture of sand, soil and compost (2:1:1). Plantlets were regularly watered and covered with polythene bags. Polythene bags were removed after a week and plantlets were successfully transferred to field conditions.

Results and Discussion

The development of first shoots were recorded after 8-9 days of culture in MS medium supplemented with various concentrations of BAP and KN either singly or in combination. Out of these MS medium with 0.5 mg/l BAP produced maximum average of 7.1 (Fig. 1A) number of shoots and a maximum shoots length of 3.6 cm with 10.3 average number of leaves (Fig. 1B) per explants after 60 days. The number of leaves was seen to increase with the increase of the length of shoots. KN (0.5 mg/l) showed maximum average of 2.0 numbers of shoots with a maximum value of 2.1 cm of shoot length. The control showed development of first shoots after 47 days of culture, but no multiple shoot even after 60 days of culture. The combinations of BAP + KN (both in varying concentrations) were also studied, but it proves less responsive than 0.5 mg/l BAP and KN used singly. So, total number of shoots per explants reached a maximum value of 7.1 at 0.5 mg/l BAP (Table-1). Earlier Bhojwani (1981) also reported that BAP alone or in combination with other hormone favours multiple shoots proliferation in several species.

For root induction, IAA and IBA were used either singly or in combination. Initiation of root was observed after 9 days in MS medium containing IBA, whereas IAA containing medium showed root imitation after 12 days of culture of the microshoots. No root initiation was observed in control even after 35 days of culture.

The best rooting was observed in MS medium containing 0.2 mg/l IBA with maximum average of 9.6 numbers of roots per explants (Fig. 1C) with a mean length of 2.7 cm after 5 weeks of culture (Table-2). Similar results were reported by Yadav *et al.* (1990) in *Morua nigra* and Das and Handique (2002) in *Polygonum microcephallum*. Well-rooted shoots were allowed to grow in hormone free full strength MS medium. After 12 days, they were transferred to SDDW for 1 day. Then the plantlets were transferred to pots containing sterilised mixture of sand, soil and compost in 2:1:1 ratio (Fig. 1D). The potted plantlets were watered regularly and kept under covering with polythene bags for 5-6 days to avoid desiccation. Thereafter plantlets were successfully transferred to field condition with 75% surivability.

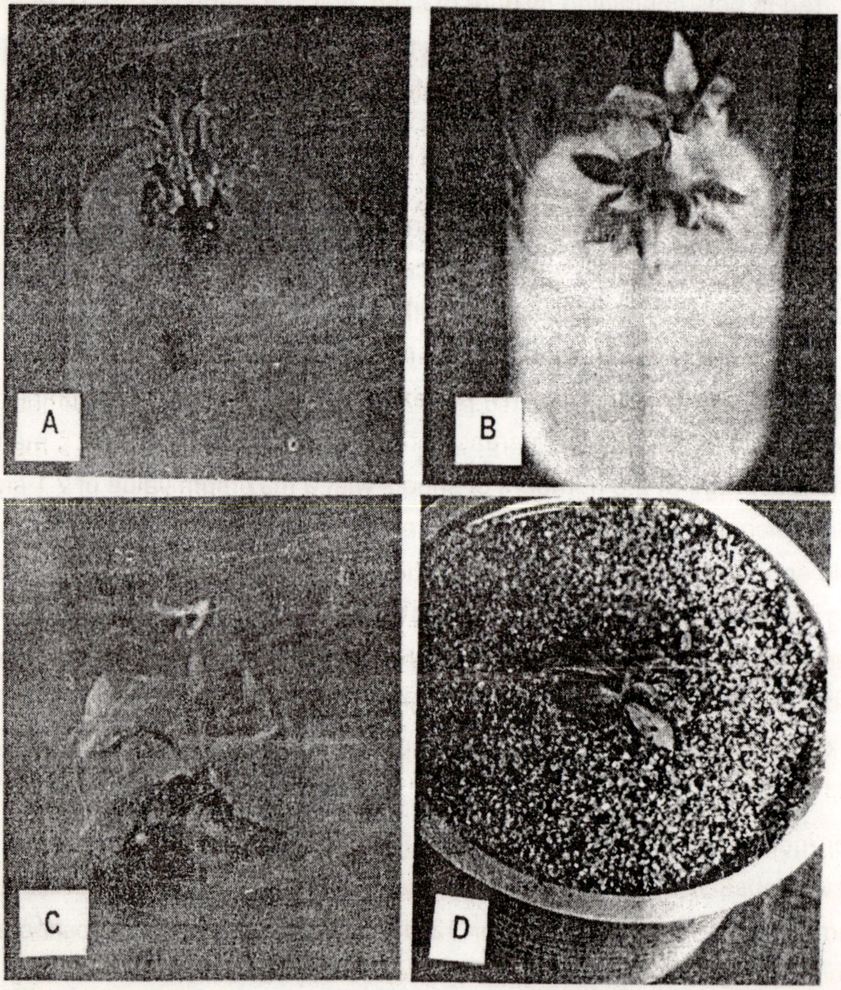

Fig. 1: Stages of *in vitro* regeneration of *Lawsonia inermis* L. from nodal explants:

A. Multiple shoots formation on nodal explants in MS medium containing BAP.

B. Growth of individual shoot with proliferation of leaves.

C. Rooted plantlets in MS Medium containing IBA.

D. Potted plants in the mixture of sand, soil and compost (2:1:1).

Table-1: Effect of BAP and KN on shoot multiplication of *L. inermis* from nodal explants

Growth Regulator (mg/l)		After 60 Days of Culture		
BAP	KN	No. of shoots/explants	Length of shoots/ explants (cm)	No. of leaves/explants
-	-	1.0	0.5	2.0
0.2	-	5.3	3.2	10.1
0.5	-	7.1	3.6	10.3
1.0	-	4.8	2.8	7.1
1.5	-	3.0	2.0	6.3
2.0	-	2.1	1.7	6.1
2.5	-	1.7	1.1	4.6
-	0.2	1.8	1.5	8.2
-	0.5	2.0	2.1	9.1
-	1.0	1.6	1.1	7.1
-	1.5	1.4	0.9	6.0
-	2.0	1.3	0.8	3.6
0.2	2.5	2.7	1.8	5.3
0.5	2.0	3.2	1.9	5.5
1.0	1.5	1.8	0.9	4.8
1.5	1.0	1.6	1.0	4.6
2.0	0.5	2.8	1.2	5.1
2.5	0.2	2.1	1.0	3.0

* Values represented on the table are mean of 7 replications.

Table 2: Responses of root induction from regenerated shoots after 35 days of culture

Growth regulator (mg/l)		No. of root/explants	Length of roots (cm)
IBA	IAA		
-	-	0.0	0.0
0.2	-	9.6	2.7
0.5	-	7.8	2.1
1.0	-	4.9	1.3
1.5	-	2.3	0.9
2.0	-	1.9	0.5
-	0.2	7.8	2.2
-	0.5	5.7	1.4
-	1.0	2.1	1.0
-	1.5	1.7	0.7
-	2.0	1.2	0.4

References

Anonymous, 1962. *The Wealth of India -Raw Materials*, Vol. 6., CSIR, New Delhi, pp. **47-50.**

Bhojwani, S.S. 1981. *In vitro* propagation of garlic by shoot proliferation. *Scientia Horticulturae*, **13: 47-52.**

Das, Jhumur and Handique, P.J. 2002. Micropropagation of a rare medicinal plant species—*Polyfonum microcephallum* D. don. through high frequency shoot multiplication, *J. Phytol. Res* **(2): 197-200.**

Handique, P.J. 1994. Ethno-medico-botanical studies of the indigenous medicinal plants of Golaghat and Jorhat districts of Assam. *ASTEC (Guwahati) Project report*. pp. 89.

Kirtikar, K.R. and Basu, B.D., 1981. *Indian Medicinal Plants*, Vol. **2 : 1076-1080.**

Murashige, T. and Skoog, F. 1962. A revised medium for the rapid growth and bio-assays with tobacco tissue culture, *Physiol. Plant.* **15: 473-497.**

Prain, D. 1963. *Bengal Plants*, Botanical Survey of India, Kolkata, **1: 449.**

Yadav, U.; Lal, M. and Jaiswal, V.S. 1990, Micropropagation of *Morus nigra* L From shoot tip and nodal explants of mature trees. *Scientia Horticulturae*, **44: 61-67.**

❑❑❑

10

Changes in Water Parameters on Oyster Habitat in Bhatye Estuary at Ratnagiri Coast of Maharashtra

G.D. Suryawanshi, K.B. Shejule and Z.A. Langekar*
Department of Zoology, Dr. B.A.M. University, Aurangabad
*Naik Sea Food Ltd., Ratnagiri

Introduction

In India, estuaries like any other country have been a focal point of a lot of human activities. These activities include development of port and harbours. The fishing activity, however, gets considerably reduced because of turbulent conditions, or even suspended (Qasim, 2004). Like open sea, overall the creeks, esturaries and sheltered waters of near shore play an important role in the production, maintenance and protection of potential biological resources like shellfishes and wild life. Although, some efforts for mitigation of deleterious impact due to environmental stress have been carried out, in the case of commercially important species especially those, which are non-migrants and can't escape stress there is threat to their very existence under such circumstances. However, the protection and management of coastal waters have still not been approached on an expected scale. An extensive literature is available on various physicochemical parameters of the creeks, estuaries and offshore waters along the west coast of India (Gajbhiye *et al.*, 1995, Yennawar, 1997, Senquopta and Deshmukh, 2000 and Mohite, 2002). In the present study an attempt was made with an objective of providing existing knowledge about the ecology of near shore, especially from the oysters bed of (*Crassostrea cattuckensis*) in Bhatye estuary at Ratnagiri. Incorporated in this profile is the summary of the principle water quality parameters.

Materials and Methods

Ratnagiri is southern district of Maharashtra state bearing productive estuaries and creeks. Bhatye estuary [73°15 East and 16°′51 North] is one of most productive region in south of Ratnagiri coast where Kajvi River meets the sea. Along the banks of the river, far interior in the estuary and along the estuary the human population of traditional fishermen community resides. A number of mechanised boats for fishing in the sea pass regularly through this estuary. During such fishing operations, the hydro-chemical parameters of the seawater from oyster habitat were analysed. Temperature was recorded with standard mercury thermometer, while salinity using salinometer (Hanna Instruments, Italy), pH by the meter (pH 600 Milwaukee, Portugal) and for the estimation of dissolved oxygen the samples of sea water samples were fixed on the field and brought to the laboratory to determine the oxygen by winklers azide modified method.

Results and Discussion

In marine environment benthic invertebrates have to constantly make physiological adjustment in relation to environment variables, especially in estuarine tides river discharge, density differences between water masses in the estuary caused by differences in salinity and temperature. Among the water parameters, temperature is one of the most important factors having profound influence on the biotic communities. In particular planktonic communities constituting the very base of the energy cycle are prone to temperature influence. Kavern and Bell (1965) reported an increase in gross productivity with increase in temperature. Temperature conditions in estuaries depends upon bathymetry of the environment, water exchange between estuary and sea, and seasons of variations (Borego and Borego, 1982). Nair (1984) studied the temperature variation in different habitats in relation to meteorological parameters. In the present study on the oyster bed during monsoon (July-August) showed temperature of 30.2 to 28.2°C (Table 1). The influx of freshwater through various sources and the intrusion of seawater through the lower reaches during monsoon have profound influences on the distribution of temperature in estuary or backwater regions. During winter season (December), while minimum temperature was 27.5°C and maximum 28.0°C recorded, in summer (May) minimum and maximum temperature values were 30.5°C and 32.7°C respectively. Similar observation made by Solunkhe (1999) on Achara creek, west coast of Maharashtra.

Table-1: Physico-chemical characteristics of water in Bhatye estuary Ratnagiri

Period		Temperature (°C)	pH	Salinity (mg/lit.)	Oxygen Content (ml/lit.)
Monsoon	Minimum	28.2	7.1	16.22	4.0
(August)	Maximum	30.2	7.4	18.25	4.3
Winter	Minimum	27.5	7.6	30.55	3.3
(December)	Maximum	28.0	7.8	30.76	3.5
Summer	Minimum	30.5	7.5	33.00	3.2
(May)	Maximum	32.7	7.9	35.30	3.7

Generally, pH of water sample is dependent on the relative quantities of calcium, carbonates and bicarbonates (Zafar 1966). The pH change is also governed by the amount of free CO_2 and HCO_3 and is directly related to the pH value (Kant and Raina, 1990). They have further stated that any change in the former complex has a direct bearing on the maxima and minima of pH. In addition to the heavy rainfall by causing considerable dilution of seawater, the buffering effects cause fluctuations in the pH. In the present study the pH of water over the oyster bed in Bhatye estuary during monsoon during between 7.1 to 7.4, during winter it was 7.6 to 7.8, while in summer 7.5 to 7.9. The observed variations in pH did not show seasonal pattern. However, the pH values were low during monsoon due to heavy influx of freshwater. Similar observation was made by Mohite (2002) in kalbadevi creek at Ratnagiri.

Salinity is considered the key factor amongst all the environmental characteristics in estuarine habitats. In the present study the salinity of water over the oyster bed during monsoon ranged between 16.22‰ to 18.25‰ during winter it was 20.55‰ to 30.76‰, while between it was during winter and in summer it was 33.0 to 35.30‰. The greater influx of river water in the inter tidal region in Bhatye estuary might have accounted for this low salinity existence and this feature was obviously dependent on the time and extent of freshwater discharge from the rivers, local precipitation, water evaporation and also on the ebb tides. Nair and Ganpathy (1983) found that increase in salinity values in summer was possibly mainly due to continual evaporation of water body which remains almost in a land locked state during this period and the bar mouth has remained open during most of December and this result in significant increase in salinity in the backwaters. However, no such bar exists in Bhatye estuary. Lower salinity values were observed in Bhatye estuary during the periods of heavy rainfall due to southwest monsoon. The salinity variations in other seasons are possibly controlled by actions of tides and influence of the river water (Padmavati and Goswami, 1996). The Bhatye is situated near to the river mouth, some two Kms. interior and the broadness and shallowness of the estuary creates typical tropical estuarine conditions. In present study the maximum salinity was recorded in May when the temperature of water was quite high and water level in the reservoir decreased due to evaporation.

Dissolved oxygen (DO) is one of the most important abiotic parameters influencing the life in the coastal environment. Normally high dissolved oxygen is encountered in unpolluted areas, while at relatively polluted lower levels of DO is very low. Further, depletion of dissolved oxygen to the level of anaerobia is the most critical manifestation of pollution. Lester (1975) suggested the usefulness of dissolve oxyfen as an indicator parameter for organic pollution. In the present study the oxygen saturation varied seasonally. In general during monsoon, water is well oxygenated as compared those of summer and winter. In the present investigation DO ranged between 4.0 to 4.3 ml/l. in monsoon, but declined to the range of 3.3 to 3.5 ml/l. in winter and in summer 3.2 to 3.7 ml/l. Wide fluctuations in the oxygen contents observed during the present study can be attributed to turbulence and turbidity, freshwater influx and temperature changes. More or less similar trends have been reported by Qasim (2004) in Indian estuaries from season to season and from one region to the other in the same estuary. Generally, low oxygen content to the other in the same estuary. Generally, low oxygen content was associated with a high temperature and salinity. Many investigators found an inverse relationship of the oxygen and salinity, high values of ebb tides and low values at flood tides. Thus, the precisely timed physiological processes help estuarine animals for diverse activities especially like survival, growth and reproduction, within the required limits of temperature, salinity and oxygen contents in a given habitat.

References

Borrego, A.S. and Borrego, S.A. 1982: Temporal and spatial variability of temperature in two coastal lagoons. *Cal Coti. Rep.* XXIII: **185-197.**

Gajbhiye, S.N., Mustafa, S., Mehta, P.M. and Nair, V.R. 1995: Assessment of biological characteristics on coastal environment of Murud (Maharashtra) during the oil spill (17 May 1993). *Ind. J. Mar. Sci.*, **24: 196-202.**

Kant, S. and Raina, A.K. 1990: Limnological studies of two ponds in Jammu, II. Physiochemical Parameters. *J. Environ. Biol.*, **11: 137-144.**

Kavern, N. and Bell, R.C. 1965: Primary productivity and energy relationship in artificial streams. *Limnol. Oceanogr.* **10: 78-87.**

Lester, W.F. 1975: Polluted river, River Trent. England, In: *River Ecology* (Ed. Withion, B.A.) Blackwen Scientific Publications, London.

Mohite V.T. 2002: Base levels of heavy metals and their detoxification mechanism in green mussles *P. viridis* from coast of Maharashtra. Ph.D. Thesis, Institute of Science, Mumbai. pp. **1-250.**

Nair, N. Unnikrishnan 1984: Studies on the backwater oyster, *Crassostrea madrasensis* (Preston) of the Cochin, Harbours, Ph.D. Thesis, University of Cochin, **1-355.**

Nair, K.V.K. and Ganpathy, S. 1983: Baseline ecology of Edaiyur-Sadras estuartine system at Kalpakam part-I General hydrography and chemical feature. Mahasagar. *Bull. Nat. Inst. Oceangr.* **16(2): 143-151.**

Padmavati, G. and Goswami, S.C. 1996: Zooplankton ecology in the Mandavi-zuari eatuarine system of Goa, west coast of India. *Ind. J. Mar. Sci.* **25: 268-273** pp.

Qasim S.Z. 2004: *Handbook of Tropical Ertuarine Biology*. pp. **1-131.**

Salunkhe, S.V. 1999: Study on the reproduction of edible bivalve molluscs from Sindhudurg District Maharashtra State. Ph.D. thesis, Dr. Babasaheb Ambedkar Marathwada University, Aurangabad, S.M., India. **1-318.**

Sengupta, R. and Deshmukh, G. 2000: *Coastal and Marine Environments of Gujarat Ecology and Economics*, pp.**150.**

Yennawar, P.L. 1997: Some aspects on reproductive biology of west bivalve molluscs. Ph.D. Thesis. Dr. Babasaheb Ambedkar Marathwada Univesity, Aurangabad (M. S.), India, pp. **1-394.**

Zafar, A.R. 1966: Limnology of Hussain-sagar lake of Hyderabad. *Phykos*. **6: 115-126.**

❑❑❑

11

A Study on the Selection of Habitat by the Post Larval and Juvenile Banana Shrimp, *Penaeus merguiensis* in and around Kali Estuary, Karwar

Sampath Kumar S. and Kusuma Neelakantan
School of Ocean Sciences, Kodibag, Karwar–581 303

Introduction

The influence of environmental factors on the process of habitat selection by pinheads in general and *Penaeus merguiensis* in particular is not well understood. Most species ultilise habitats, which encompass ecologically hospitable abiotic and biotic parameters, guided by their physiological limits of tolerance The habitat selection is essentially the relationship between behaviour and environment which determines the local distribution (Meadows and Campbell, 1972).

P. merguiensis like most penaeids make use of magrove lined estuary as the nursery feeding ground (Gunter, 1967; Staples and Vance, 1979; Staples *et al.* 1985; Dall *et al.*, 1990; Vance *et al.*, 1990). Studies have been carried out in the Kali estuary Karwar to assess the seasonality in the distribution and abundance of penaeid prawns in general (Gunaga *et al.*, 1989) and *P. merguiensis* in particular (Sampath Kumar, 1996). Some information is available on the habitat requirement of *P. merguiensis* from Australian waters (Staples *et al.*, 1985; Vance *et al.*, 1990; Robertson, 1988) and also from the waters of Goa (Achuthankutty, 1988).

The present study was an extension of a work on the seed abundance of *P. merguiensis* in relation to some abiotic and biotic factors in Kali estuary for a period of two years (Sampath Kumar, 1996). During the study it was noted that even though immigration was occurring into all the creeks during the flood tides, the density of post larvae and juveniles in different creeks adjoining the estuary during low tide were different. As a result of this observation, this particular study was planned and carried out, during the peak immigration period, to know the environmental conditions prevailing in the adjoining locations and the factors limiting the survival in these area, which in turn might help in elucidating the habitat preference of *P. merguiensis* and to identify the natural causes of variation in density of post larval and juveniles of *P. merguiensis* in the Kali estuary.

Materials and Methods

The habitat preference and selection were calculated on the basis of the number of individuals obtained per unit area and is expressed as no/m^2.

The study was carried out during pre-monson, when the immigration was frequent and the abundance of the post larvae and juvenile stages of *P. merguiensis* in the estuary was at its peak (Sampath Kumar, 1996) at four study stations, referred to as Habitat 1, Habitat 2, Habitat 3 and Habitat 4.

Weekly sampling was carried out at all the four habitats, during low tide for a period of one lunar month. The sampling was done using a standard velon screen hand held drag net, having an area of 3 × 1 m and 2 mm mesh size. The sampling was done at each sampling area, by dragging the net over a distance of 10 mts, near the edge of the water, by holding the two ends of the net by two people and counting the number of post larvae and juveniles per drag. The next was dragged 3 times and the average number was calculated. Concomitantly, surface water and sediment (core) samples were collected for analysis.

Temperature was measured using a centigrade thermometer. Salinity was determined by Mohrknudson's method (Strickland and Parson, 1956), dissolved oxygen was estimated by Winkler's method (Strickland and Parson, 1956), and nutrients like Nitrate, Nitrite, Phosphate and silicate were determined by following the methods of Strickland and Parson (1956). Organic matter in sediments was estimated by the method of El Wakeel and Riley (1956) and pH was measured using pH meter (Systronic make, model 7024). Size fraction was determined by sieve analysis and pipette analysis (Holme and Intyre, 1984). Suspended load was determined by filtering 1 liter of water sample collected from the sampling site, through preheated weighed G. F/C filter paper and weighing it after drying till constant weight was attained and is expressed as mg/l.

Study Stations:

Habitat 1, Kodibag Creek

This creek joins the Kali estuary along the southern bank at a distance of 0.3 km from the mouth. Even though only about 1 km long, the link with the estuary is lost during low tide as major portion of the creek bottom lies exposed during low tide, except the far end where the water depth is around 1.5 m. It appears like a pond, when the connection with the estuary is lost, during low tide. The area is bordered by a few magrove shrubs. The initial survey conducted revealed that immigration of post larvae of *P. merguiensis* does occur during the flood tide, but sometimes no shrimps were caught during the ebb tide.

Habitat 2, Mavinahole Creek

This creek joins the estuary along the northern side, nearly opposite to Kodibag creek and is approximately 6 kms long. The creek near the study area is bordered by fringing mangroves. The creek bottom is covered sea grass and water is shallow and sediment is sandy.

Habitat 3, Sandy backwater area

This is located on the southern side of the Kharland bund near the mangrove free area of the Kadwad backwaters of the Kali estuary. It is approximately 6 kms from the mouth of the estuary and has a sandy bottom with no vegetation cover, which gets exposed during the low tide.

Habitat 4, Muddy managrove area

This is located in the backwater area of Kadwad, near to Habitat 3 on the northern side of the Kharland bund. This area includes a steep intertidal bank adjoining the fringing mangrove forest.

Statistical Analysis:

The data acquired from the investigation was subjected to correlation analysis (Sendecor and Cochran, 1967) to test the significance of the relationship.

Result and Discussion

From Table 1, it can be seen that only post larvae and early stages of juveniles were seen at Hab. 1 and Table 2 shows that during 2nd and 4th collection they were totally absent at Hab. 1, thus suggesting that even though post larvae entered the creek, they did not remain or survive there as some of them were seen to move out during the following ebb tide along with the water and were collected in the drag net tide against the receding waters. This may be attributed to the slightly low pH of the water.

Table-1: Percentage availability of different size groups collected from each habitat

Size Groups	Habitat 1	Habitat 2	Habitat 3	Habitat 4
60	0	0	0	21
50 mm	0	4	0	42
40 mm	0	33	0	92
30 mm	0	21	0	105
20 mm	18	40	0	147
10 mm	7	6	0	12

Table-2: Variation in the density of post larval and juveniles of *P. merguiensis* at different habitats

Collections	Post larval and Juvenile Density (no/m^2)			
	Habitat 1	Habitat 2	Habitat 3	Habitat 4
1	19	83	0	234
2	0	121	0	583
3	84	144	0	302
4	0	117	0	148
5	21	56	0	827

Williams had indicated that hydrogen ion concentration of the substrate may influence attraction to a substrate, but the present observation suggests that slightly low pH can induce repulsion from a particular substrate type, when it is not tolerable. Pedini (1981) has shown that acidic sediments can decrease growth rates and increase mortality in aquaculture ponds.

The results of hydrological analysis shows that, the pH of water of the sediment and dissolved oxygen was less in comparison to all other habitats (Table 3). Even though there was no significant relationship existing between seed abundance and any particular environmental factors statistically, at Habitat 1 there existed significant relationship between water temperature and sediment temperature (at $P < 0.05$), water temperature and dissolved oxygen (at $P < 0.01$), which might suggest that as the soil is exposed to the sun during day time at low tide it gets heated up and in turn increasing the temperature of the inflowing flood tide waters, which result in decreasing the oxygen content of water.

At Hab. 2 the density was (56-144 no/m^2) greater than Hab. 1 (0-84 no/m^2). This habitat recorded almost all the size groups except the larger size group (Table 1), even though the density was low in comparison to Hab. 4 (148-827 no/m^2). The continued presence of early stages suggests regular immigration during flood tides (Staples and Vance, 1979). The amount of organic content of the sediment implies richness in terms of food availability for the shrimps. The post larvae appear to surivie even though the substrate was predominated by sand (Table 3). The lack of muddy substrate, which is its preferred habitat (Robertson and Duke, 1987) may be outweighed by the protection rendered by vegetation cover and the organic richness as a food source of food.

Hab. 3 was sandy and without any vegetation cover and no post larvae were collected. This might be attributed to the lack of food (absence of organic matter) and lack of protection by way of the absence of vegetation cover and no soft muddy substratum. The presence of coarse sand makes it difficult to bury quickly, unlike muddy substrate. Somers *et al.* (1987) inferred from their study that low organic carbon becomes a limiting factor in the distribution of the commerical species, during their estuarine phase. Vatiation in shrimp abundance has been linked to the organic carbon content of the sediment both in Gulf of Mexico (Gardy, 1971) and in the Sudanese Red Sea (Branford, 1981).

Hab. 4 recorded very high densities (148-827 no/m^2) and all size groups were represented. This may be due to factors like, increased oxygen content, high organic detritus (Table 3), floral and faunal abundance in the water and sediments (Sampath Kumar, 1996) and protection offered by the soft muddy substratum.

Correlation analysis did not reveal any significant relationship between seed abudance and environmental factors at any Habitat and hence the data is not presented.

Table-3: Variation in the abiotic factors and Seed Abundance at different habitats during the study period

Parameters	Hab. 1	Hab. 2	Hab. 3	Hab. 4
Seed Abundance (no/m^2)	124	521	0	2094
Water Temperature (°C)	30.1	31.3	30.4	30.4
Salinity (‰)	27.5	28.6	19.6	20.2
pH	6.4	7.3	7.8	7.8
Suspended Load (mg/g)	0.009	0.009	0.007	0.007
Dissolved Oxygen (ml/l)	2.9	3.9	4.6	5.2
Sediment Temperature (°C)	29.7	31.2	31.1	30
pH	6.6	7.6	7.7	7.7
Orgamic Carbon (mg/ g)	4.9	5.4	0.4	6.3
Sand (%)	99.5	86.3	99.5	11.8
Silt (%)	0.5	10.5	0.5	33.7
Clay (%)	0	3.2	0	54.5

Sick *et al.* (1972) have inferred that a suitable substratum is not only important for prawn species to maintain their normal burrowing habits but also in promoting high survival and growth rates in culture systems. The present result apparently agrees with that if Young and Carpenter (1977) who determined that the final selection of a nursery ground probably depends on habitat features such as depth, substratum and types of plant or other cover for early juveniles of various species.

The greater densities in muddy mangrove area are mainly due to the aggregation of immigrating population (Sampath Kumar, 1996). Vance *et al.*, (1996) observed during their study that large catch of *P. merguiensis* post larvae were only recorded from mangrove lined banks of the estuary, although small numbers of post larvae settled in all parts of the estuary.

The present observation at Hab. 4 provides further evidence for the hypothesis suggested by Robertson and Duke (1987) that mangrove habitat is more productive and has greater food resource for consumers. However, they further added that the structural hererogeneity provided by mangrove tree, roots and mangrove snags in creeks afford greater protection from predators. In the present study, it was observed that during the low tide the water records to such an extent that it reaches below the level of any growing mangrove trees, and has no prop roots or stilt roots to provide protection to the inhabiting organism. The only way of self defense in this habitat of high density seems to be burying themselves in the muddy substratum and disappearing from its predators. Hence it shows that the degree of protection that the shrimps from the mangrove roots is less significant in comparison to the muddy substratum.

The observation recorded at Hab. 1 and Hab. 2 similar to the influence drawn by Vance *et al.*, 1990. From their study in the Embley estuary they concluded that, if the habitat in which the post larvae and small juveniles settles, bears less than optimal environmental conditions, then either they move back into the main estuary to be carried by subsequent flood tides (Hab. 1) or remain where they first settle and be subjected to increased predation and decreased growth (Hab. 2). The difference in the densities of post larvae and juveniles of *P. merguiensis* recorded at Hab. 2 and Hab. 4 might help us in suggesting that Hab. 2 does not bear very optimal environmental condition for the species, but remains there as it gets food and protection. In addition the temperature recorded is relatively higher than all other habitats.

As the environmental condition of the estuary during the study in general was appropriate for immigration and proper growth (Gunaga *et al.*, 1989, Sampath Kumar, 1996), the differences in densities at each habitat was due to the differences in the prevailing environmental conditions (Table 4). Thus it is possible to conclude that muddy areas provide the most protection to these prawns and it may be their most preferred

Table-4: Probable Abiotic factors that might be having some influence on Seed Abundance at different study sites

Factors	Hab. 1	Hab. 2	Hab. 3	Hab. 4
Seed Abundance (no/m^2)	124	521	0	2094
pH of Water	6.4	7.3	7.8	7.8
Dissolved Oxygen (ml/l)	2.9	3.9	4.6	5.2
Sediment pH	6.6	7.6	7.7	7.7
Organic Carbon (mg/g)	4.9	5.4	0.4	6.3
Sand (%)	99.5	86.3	99.5	11.8
Clay (%)	0	3.2	0	54.5

habitat and if they land up in a sandy habitat, they might remain there if the area provides protection by means of vegetation cover. If the inundating waters take the post larvae to an area with unfavourable pH, then they do not remain there and more out along with the receding waters.

References

Achuthankutty, C.T. 1988. Biology of the banana prawn, *Penaeus merguiensis* de Man from Goa waters. *Journal of Aquatic Organisms*., **2.1: 1-10.**

Branford, J.R. 1981. Sediment and distribution of penaeid shrimp in Sudanese Red Sea. *Estuar. Coast. Shelf Sci.*, **13: 349-354.**

Dall, W., Hill. B.J., Rothlishberg, P.C. and Staples, D.J. 1990. The Biology of the Penaeidae. *Adv. Mar. Biol* . **27: 1-489.**

El-Wakeel, S.K. and Riley, J.P., 1956. Determination of Organic Carbon in marine muds. *J. Cons. Perm. Int. Explor. Mer.,* **22: 180-183.**

Gardy, J.R. 1971. The distribution of sediment properties and shrimp catch on two shrimp shrimping grounds on the continental shelf of Gulf of Mexico. *Proc. Gulf and Caribean Fish. Inst.* 23 (Ann. Sess.): **139-148.**

Gunaga, V.R., Neelakantan, K and Neelakantan, B. 1989. Seasonal abundance of penaeid post larvae in the Kali estuary, Karwar. Mahasagar, **22(4): 193-197.**

Gunter, G. 1967. Some relationships of esturaries to the fisheries of the Gulf of Mexico. In: *Estuaries*, edited by G.H. Lauff. American Association for the Advancement of Science, Washington DC, **621.**

Holme, N.A. and McIntyre, A.D., 1984. *Methods for the Study of Marine Benthos*, Blackwell Scientific Publication, Oxford. p. **172.**

Meadows, P.S. and Campbell, J.I. 1981. Habitat selection by aquatic invertebrates. In: Russel, F.S., Young, M. (eds). *Adv. Mar Biol.*, (Academic Press, London). pp. **10: 271.**

Pedni, M. 1981. Penaeid Shrimp Culture in Tropical developing Countries. *FAO. Fish Cir.* p. **732.**

Robertson, A.I. 1988. Abundance, diet and predators of juvenile banana prawn, *Penaeus merguiensis* in a tropical mangrove estuary., *Aust. J. Mar. Freswat. Res.* **39: 467-478.**

Robertson, A.I. and Duke, N.C. 1987. Mangroves are nursery sites. Comparison of the abundance and species composition of fish and crustaceans in mangroves and other near shore habitats in tropical Australia. *Mar. Biol (Berl)* **96: 193-206.**

Sampath Kumar, S. 1996. Some eco-physiological aspects of *Penseus merguiensis* from karwar waters, *Ph.D. thesis*, Karnatak Univesity. p. **265.**

Sick, L.V., Andrews, J.W. and White, D.B. 1972. Preliminary studies of selected environmental and introital requirements for the culture of penaeid shrimps. *Fish. Bull. US Fish. Wildife Ser.*, **70: 101.**

Sivabalan, A. 1993. Studies on the ecology of the mangrove of kali estuary, Karwar. *Ph.D. thesis*, Karnatak University. p. **390.**

Sendecor, G.W. and Cochran, W.C. 1967. Statistical Methods, 6th edition. Oxford and IBH Publishing Co., New Delhi. p. **172.**

Somers, I.F., Croaces, P.J. and Hill, B.J. 1987. Distribution and abundance of tiger prawns *Penaeus esculentus* and *Penaeus semisulactus Aust. J. Freshwat. Res.* **38: 63.**

Staples, D.J. and Vance, D.J. 1979. effects of changes in catchability on sampling of juvenile and adolescent banana prawns *Penaeus merguiensis* de man. *Aust. J. mar. Freshwat Res.* **30: 511-519.**

Staples, D.J., Vance, D.J. and Heals, D.S. 1985. Habitat requirement of juvenile penaeid prawn and their relationship to offshore fisheries, In: *Sec. Aust. Prawn Sem.*, Rothlisberg, P.C., Hill, B.J. and Dtaples, D.J. (Eds) NPS2 Clevaland, Australia, **47.**

Strickland, J.D.H. and Parson, T.R. 1956. A practical handbook of seawater analysis. *Bull. Fish. Res. Bd. Canada.*, pp. **167: 310.**

Vance, D.I., Haywood, M.D.E. and Staples, D.J. 1990. Use of a mangrove estuary as a nursery area by postlarval prawns, *Peneus merguiensis* de Man, in northern Australia. *Estuar. Coastal Shelf Sci.*, **31: 689-701.**

Williams, A.B. 1958. Substrate as a factor in shrimp distribution. *Limnol Oceanogr,* **3: 283-291.**

Young, P.C. and Carpenter, S.M. 1977. recruitment of postlarval prawns to nursery areas in Moreton Bay, Qld. Australia. *Aust. J. Mar. Fresahwat. Res.* **28: 745-773.**

□□□

12

The Role of Abiotic Factors in the Distributional Patterns of Acarina and Collembola in the Sub-tropical Forest Ecosystem of Manipur

Ch. Chitrapati and Th. Biony Singh
Department of Life Sciences, Manipur University,
Canchipur–795 003, Imphal

Introduction

The soil is a physically and chemically diverse environment that is used by a wide variety of invertebrate animals and micro-organisms. The species differ in the distribution of their activities in space, time and biological function. The mites and collembolas constitute 72-97 per cent of the total arthropod fauna of Indian soil (Singh and Mukharji, 1973; Prabhoo, 1976; Roy *et al.,* 1998). In the present study soil microarthropod would mean the microarthropods fauna in soil in general with Acarina and Collembola in particular. Soil microarthropods play a key role in regulation of microbial population, decomposition of organic matter and cycling of nutrients within the soils (Coleman and Crossely, 1996; Brady and Weil; 1996; Tian *et al.*, 1998). The abiotic factors which are physical and chemical in nature are important in determining the population pattern and species composition. Fluctuations in temperature and moisture are less secure than they are at and above and ground surface and the amplitude of these fluctuations decreases with depth in the soil (Geiger, 1959). Moreover soil organic matter play an important role in determining the character of the soil.

The soil organic carbon is the major constituent of organic component of which the plant are made and they were concerned with different metabolic reaction which are vital for growth of plants.

In north-east India soil microarthropods communities have been investigated (Vatsauliya and Alfred, 1980; Darlong and Alfred, 1981). However, under Manipur climatic conditions except studies made by Singh *et al.*, (1986), Singh and Yadava (1998), no work has been reported so far. gradual soil degradation particularly due to slash and burnt method has led to the drastic changes in the population dynamics of the soil microarthropods and therby either directly or indirectly affecting the soil ecosystem with this objective. The present study carried out at two sites in the substropical forest ecosystem of Manipur in order to know the distributional patterns of soil microarthropods under the effect of abiotic factors.

Materials and Methods

In the present investigation three plots having a size of 10m × 10m each were demaracted in both the protected forest and degraded sites. Each plot was again divided into subplot I, II and III in both the sites. A total of 234 soil samples were drawn at the rate of 27 samples from each study site per month during the study period from March, 2000 to February, 2001. The soil samples were collected from both the sites of forest with the help of cylindrical soil cores of 19.64 cm^2 in surface area at three layers *i.e.* 0-5, 5-10 and 10-15 cm. The soil microarthropods were extracted using modified Tullgren funnel at a constant temperature of 40° (± 2°C) for 6-7 days depending on the moisture content of soil samples and they were collected in vials containing 80 per cent alcohol. The soil microarthropods were identified and counted under a binocular microscope and were classified upto order level. In both the study sites the measurement of soil temperature were recorded with the help of soil thermometer for each layers *i.e.* 0-5 cm, 5-10 cm and 10-15 cm during the sampling period which was set at few cm. depth.

The soil moisture contents were determined by gravimetric method as given by Misra (1968) and Wilde *et al.* (1985). The soil organic carbon were determined by oxidation calorimetric method after modified Walkley and Black method as given by Anderson and Ingran (1993). The Collembola and Oribatid communities of the protected forest and degraded sites were analysed to study the Community Similarly using the following statistical formula.

Community Similarity

The similarities or dissimilarities of Collembola and Oribatid communities in the protected forest and degraded sites were worked out using the following index.

The Coefficient of Similarity (CCS) of Sorensen (1948) also known as Quotient of Similarity (Q/S) was calculated as

$$\frac{Q}{S} = \frac{2C}{S_1 + S_2} \times 100$$

Where, C = number of species common to both the communities.

S_1 = number of species in community 1 (protected site).

S_2 = number of species in community 2 (degraded site).

Results and Discussion

The vertical distribution pattern and the total population density of Acarina and Collembola in the protected forest and degraded study sites were shown in Tables 1 and 2 respectively. The total population density of Acarina and Collembola recorded higher in the protected forest site as compared to the degraded study site both annually and vertically. This may be due to high soil moisture content, high organic carbon content and close canopy in the study site I Fig. 1a and 2a. Vegetation itself does not directly influences Collembola but may affect them indirectly through microclimate, soil fertility, soil profile, microflora and soil porosity (Hagvar, 1982). The population density of Acarina exhibited highest aggregation of 38 per cent and 41 per cent in the uppermost soil layer *i.e.* at 0-5 cm in the protected forest and degraded sites respectively.

Table-1: Showing the total population density of Acarina in the protected forest and degraded ecosystem (Number ± S. E.) × 10^2 m^{-2}

Protected forest		Degraded ecosystem	
Soil layers (cm)	**Number ± S.E.**	**Soil layers (cm)**	**Number ± S.E.**
0-5	718 ± 36.05	0-5	495 ± 19.20
5-10	657 ± 20.05	5-10	372 ± 20.33
10-15	530 ± 34.97	10-15	339 ± 25.69
Total	1905 ± 91.07	Total	1206 ± 65.22

The least population density of Acarina was found at 10-15 cm soil layer contributing only 28 per cent in both the study sites. Similarly maximum population density of Collembola was recorded at the upper soil layer and decreased with depths. Probably may be due to adequate living space, favourable moisture conditions, aeration rates and rich accumulation of human soil in the upper soil layer (Hale, 1967; Wallwork 1970). Seasted and Crossely (1981), Darlong (1984) also reported maximum confinement of soil microarthropod in the upper soil layers.

Table-2: Showing to total population density of Collembola in the protected forest and degraded ecosystem (Number ± S. E.) × 10^2 m^{-2}

Protected forest		Degraded ecosystem	
Soil layers (cm)	**Number ± S.E.**	**Soil layers (cm)**	**Number ± S.E.**
0-5	241 ± 12.68	0-5	166 ± 4.95
5-10	190 ± 3.32	5-10	148 ± 3.8
10-15	171 ± 7.66	10-15	143 ± 6.20
Total	602 ± 23.66	Total	457 ± 14.15

In the present investigation Acarina has been classified into four suborders *viz.* Cryptostigmata, Mesostigmata, Prostigmata and Astigmata. Among the different suborder of Acarina, Cryptostigmata exhibited highest population density and Astigmata occupied lowest population density in both the study sites. The Acarina and Collembola exhibited maximum population density during rainy season and gradual decreasing trend towards the dry winter season in both the study sites showing distinct seasonal trend.

This may be due to congenial microclimatic condition during this season particularly high soil moisture content had a favourable effect in the decomposition rate which was reflected in the higher organic carbon content of the soil thus making suitable for multiplication of soil microarthropods. Loots and Ryke (1966) working in the tropical forest of Africa found a peak density of microarthropods during late summer and minimum during dry winter season. Yadava and Singh (1988) while working in the substropical forest ecosystem of Shiroy Hill of Manipur in North Eastern India also exhibited higher population density of soil microathropods during rainy season which gradually decreased with the approach of winter season. These findings are in agreement

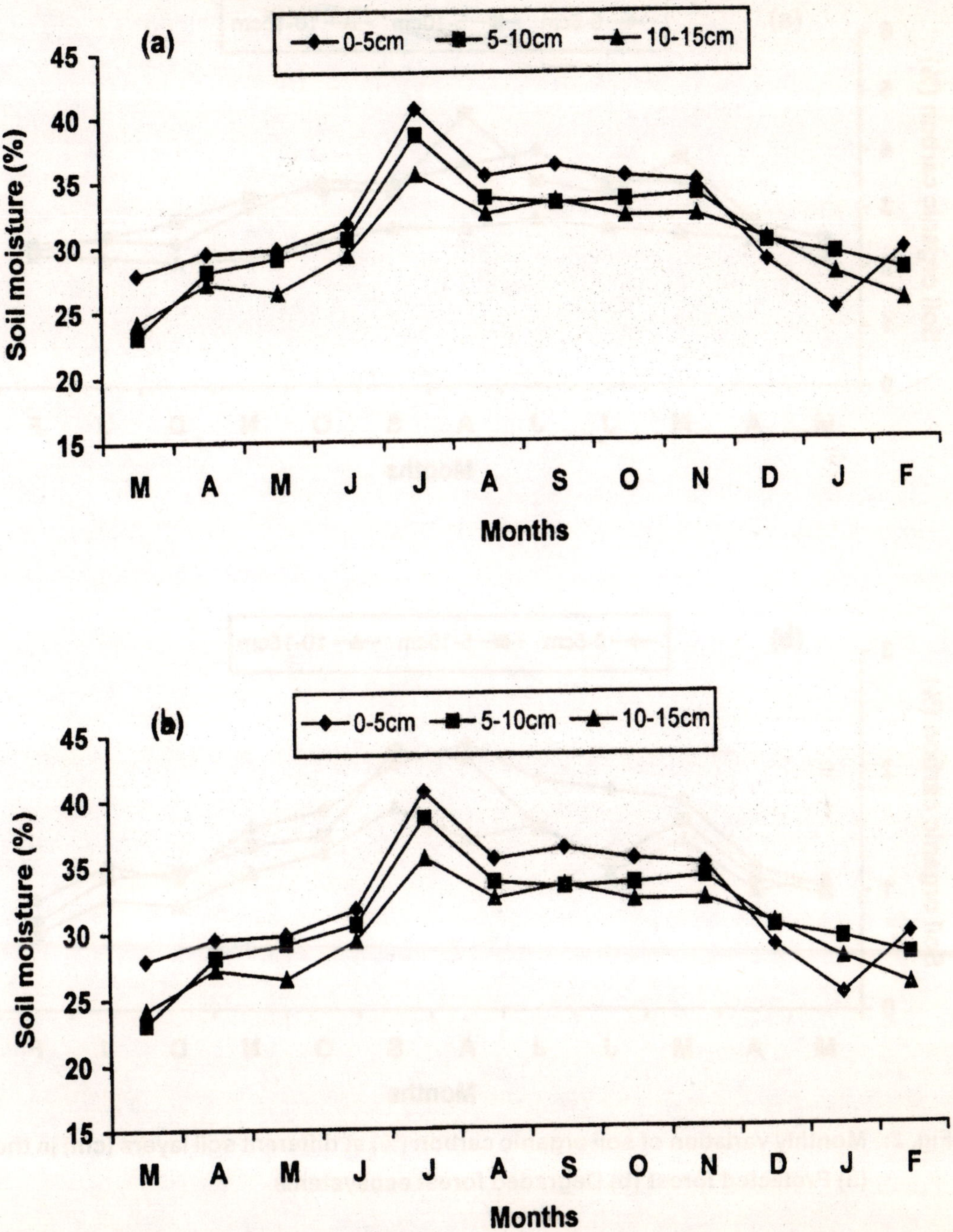

Fig. 1: Monthly variation of soil moisture (%) at different soil layers (cm) in the (a) Protected forest (b) Degraded forest ecosystems

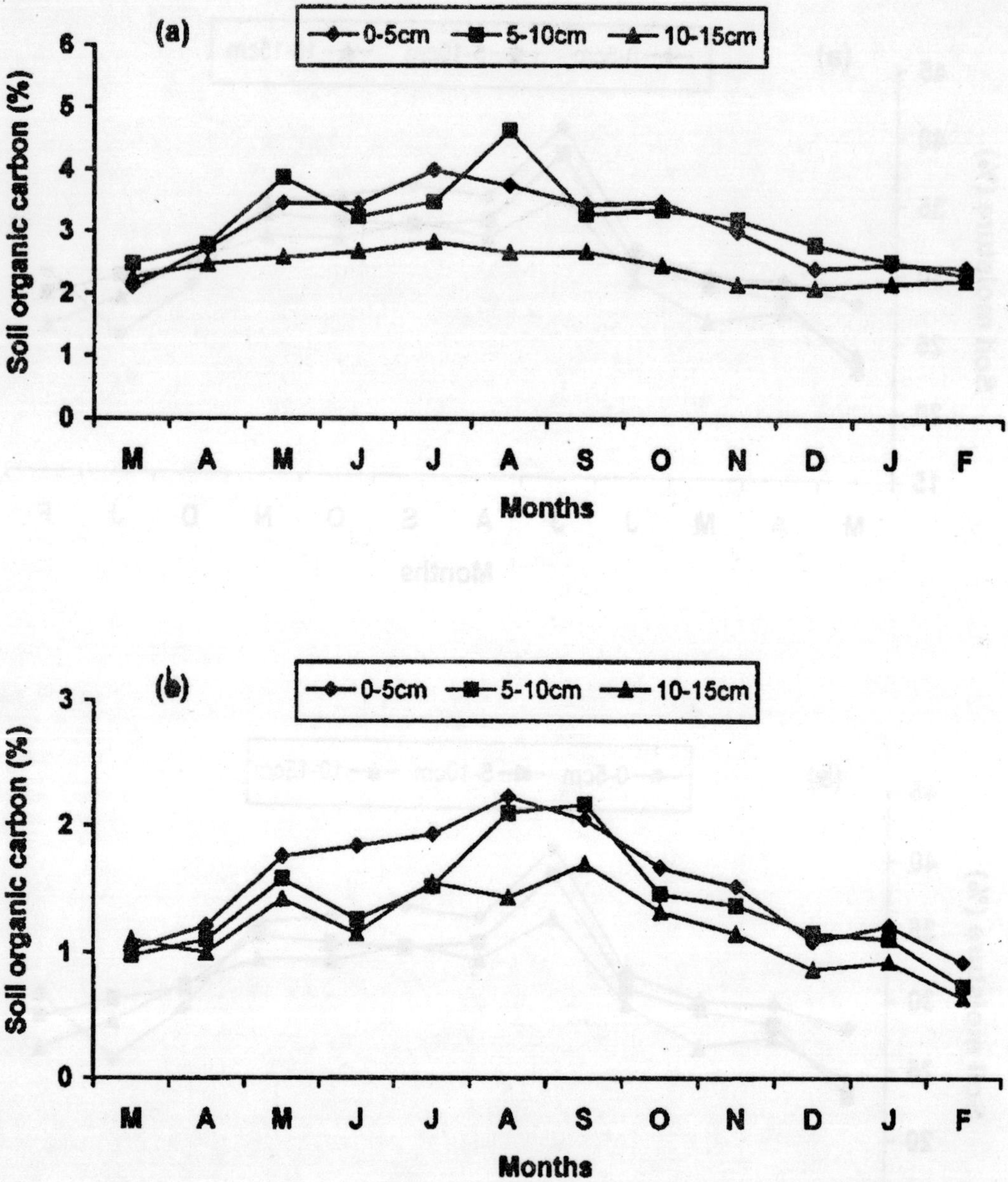

Fig. 2: Monthly variation of soil organic carbon (%) at different soil layers (cm) in the (a) Protected forest (b) Degraded forest ecosystems

with the present investigation. Sharon *et al.* (2001) while working in an Oak forest reported that abundance of soil fauna is influenced by season and not only the upper leaf layer as shown by Argyropoulou *et al.* (1993).

Comparatively higher values of soil moisture and organic carbon content in he protected study site (Fig. 1a and 2a) than that of degraded open area (Fig. 1b and 2b) have found to be important in affecting aggregation behaviour as well as fluctuation pattern of both Acarina and Collembola population. Thus in the degraded and open study site (sit-II) the population of both of these soil microarthropods group resulted in the dispersed and less population activity. The correlation analysis between Acarina and Collembola population with soil moisture content, soil temperature and organic carbon showed a positive significant relationship in both the study sites (Tables 3 and 4). Badejo *et al.*, (1997) reported that Mites and Collembola were generally aggregated under high moisture conditions in the wet season which was also in accordance with the present finding.

Table-3: Correlationship between Acarina with soil moisture and temperature in site I and II

Variables	Soil layers (cm)	r	df	Y	t	P
Protected Forest						
Soil moisture (%)	0-5	0.8660	10	-8.41+0.45x	5.4732	<0.01
	5-10	0.6719	10	-6.81+0.39x	2.8670	>0.05
	10-15	0.7040	10	-6.39+0.36x	3.32	<0.01
Soil temperature (°C)	0-5	0.8440	10	-1.03+0.35x	4.976	<0.01
	5-10	0.807	10	-1.54+0.36x	4.315	<0.01
	10-15	0.785	10	-11.94+5.48x	3.937	<0.01
Soil organic carbon (%)	0-5	0.801	10	-3.13+2.93x	7.062	<0.01
	5-10	0.860	10	-8.55+4.69x	5.318	<0.01
	10-15	0.789	10	-8.08+14.12x	4.060	<0.01
Degraded Ecosystem						
Soil moisture (%)	0-5	0.828	10	-3.79+0.27x	4.67	<0.01
	5-10	0.673	10	-3.42+0.25x	2.875	<0.01
	10-15	0.559	10	-2.34+0.20x	2.133	<0.05
Soil temperature (°C)	0-5	0.879	10	-0.49+0.23x	8.005	<0.01
	5-10	0.915	10	-1.45+0.23x	7.197	<0.01
	10-15	0.911	10	-1.38+0.21x	6.954	<0.01
Soil organic carbon (%)	0-5	0.799	10	-0.005+2.68x	4.198	<0.01
	5-10	0.500	10	1.71+0.93x	1.824	>0.05
	10-15	0.649	10	0.12+2.50x	2.697	<0.05

Table-4: Correlationship between Collembola with soil moisture and temperature in site I and II

Variables	Soil layers (cm)	r	df	Y	t	P
Protected Forest						
Soil moisture (%)	0-5	0.596	10	-2.81+0.15x	2.339	<0.05
	5-10	0.514	10	-1.29+0.09x	1.897	>0.05
	10-15	0.627	10	-1.91+0.11x	2.859	<0.05
Soil temperature (°C)	0-5	0.655	10	-0.64+0.14x	2.738	<0.05
	5-10	0.628	10	-0.083+0.085	2.551	<0.05
	10-15	0.627	10	-0.17+0.07x	2.543	<0.05
Soil organic carbon (%)	0-5	0.678	10	-1.76+1.21x	2.914	<0.05
	5-10	0.539	10	0.14+0.53x	2.022	>0.05
	10-15	0.629	10	-2.19+1.48x	2.556	<0.05
Degraded Fecosystem						
Soil moisture (%)	0-5	0.880	10	-1.35+0.09x	5.793	<0.01
	5-10	0.858	10	-1.52+0.103x	5.291	<0.01
	10-15	0.580	10	-0.55+0.066x	3.165	<0.05
Soil temperature (°C)	0-5	0.766	10	0.08+0.06x	3.765	<0.01
	5-10	0.819	10	-0.15+0.68x	4.518	<0.01
	10-15	0.576	10	0.17+0.05x	6.954	<0.01
Soil organic carbon (%)	0-5	0.670	10	-0.12+0.92x	2.853	<0.05
	5-10	0.711	10	0.25+0.72x	3.196	<0.01
	10-15	0.440	10	0.61+0.49x	1.550	>0.05

Table-5: Showing community similarity of Oribatid and Collembola in different seasons and soil layers

Soil layer (cm)	Season	Sorensen's coefficient Q/S (%)	
		Oribatid	Collembola
0-5	Summer	66.70	54.54
	Rainy	80.00	54.54
	Winter	72.72	66.67
5-10	Summer	54.54	60.00
	Rainy	66.70	60.00
	Winter	25.00	50.00
10-15	Summer	75.00	57.14
	Rainy	88.80	50.00
	Winter	40.00	28.57

In the present investigation the community analysis have been confined to the individuals belonging to only nine genera each of Cryptostimata (Acarina) and Collembola. Oribatid belonging to nine genera *viz. Chaunoproctus* sp., *Scheloribates* sp., *Zetorchestes* sp., *Rhysotritia* sp., *Haplophorella* sp., *Oppia* sp., *Xylobates* sp., *Galumna* sp., and *Eupthiracus* sp. and nine genera of Collembola *viz. Sminthurnus* sp., *Isotoma* sp., *Sinella* sp., *Tullbergia* sp., *Folsomia* sp., *Lepidocyrtus* sp., *Onychiurus* sp., *Xenylla* sp. and *Tullbergia* sp. were analysed to show the similarity between the two study sites.

The community similarity of Acarina and Collembloa between the two sites were shown in table 5. The value of Sorensen Quotient of Oribatid and Collembola showed high similarity between the two study sites which was regulated by different abiotic factors. Some genera such as *Chaunopoctus* sp., *Scheloribates* sp., *Oppia* sp. and *Xenylla* sp. were commonly available throughout the seasons and soil layers and exhibited high similarities between the two study sites. Lamoncha and Crossley (1998) and Sharon *et al.* (2001) pointed out that every taxon has a geographical range which fluctuates through times and becomes discontinuous and though individuals were distributed within the range of various points at any given time, they may not be available at the same points. But Addison *et al.* (2003) reported that the community structure of Collembola in old-growth forests differed significantly from those of younger forests and could still be differentiated from those of mature forests 80-102 years after clearcutting. However, the results of the present study showed that Collembola and Acarina population tends to be similar in diversity and richness so long the microhabitat factors are similar and least difference in the elevation of the study site could not influence much in their distribution pattern. The population response of Collembola and Acarina significantly showed high correlationship with soil moisture content and temperature in both the study sites. As it was revealed from the investigation in the protected site with congenial abiotic factors particularly soil moisture played important role in determining population distribution both horizontally as well as vertically protected site were optimum soil moisture and temperature condition for the population growth of soil arthropods. In the degraded site, population response, associated with clear cutting and fire, resulted in a low population density as compared to the protected study site. Both Collembola and Acarina abundance preferred humus rich soils as well as soil moisture content which was clearly indicated from the high significant correlations analysis of the present study. A similar pattern was observed by Usher, (1976), and

Verhoef *et al.* (1988) and they reported soil moisture as the most important environmental factor acting upon soil arthropods distribution. Thus the present study revealed that the microhabitat factors particularly soil moisture and soil organic matter content tended to respond more rather than the microhabitat factors in the distribution pattern of Acarina and Collembola population both horizontally as well as vertically.

References

Addison, J.A. Trofymow, J.A. and Marshall, V.G. 2003. Abundance, species diversity and community structure of Collembola in successional coastal temperate forests on Vancouver Island, Canda. *Applied Soil Ecology*, **24: 233-246.**

Anderson, J.H. and J.S. Ingram leds. 1993. *Tropical Soil biology and Fertility*: A Handbook of Method, U.K.C.A.B.—International.

Argyropoulou, M.D., M.D. Asikidis, G.D. Latrou, G.P. Stamou. 1993. Colonisation patterns of decomposing litter in a maquis ecosystem, *European Journal of Soil Biology*. **29: 183-191.**

Badejo, M.A., T.O. Obilade and B.A. Oblubakin. 1997. Spatial distribution and abundance of mites and springtails under different temperature and moisture regimes in a tropical rain forest floor. *Tropical Ecology*. **38(1): 31-38.**

Brady, N.C. and R.R. Weil. 1996. *The Nature and Properties of Solid*, 11th Edition, Prentice Hall-upper saddle River, N.J.

Coleman, D.C. and D.A. Crossley. 1996. *Fundamentals of Soil Ecology*, Academic Press, New York, USA.

Barlong, V.T. 1984. Ecological studies of soil arthropods in forest and Jhum system of Laitkov, Meghalaya: In Ph. D. Thesis, Nehu, Shillong.

Darlong, V.T. and Alfred, J.R.B. 1981. Differences in arthropods population structure in soils of forest and Jhum sites of North Eastern. India. *Pedobiologia*. **23: 112-119.**

Geiger, R. 1959. *The Climate Near the Ground*, Harvard Univ. Press, Cambridge, Mass, USA.

Hagvar, S. 1982. Collembola in Norwegian coniferous forest soils. Relations to plant communities and soil fertility. *Pedobiologia*. **24: 255-296.**

Hale, W.G. 1967. *Collembola in "Soil biology"* (Burges, N. and Raw, Feds). Academic Press London and New York. pp. **397-411.**

Lamoncha; K.L. and D.A. Crossely, Jr. 1998. Oribatid mite diversity along an elevation gradient in a south eastern Appalachian forest. *Pedobiologia*. **42: 43-45.**

Loots, G.C. and P.A.J. Ryke. 1966. A comparative quantitative study of the microarthropods in different types of pasture soils. *Zool. Afr.* **21: 157-197.**

Misra, R. 1968. *Ecology Work Book*, Oxford and IBH Pub. Co. Kolkata, India.

Prabhoo, N.R. 1976. Soil microarthropods of virgin forest and adjoining tea fields in the Western Ghats in Kerala. *Oriental insects.* **10(33): 435-442.**

Roy, S., A.K. Shrivasav and M.M. Roy. 1998. Abundance of collembolans and Acarines in natural grassland of Bhundelkhand, *Flora and Fauna.* **4(2): 55-58.**

Seastedt, T.R., D.A. Crossely Jr. 1981. Microarthropods response to cable logging and clear cutting in the southern Appalachians, *Ecol.* **62: 126-135.**

Sharon, R., G. Degani and M. Warburg 2001. Comparing the soil macrofauna in two Oak wood forests: does community structure differ under similar ambient conditions?

Singh, J. and S.P. Mukharji. 1973. Population size of soil arthropods in some field at Varanasi. *Sci. and Culture.* **39: 498-500.**

Singh, T.B. and P.S. Yadava and E.J. Singh. 1986. Population density of soil arthropods in the subtropical forest ecosystem at Shiroy hill. Manipur. *Proc. Symp. Pest, Resid & Env. Pollu.* pp. **272-288.**

Singh, T.B. and P.S. Yadava. 1998. Seasonal fluctuations of oribatid mites in a subtropical forest ecosystem of Manipur, North Eastern Inda. *International Journal of Ecology and Environmental Sciences.* **24: 923-129.**

Sorensen, T. 1948. A method of establishing groups of equal amplitude in plant sociology based on the similarity of species content and its application to analysis of the vegetation Danish commons. *Biol. Skr.* **5: 1-34.**

Tain, G., C.O. Adejuyigbe, G.O. Adeoye and B.T. Khang. 1998. Role of soil microarthropods in leaf decomposition and N release under various land-use practices in the humid tropics. *Pedobiologia.* **42: 33-42.**

Usher, M.B. 1976. Aggregation responses of soil arthropods in relation to the soil environment. In: J.M. Anderson and A. Macfadyen (Editors). *The Role of Terrestrial and Aquatic Organisms in Decomposition Processes.* Blackwell Scientific Oxford, pp. **61-94.**

Vatsauliya, P.K. and J.R.B. Alfred. 1980. Qualitative study of soil arthropods in Jhum ecosystems of North eastern India. *Indian Zoologist*, **4: 153-160.**

Verhoef, H.A., Prast, J.E. and Verweij, R.A. 1988. Relative importance of fungi and algae in the diet and nitrogen nutrition of *Orchesella cinta* (L.) and *Tomocerus minor* (Lubbock). *Funct. Ecol.* **2: 195-201.**

Wallwork, J.A. 1970. *Ecology of Soil Animals.* McGraw Hill Publishing Company Limited.

Wilde, S.A., R.B. Cosey, J.G. Iyer and G.K. Voigt. 1985. *Soil and Plant Analysis for Tree Culture*, Oxford and IBH Pub. Co. New Delhi.

Yadava, P.S. and E.J. Singh. 1988. Some aspects of ecology of oak forest in Shiroy Hills, Manipur (North-Eastern India). *International Journal of Ecology and Environmental Sciences.* **14: pp. 111.**

❑❑❑

13

Isolation of Bacteriophages Specific to *Pseudomonas aeuroginosa*

P. Dhevagi, M.A. Shibu*, M. Rabeeth* and V. Bharath Kumar
Department of Environmental Sciences, Tamil Nadu Agricultural University
*Kongunadu Arts and Science College, Coimbatore–641 029

Introduction

Over evolutionary time, bacteria becoming resistance to antibiotics by acquiring genes for proteins that either destroy the antibiotic molecule or expel it from the cell. In addition to it, increasing concern is about the impact of antibiotics on food products. Bacteria such as *Staphylococcus aureus, Mycobacterium tuberculosis, Klebsiella pneumoniae*, *Pseudomonas aeuroginosa* are some of the dreadful pathogens resistant to antibiotics. *Pseudomonas aeuroginosa* are gram negative, fermentative, motile and opportunistic human pathogen. It is responsible for ventilator-associated pneumonia in incubated patients reaching the death rate of about 38 per cent. It is also responsible for urinary tract infections, surgical wound infections, burns etc. (Brewer *et al.,* 1996). This oppurtunitic pathogen can be eliminated by a prehistoric era usage-Bacteriophages. Recent studies and decade long use by Georgians have brought renewed interest in the possibilities of bacteriophage therapy, the use of specific kind of viruses that attack only bacteria (Carlton 1999, Sulakvelidse *et al.,* 2001). Their specificity and harmless nature to eukaryotes makes them ideal as therapeutic agents (Merill *et al.,* 2003). In recent years, well controlled animal models have been demonstrated that phages can rescue animals from a variety of fatal infections, while non-controlled clinical reports have shown that phages can be effective in treating multidrug resistant infections in human. The potential of bacteriophages as therapeutic agents against infection has yet to be fully explored. Hence the present investigation was planned on the possibility of using phages to *Pseudomonas aeuroginosa* as *invitro*-therapy for infection.

Materials and Methods

This study was carried out in four phases: (1) Isolation of *P. auroginosa* and phages specific to *P. aeuroginosa*. (2) Establishing the MLD in the animals model. (3) Test for phage toxicity. (4) Bacteriophage therapy.

(1) Isolation of *P. aeuroginosa* and phages specific to *P. aeuroginosa*

(a) Isolation of pathogen— *Pseudomonas aeuroginosa* was obtained from the blood samples collected from the hospitals located in and around Coimbatore. Collected samples were transported in sterile peptone water to laboratory and used for further studies. Isolation and characterisation of the pathogen was done by the method suggested by Johnson and Case (1995) and identification was done by the method suggested by Holt *et al.,* (1994).

(b) Enrichment and Isolation of bacteriophages— Bacteriophage was isolated from sewage samples at three different places by the method suggested by Nuttall *et al.* (1987). Sewage water was collected in sterile container. Five ml of Decca strength phage broth (DSPB) media and five ml 24 hrs old broth culture of *Pseudomonas aeuroginosa* was added to forty-five ml of sewage and incubated at 37°C for 24 hrs. After 24 hrs the suspension was centrifuged at 10,000 rpm for 10 min. Supernatant was filtered using Millipore filter (0.2μ) and filtrate was collected and stored at 40°C. Phages in the filtrate were determined by seeding-agar overlay method (Sambrook and Russell, 2001). The plaques appeared on the *P. aeuroginosa* lawn was individually isolated, and used as a phage stock. The plaques appeared on the double layer plating method was individually confirmed for the specificity against *P. aeurognosa*.

(2) Establishing the MLD in the animal model

Germ free Swiss albino mice of the same race was used for the study. *P. aeuroginosa* was grown in Kings B Media at 37°C for 24 hrs. Log phase cultures of *P. aeuroginosa* were used at different concentrations in order to fix the minimal lethal dose (MLD). The treatment used was 0 (T_1), 0.1 (T_2) 0.25 (T_3), 0.50 (T_4), 1.00 (T_5) ml of 99 × 10^9 CFU, where T_1 was used as control. All the treatments were replicated four times. The pathogen was injected intraperitoneally to mice. The animals were observed for symptoms within 24 hrs (Behavioural study of mice *i.e.* temperature and bodyweight was done to check the efficacy of the bacteria and bacteriophages). Mice inoculated with bacteria were scored for their state of health on a scale of 5 to 0, based on progressive disease states reflected by several clinical signs. A normal and unremarkable condition was scored as 5, slight illness defined as lethargy, as ruffled fur was secured as 4, moderate illness defined as severe lethargy, ruffled fur and hunched back was scored as 3, severe illness with above sign plus exudative accumulation around partially closed eyes was scored as 2, a moribund state was scored as 1, and death was scored as 0. To verify if the cause of death was really due to *P. aeuroginosa,* the serum was collected from the dead mice and streaked on *P. aeuroginosa* plate. Colonies growing on plates were then compared to the original *P. aeuroginosa* culture (Biswas *et al.,* 2002).

(3) Test for phage toxicity

The efficacy of phages was evaluated in the animals model with four different treatments to find out whether phages were safe for injection and inoculation into the mice. Dilutions of 0 (T_1), 0.25 (T_2), 0.50 (T_3), 1.00 (T_4) ml of 800 × 10^{14} PFU of phages were prepared in PBS and injected intra-peritoneally, where T_1 was used as control. All the treatments were replicated four times.

(4) Bacteriophage therapy

The MLD for *P. aeuroginosa* was found to be 0.1 ml and selected for inoculation. It was injected pertioneally followed by phages after 15 minutes were administrated with different concentrations 0 (T_1), 0.25 (T_2), 0.50 (T_3), 1.00 (T_4) ml, where T_1 was used as control mice. Behavioural changes and survival of mice were observed.

Result and Discussion

P. aeuroginosa were isolated from Hospital sample and identified as *P. aeuroginosa* using, morphological and biochemical characterisation (Tables 1a and 1b). The isolated *P. aeuroginosa* was designated as KDB-1. Isolated strain showed resistance to most of he antibiotics indicates the multi drug resistance of the pathogens (Ahmad, 2002). Infection by the ubiquitous multidrug-resistant *Pseudomonas* spp creates no additional complexity to the infected patients. Levin and Bull (1996), Lederberg (1996), Barrow and Soothil (1997) also isolated the multidrug resistant bacteria.

Table-1 (a): Morphological, Biochemical characterisation of *P. aeuroginosa* KDB-1

Tests performed	Result
Shape	-
Gram reaction	Positive
Motility	Motile
Gelatin utilisation	+
Citrate utilisation	+
Methy red	+
Voges proskeur	-
Acid from glucose	+
Gas from glucose	+
Urease production	+
Indole production	+

+ → Positive result

— → Negative results

Table-1 (b): Antibiotic sensitivity test of *P. aeuroginosa* KDB-1

Antibiotics	Result
Gentamycin	R
Ciproflaxin	I
Chloramphenical	R
Teracycline	R
Streptomycin	R
Kanamycin	R
Ampicillin	R
Erythromycin	R
Peniclin	R
Cephalosporin	R
Rifampicin	R

R - Resistance

I- Intermediate

Bacteriophages specific to *P. aeuroginosa* were isolated from sewage using chloroform technique. The isolated phage was quantified using double overlay technique for plaque assay and designated as PAD-1. The result on plaque assay was given in Table. 2. Isolation of bacteriophages from sewage has been demonstrated by Nutall *et al.* (1987) in which a total of 27-INCH plasmid dependent phages were isolated and all were shown to produce turbid plaques. Gowrishankar *et al.* (1998) also isolated phages from sewage.

The concentration 0.1, 0.25, 0.50, 1.00 ml of 99×10^{9} CFU/ml of *P. aeuroginosa* (Table 3) was tested in mice to check the MLD. Mice were found to be dead at the minimum dosage of 0.1 ml. The state of mice health was observed for 24 hrs and the results given in Fig. 1. No mortality was observed in the uninfected animals, on contrary the inoculated mice showed the symptoms of illness and had a very short survival period of half an hour. Difference in progression of disease can be ascribed to the innate variations in their immunity and defense in response to bacterial colonisation, and for the varied transition of the disease from bacterial colonisation to septic shock (Baron *et al.*, 1994).

Table-2: No of plaque forming units (PFU/ml of lysate (PAD-1)

S. No.	Dilution	Number of plaques formed per ml of lysate
1	10^{-6}	800×10^{8}
2	10^{-8}	737×10^{10}
3	10^{-10}	529×10^{12}
4	10^{-12}	412×10^{16}
5	10^{-14}	200×10^{18}

Values represent mean of three replications.

Table 3: Cell count of *P. aeuroginosa* (CFU/ml) in different dilution (KDB-1)

S. No.	Dilution	CFU/ml of sample
1	10^{-1}	300×10^{2}
2	10^{-2}	203×10^{4}
3	10^{-3}	99×10^{6}
4	10^{-4}	99×10^{6}
5	10^{-5}	9×10^{10}

Value represent mean of three replications.

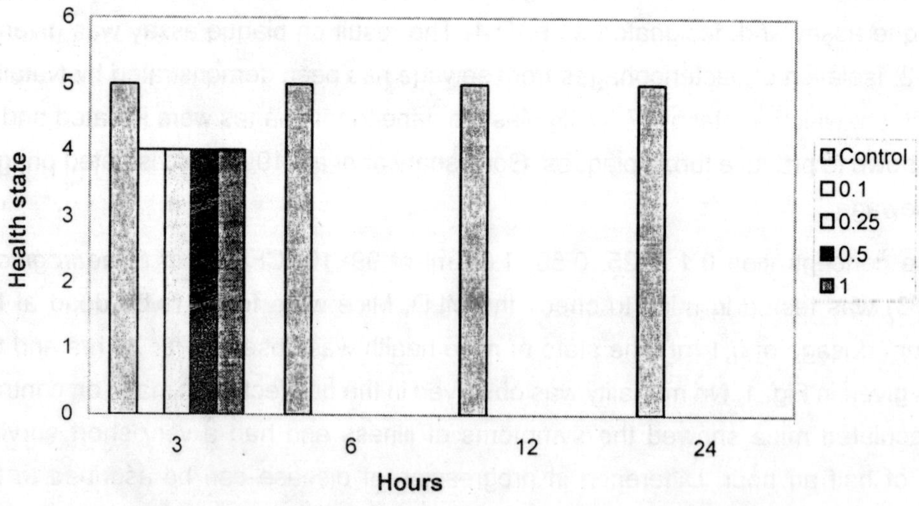

Fig. 1: *Pseudomonas aeuroginosa* treated mice

Bacteriophages for *Pseudomonas aeuroginosa* were tested in experimental infections of mice to investigate their potential for the treatment of infections of man. The phage lysates of 0.25, 0.50, 1.00 ml of 800 × 10^{14} PFU/ml were not toxic as they induced only slight ill condition in mice (Fig. 2). Mice showed symptoms of slight illness at about half an hour after phase injection and this persisted until the 12th hour, after which all mice recovered their normal health. This decline in the health status of mice may be attributed to the endotoxins that are always present in crude phage lysates (Merril *et al.,* 2003). Studies by Soothill (1992) support the view that bacteriophages could be useful in the treatment of human infections caused by antibiotic-resistant strains of bacteria. Soothill (1994) reported that bacteriophages prevents destruction of skin grafts by *Pseudomonas aeuroginosa.*

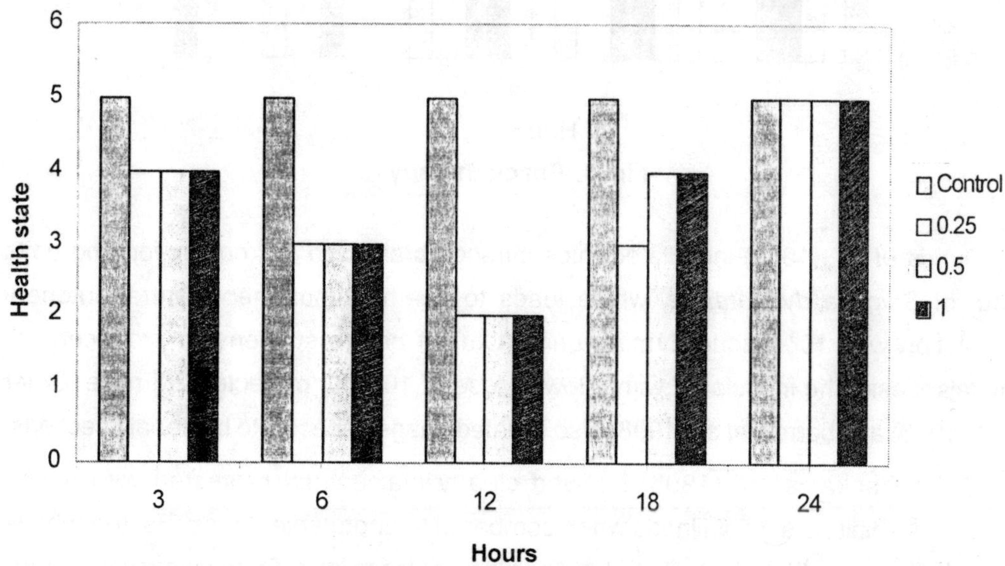

Fig. 2: Lysate treated mice

Mice treated with lower concentration of lysate lead to expression of lethargy, reduced appetite, hunched back and severe illness. As the phage dose increases from 200 to 800×10^{14} PFU/ml resulting in increase multiplicity of bacteriophage leads to surivial of animal. After 24 hrs all the mice showed no sign of illness and were comparable with that of healthy mice. At lower concentration of lysate (0.25) the expression of symptoms were ruffled, severe lethargy and hunched back. Mice showed symptoms of slight illness at about 6 hrs after phage injection and after which all mice recovered their normal health (Fig. 3). Further analysis of animals tissue and serum indicates the absence of bacterial load in response to phage administration, which proves the potential of phage for treatment of *P. aeuroginosa* infection.

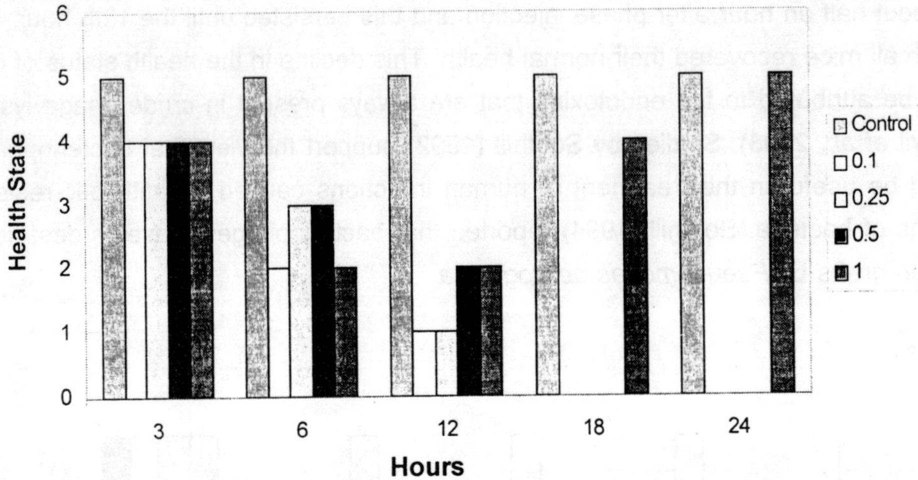

Fig. 3: Phage therapy

Dubos *et al.*, (1943) inoculated mice intracerebrally with 10^6 colony-forming units (Cfu) of *Shigella dysenteriae*, which leads to death. Simultaneous intraperitoneal inoculation with 10^9 plaque-forming units (Cfu) of phage suspension produced 6/8 animals, beside he inoculated with a lower dose of 10^5 pfu, protected 2/8 mice. Geier *et al.* (1973) and barrow *et al.* (1998) also isolated phages specific to bacterial infections.

Gowrishankar *et al.* (1998) showed clearly that animal protected with phage ingestion exhibit no sign of illness when compared to unprotected ones. Bacteriophage for *Acinetobacter baumanii, Pseudomonas aeuroginosa* and *Staphylococcus aureus* were tested in experimental infections of mice to investigate their potential for the treatment of infections of man (Soothill, 1992). Kozminska *et al.* (1997) examined bacteriophages for the treatment of bacteriology infection by influencing bacteriophages in producing cytokines by leukocytes of human peripheral blood.

In case of acute human infection, phage therapy have been considered valuable adjunct when compared with antibiotics. This present experiment reveals that lytic phages are effective therapy for *P. aeuroginosa* infection and indicates clearly the relevance of phage therapy, to reduce animals infected with *P. aeuroginosa* with no sign of illness. The ability of phage to reduce animals infected with these high titers of bacteria is especially putting forth the use of bacteriophages as therapeutic agent. This study further shows that a single IP dose of phage lysates was enough to bring about full recovery of mice infected by lethal concentrations of *P. aeuroginosa* (3×10^3 CFU/ ml). The potential of phage therapy has been given by several persons (Alisky *et al.*, 1998, Carlton 1999, Pirisi 2000, Soothill and Barrow 1997) and the present study reinforces the view that this study is worth exploring. In addition to its application in defense against biological warfare agents, it would also provide an inexpensive oral vaccine for the developing world.

References

Ahmad, S.I. 2002. Treatment of post-burns bacterial infections by bacteriophages, specifically ubiquitous *Pseudomonas* spp. notoriously resistant to antibiotics. *Med Hypotheses.* **58: 327-331.**

Alisky, J., Iczkowski, K., Rapoport, A., and Trositsky, N. 1998. Bacteriophages show promise as antimicrobial agents, *J. Infect.* **36: 5-15.**

Barrow, P., Lovell, M., and Berchieri, A.J.R. 1998. Use of lytic bacteriophage for control of experimental *Escherichia coli* septicemia and meningitis in chickens and calves. *Clin. Diag. Lab. Immunol.* **5: 294-298.**

Barrow, P.A. and Soothill, J.S. 1997. Bacteriophage therapy and prophylaxis: rediscovery and renewed assessment of the potential. *Trends Microbiol.* **5: 268-271.**

Baron, E.J., Peterson, L.R. and Finegold, S.M. 1994. *Escherichia coli. Bailey and Scott's Diagnostic Microbiology*, 9th. Moseby-year Book. Inc.

Biswas, B., Adhya, S., Washart, P., Paul, B., Trostel, A.N., Powel, B., Carlton, R., and Merril, C.R. 2002. Bacteriophage therapy rescue mice bacteremic from a clinical isolate of Vancomycin resistant *Enterococcus faecium. Infection and Immunity.* **70: 204-210.**

Brewer, S.C., Wunderink, R.G., Jones, C.B., Leeper, K.V.J. 1996. Ventilator-associated penumonia due to *Pseudomonas aeuroginosa. Chest.* **109:1019-29.**

Carlton, R.M. 1999. Phage therapy: past history and future prospectous. *Archivum Immunologii et therapiae Experiements.* **47: 267-274.**

Dubos, R.J., Straus, J.H. and Pierce, C. 1943. The multiplication of bacteriophage *in vivo* and its protective effects against an experimental infection with *Shigella dysenteriase. J. Exp. Med.* **78(20?): 161-168.**

Geier, M.R., Trigg, M.E. and Mereil, C.R. 0 1973. The fate of Bacteriophage lambda in non-immune germ free mice. *Nature* **246: 221-223.**

Gowri Sankar, R., Madhusudhan, V. and Palaniappan, R. 1998. Evaluation of phage therapy to treat experimental infections in mice. *Indian Journal of Microbiology*. **38: 101-103.**

Holt, G.J., Keieg, R.N., Sneath, A.H.P., Stately, T.J. and Williams, T.S. 1994. *Bergey's Manual Determinative Bacteriology* 9th edition. Lippincott Williams and Wilkins Company.

Johnson, L.E., Case, C.L., 1992. *Laboratory Experiments in Microbiology*. pp. **125-129.** The Benjamin/Cummings publishing company, inc.

Kluytmans, J. 1997, Surgical infections including burns. In: Wenzel RP, editor. *Prevention and Control of Nosocomial Infections*. 3rd ed. Baltimore: Williams and Wilkins; 1997. pp. **841-65.**

Kozminska, J., Weber-Dabrowska, B., Mulczyk, M. 1997. The study on biology of bacteriophages and their usage in the treatment of bacterial diseases and on the influence of different bacteriophages on cytokine production by leukocytes in human peripheral blood. *OTOLARYNGOLOGIA POLSKA* 51 Suppl. **25: 195-198.**

Lederberg. 1996. Samller fleas ....ad infinitum. *Therapeutic bacteriophage*. PNAS **93: 3167-3168.**

Levin, B. and Bull, J.J. 1996. Phage therapy revisited; The population Biology of a Bacterial infection and its treatment with Bacteriophage and antibiotics. *The American Naturalist* **147: 881-898.**

Merril, C.R., Scholl, D., Adhya, S.L. 2003. The prospect for bacteriophage therapy in Western Medicine. *Nat. Rev.* **2: 489-497.**

Nuttall, D., Maher, D. and Collerean., E. 1987. A method for direct isolation of INcH plasmid dependent bacteriophage. *Applied Microbiology*. **5: 37-40.**

Pirisi, A. 2000. Phage therapy-advantages over antiobiotics? *Lancet* 356: 1418. Sambrook, J. and Russell, R.W.D. 2001, *Molecular Cloning: A Laboratory Manual*, 2nd ed., vol. 1, p. 2.25 to 2.33. Cod spring harbor laboratory press, cold spring harbor, NY.

Soothill, J.S. 1992. Treatment of experimental infections of mice with bacteriophages. *J. Med. Microbiol.* **37: 258-262.**

Soothill, J.S. 1994. Bacteriophage prevents destruction of skin grafts by *Pseudomonas aeuroginosa. Burns.* **20: 209-211.**

Soothill, J.S. and barrow, P.A. 1997. Bacteriophage therapy and prophylaxis: rediscovery and renewed assessment of potential. *Trends Microbial.* **5: 268-271.**

Sulakvelidze, A., Alvidze, A.Z., and Morris, J.G. 2001. Bacteriophage therapy Mini review. *Animicrobial Agents and Chemotherapy*. **45(3): 649-659.**

❑❑❑

14

Seed Germination in Four Species of *Cassia* L. in Response to Different Temperature Conditions

Kiran Bargali

Department of Botany, K.U. Nainital–263 002 (Uttaranchal)

Introduction

The ability of seeds to either germinate when the set of environmental factor is favourable or to remain dormant during conditions unfavourable for germination determines the likelihood of survival of a given species. Seed germination is strongly influenced by variation in temperature, water stress and light requirements, and these factors often show significant interaction in their effects on germination (Bokhari *et al.*, 1975, Erasmus and Staden, 1986). Among these factors temperature is the most important to which the seeds are exposed. Temperature regulates germination in the field by determining the capacity and rate of germination, by removing primary and/or secondary dormancy (Bewley and Black, 1994).

Genus *Cassia* L. is a member of family Caesalpiniaceae. It is represented by about 500-600 species in tropics and warm temperature region. Of these, 35 species are reported from India. Most of the *Cassia* species are considered as an important medicinal plant not only in India but in other countries as well (Arjariya and Khare, 2003). Germination is one of the most critical phases in the plants life cycles, particularly in natural conditions where seed has to face several environmental constraints. Ecology of seed germination probes into adaptation strategy in the germination phase. A series of experiments were, therefore, conducted on seed germination ecology of four species of *Cassia viz. C. floribunda, C. mimosoides, C. occidentalis* and *C. tora* in relation to temperature regimes. Results of these experiments are being described and discussed in this paper.

Materials and Methods

The list of species studied with their habit and habitat condition/ecology is presented in Table 1. The seeds of four legume species were collected from different sites at National and adjacent area from mono specific patches of plant species growing naturally. Mature disease and insect free pods were collected and brought to the laboratory. Pods were air-dried and separation of seed was done manually. Thus, collected seeds were cleaned and packed in polyethylene bags for the seed germination studies.

Table-1: List of species studied with their habit and habitat conditions/ecology

Species	Habit	Distribution	Habitat	Flowering time	Fruiting time	Local uses
Cassia floribunda Cav. Descr.	Shruby herb	Upto 1800	Throughout the hills in wastelands	July-Sept	Oct-Dec	Planted to form hedge; pods are cooked as vegetable
C. mimosoides Linn.	Prostrate herb	Upto 1600	In grassy open localities, roadside and wayside	Aug-Sept	Oct-Now	Used as a fodder species
C. occidentalis Linn.	Erect herb or undershrub	Upto 1300	In waste-lands, roadsides and sometimes near river banks	July-Sept	Oct-Dec	Leaves are used in skin disease; decoction of root is given in dropsy; leaf and root paste is useful in piles and ringwarms
C. tora Linn	Erect herb or undershrub	Upto 1200	In waste-lands, road-sides and agricultural field	July-Sept	Sept-Dec	Seeds and leaves are used in skin diseases, cuts and wounds

Germination Tests

Germination tests of each species were done by taking 5 replicates of 20 healthy seeds. Only seeds that sank to the bottom when immersed in water were used in these experiments. Seeds were sterilised in 0.1% $HgCl_2$ solution and kept under different temperature regimes.

Petri dishes and filter paper were sterolised in an autoclave at 120° C for half hour. The seeds were placed in the dishes between moistened filter papers and kept at room temperature and in three incubators at 15°C, 25°C and 35°C constant temperature, respectively. Before keeping Petri dishes the incubators were sterilised by keeping the temperature on maximum for 2 hours and then wiping thoroughly with alcohol. The papers were regularly moistened and observed daily up to 30 days.

The seeds were considered germinated if the radical exceeded 3 mm in length. Results wee expressed as the germination percentage, germination value, germination energy and energy period following Czabator (1962).

$$\text{Mean germination (\%)} = \frac{\text{Total no. of seeds germinated at the end of the experiment}}{\text{Total number of seeds}} \times 100$$

Germination value = Final DGS (Daily germination speed) x Peak value DGS where,

$$\text{DGS} = \frac{\text{Cumulative germination per cent}}{\text{No. of days since start of germination}}$$

$$\text{Germination energy} = \frac{\text{Peak value of daily total germination}}{\text{Total number of seeds}}$$

Energy period = Day with the maximum number of seeds germinated.

Niche breadth for each species was calculated using Levin's (1968) equation (see Bargali 1992).

$$B = \frac{1}{(\Sigma\, Pi^2)S} \quad \begin{matrix} s \\ i = 1 \end{matrix}$$

Where B is niche breadth; Pi is the proportional response of species P in the ith temperature regime and S is the number of temperature regimes. The measure ranges from 0 to 1 with 1 being a perfectly even distribution of response.

Statistical analysis of the data was carried out using SPSS (PC +) statistical package for analysis of variance (ANOVA).

Results and Discussion

In all the species, germination of seeds at the different temperatures differed significantly (ANOVA, $P<0.05$). Adaptations to environmental stresses operative at the seed germination phase are crucial for its future survival and development as the soil surface temperature vary from site-to-site and season-to-season. In this experiment conduced to find out the effects of temperature regimes on seed germination it was observed that alternate day/night temperature was most suitable for *C. floribunda* and *C. occidentalis* in terms of rate as well as germination percentage (Table 2). The promoting effect of alternate day/night temperature in comparison to constant temperature on seed germination of some plant species was earlier reported by Bokhari *et al.* (1975) and Rao (1987, 1992). This preference for alternate day/night temperature regime is expected, as species have to encounter marked diurnal variations in natural conditions and show adaptation of theses species to temperature fluctuations in nature. In case of *C. mimodides* germination was maximum at 35°C constant temperature and decreased toward higher and lower temperature (Table 2). Germination scores were high (55%) at constant temperature 25°C in case of *C.tora* (Table 2) indicating that all necessary physiological activities required for causing seed germination in these species have proceed most smoothly at this temperature. Bokhari *et al.* (1975) also reported that some species apparently respond favourably and germinate better in constant temperature. Evidently, seeds of the four species of *Cassia* are adapted to germinate in a wide range of temperature as indicated by the wide response breadths on temperature gradient for all species (0.902 for *C. floribunda*, 0.976 for *C. mimosoides*, 0.905 for *C. occidentalis* and 0.973 for *C. tora*). The results also provided evidence that seeds of the species studied have low initial dormancy, which would allow rapid germination once the seeds, are dispersed to a favourable environment. Rapid germination has been shown to be characteristics of many species with small seeds (Foster 1986, Grime *et al.* 1981). The species tested in this study have small seeds. Small seeds have smaller stored food resources, compared with large ones, and produce seedling, which must become entirely dependent upon photosynthesis earlier or at least a smaller size than large seeded species.

Table-2: Effect of temperature of seed germination of *Cassia* spp.

Species	Temperature regime	Mean germination (%)	Germination value	Germination energy	Energy period
C. floribunda	**Alternate day/night**	71.67 ± 1.67	12.36	0.35	8
	15°C	70.0 ± 3.47	12.14	0.25	10
	25°C	51.67 ± 1.67	8.24	0.14	15
	35°C	26.67 ± 2.12	5.13	0.10	16
C. mimosoides	**Alternate day/night**	40 ± 1.86	7.35	0.20	12
	15°C	40 ± 2.34	8.10	0.20	14
	25°C	45 ± 2.67	8.35	0.25	15
	35°C	60 ± 2.65	11.14	0.40	10
C. occidentalis	**Alternate day/night**	80 ± 4.12	14.25	0.45	6
	15°C	60 ± 2.44	12.05	0.25	9
	25°C	50 ± 1.12	8.15	0.25	12
	35°C	30 ± 1.23	5.12	0.15	15
C. tora	**Alternate day/night**	35 ± 2.45	4.15	0.20	15
	15°C	40 ± 1.32	6.05	0.20	13
	25°C	55 ± 2.65	8.11	0.40	8
	35°C	40 ± 1.24	7.10	0.35	10

Temperature is the most important environmental factor influencing germination in most plant species (Breadsell and Rechards 1987). It not only affects the capacity for germination but also affect the rate of germination (Bewley and Black, 1985). Each species have the capacity to germinate over a defined range of temperature, characteristic of that species. When seeds are set to ger,omate over a temperature range an optimum soon becomes apparent (Bewley and Black, 1982). Over the short time period, three cadinal temperatures for germination can bc recognised—the minimum, optimum and maximum. A preliminary study on seed germination behaviour of plant species could be helpful in selection of sites for growth and establishment of seedlings particularly at degraded lands.

References

Arjariya, A. and Khare, H.N. 2003. Antifungal activity in ten species of *Cassia. Ecoprint* **10: 75-76.**

Bargali, K. 1992. The responses of *Pinus roxburghii* and *Quercus leucotrichophora* to nutrient applications. *Oeco. Mon.* **2: 1-6.**

Dewley J.D. and Black M. 1994. *Seeds: Physiology of Development and Germination* (2nd edition). Plenum Press, New York, and London.

Bewley J.D. and Black M. 1982. *Physiology and Biochemistry of Seeds in Relation to Generation.* Springer –Verlag, Berlin.

Bewley J.D. and Black M. 1995. *Seeds: Physiology of Development and Germination.* Plenum Press, New York and London.

Bokhari, U.G., J.S. Singh and F.M. Smith 1975. Influence of temperature regimes and water stress on germination of three range grasses and its possible ecological significance to a short grass prairie. *J. App. Eco.* **12: 153-163.**

Breadsell, D. and Rechards, D. 1987. Ecological aspects of seed germination. pp. 14-19. In: P. Langkamp (ed.) *Germination of Australian Native Plant Seed.* Australian Mineral Industries Research Association Limited, Melbourne.

Czabator, F.G. 1962. Germination value: An index combing speed and completeness of pine seed germination. *Forest Science*, **8: 386-396.**

Erasmus, D.J. and Staden J. Van. 1986. Germination of *Chromolaena odorata* (L.) K and R Achenes: effect of temperature, imbibition and light. *Weed Research*, **26: 75-81.**

Foster, S.A. 1986. On the adaptive value of large seeds for tropical moist forest trees: A review and synthesis. *Bot. Rev.* **52: 260-299.**

Grime, J.P., Mason, G., Curtis, A.V. and Shaw, S. 1981. A comparative study of germination characteristics in a local flora. *J. Ecol.* **69: 1017-1059.**

Levins, R. 1968. *Evolution in Changing Environments.* Princeton University Press, Princeton, New Jersey, 120 pp.

Rao, P.B. 1992. Effect of temperature on seed germination of certain woody species of Central Himalaya. *Ad. Bios.* **11: 41-52.**

Rao, P.B. 1986. Seed germination of *Quercus leucotrichophora* A. Camus ex Bahadur and *Pinus roxburghii* Sarg. on certain single factor environmental gradients. *Proceeding of Indian Academy of Sciences* (Plant Sciences), **96: 63-69.**

❑❑❑

15

Evaluation of Sugarcane Endophytic Bacterial Antagonism against Sugarcane Red Rot Pathogen

K. Rajaduraipandi, K. Panneer Selvam, K. Ravikumar, Rajitha Radhakrishnan R. Vishwanathan*, and C. Manoharan#

Department of Microbiology, Dr. G.R. Damodaran College of Science, Coimbatore–641 014, Tamil Nadu, India

*Sugarcane Breeding Institute (Indian Council of Agricultural Research) Coimbatore–641 007

P.G. and Research Department of Botany and Microbiology, A.V.V.M. Sri Pushpam College (Autonomous), Poondi–613 503, Thanjavur District, Tamil Nadu, India

Introduction

Sugarcane belongs to the genus *Sacharum* of Graminae family. It is essentially a tropical plant, requiring hot humid climate. Sugarcane is the principal raw material in tropics for production of white sugar, gur and khandsari. India has the largest area under sugarcane in the world and sugar industry is the second largest agro-based industries in India. Sugarcane (*Saccharum officinarium*) is infected by various pathogens *viz*. fungi, bacteria, viruses and phytoplasmas. The seed materials and growing plants in sugarcane contain readily assimilable sugars in mono and disaccharide forms and this favours infection by many microorganisms. Among the various diseases affecting sugarcane (Rott *et al.,* 2000), red rot is an important disease caused by *Cholletorichum falcatum*.

Red rot is one of the major constraints responsible for loss in sugracane production in India and other countries (Viswanathan and Samiyappan, 1999; 2000). It was first described as a disease of sugarcane by Went in 1893 from Java. In India it was reported by Barber (1901) from Godaveri delta area of the then Madras Presidency. Currently the disease occurs in most of the sugarcane-growing regions in India. Once the disease is established in a new area, it is very difficult to contain the same. Management of red rot disease through biocontrol agents is considered as an alternative strategy for disease management, which is economically sound environmentally safe. Induced protection in plants against various pathogens by biotic and abiotic inducers have been reported in many crops (Baker *et al.*, 1997). During the 1970s specific rhizosphere bacteria applied to seeds were reported to colonise roots and promate plant growth and were later termed as plant growth promoting rhizobacteria (PGPR). PGPR are plant root colonisers and most of the reported strains are from *Pseudomonas* and *Bacillus* species. An endophytic strain EPI induced systemic resistance in sugracane against the pathogen (Viswanthan, 1999; Viswanathan and Samiyappan, 2002). Exploitation of bacterial communities inside the sugrcane stalk tissues for the management of red rot diseases has not been studied except the citations mentioned above. Hence detailed studies were carried out.

Materials and Methods

Healthy sugarcane stalks aged 8-10 months were taken and cane surface wiped with wet cotton. Each stalk was divided into three to five portions namely lower, middle and upper portions designated as 1, 2 and 3. Sugarcane stalk tissue cylinders were taken from the various portions using a cork borer. Cylinders were then washed with 0.1% mercuric chloride for 2 minutes and a number of times with sterile distilled water. The sugarcane tissues were finally put into 100 ml conical flasks containing 25 ml sterile distilled water. These flasks were put in a shaker for 1 hour. One ml of liquid from each conical flask was poured into sterile Petri plates using a micropipette. Melted and cooled King's B medium was poured onto these plates and allowed to solidify. Three replicas for each plate were maintained. The plates were incubated at 28 ± 2°C for 24-48 hours.

Soil was collected from the sugarcane rhizosphere. One gram of soil was mixed with 100 ml of sterile water, which served as the stock solution. Eight tubes containing 9 ml sterile water were taken. Serial dilution was carried out till 10^8 dilution. One ml from 10^{-5}, 10^{-6}, 10^{-7} dilutions were poured onto plates. Sugarcane stalk samples from 20 hybrid varieties, six *S. spontaneum* clones and three *Erianthus* clones were used for isolation of endophytes. The clones used for endophyte isolation are listed in Table 1.

Table-1: Sugarcane clones/varieties used for endophyte isolation

S.No.	Clone	Type
1.	Bc 91	Hybrid
2.	Co 1148	Hybrid
3.	Co 312	Hybrid
4.	Co 410	Hybrid
5.	Co 419	Hybrid
6.	Co 6304	Hybrid
7.	Co 6907	Hybrid
8.	Co 8371	Hybrid
9.	Co 86032	Hybrid
10.	Co 86035	Hybrid
11.	Co 93009	Hybrid
12.	CoA 71-1	Hybrid
13.	CoC 29	Hybrid
14.	CoC 671	Hybrid
15.	CoC 92061	Hybrid
16.	CoPant 84211	Hybrid
17.	CoJ 46	Hybrid
18.	CoJ 64	Hybrid
19.	CoS 109	Hybrid
20.	CoS 109	Hybrid
21.	IND 84-406	*S. spontaneum*
22.	IND 85-576	*S. spontaneum*
23.	IND 90-826	*Erithanus* sp
24.	IND 99-994	*Erithanus* sp
25.	IJ 76-66	*Erithanus* sp
26.	IJ 388	*S. spontaneum*
27.	IJ 76390	*Erianthus* sp
28.	IK 76-66	*S. spontaneum*
29.	SES 98B	*S. spontaneum*
30.	SES 151/6	*S. spontaneum*

Fungal discs of *Colletotrichum falcatum* (Cf 671) strain isolated from CoC" 671) from red rot stalk culture collections of Plant Pathology lab, was plated on oatmeal media, potato dextrose agar, Kings B' media at the centre of the plate. The fungus was allowed to grow for two days. Bacterial isolates were streaked on either sides of the fungus 2 cm away from the growing mycelium. The plates were incubated for a period of eight days at room temperature. Measurement of zone of inhibition and fungal growth was taken on 5th and 7th day.

The effective strains that showed antagonism were identified at the species level using standard biochemical tests.

Results and Discussion

A total of 51 isolates were isolated from sugarcane stalk tissues and one rhizosphere soil samples. Among the hybrid sugarcane varieties CoC 92061 and Co 419 yielded higher number of endophytic isolates of eight and four respectively. In other varieties the number of isolates varied from one to three. The *S. spontaneum* clone IND 85-576 yielded about seven isolates. The rhizosphere soil samples had three bacterial isolates. The isolates from sugarcane varieties/clones were designated with number name of the variety/clone and position of sampling of the stalk (Table 2).

Table-2: List of sugarcane endophytes isolated from different sugarcane varieties/clones

S. No.	Variety/Genotype	Isolates
1.	Bo 91	91-1b, 91-4b
2.	Co 1148	1148-1b, 1148-2a 1148-2b, 1148-3a
3.	Co 312	3122-b
4.	Co 419	419-1b, 419-2a, 419-2b, 419-3a, 419-3b, 419-4a
5.	Co 6304	6304-3a
6.	CoA 71-7	71-7-1a
7.	CoC 671	6711-a, 6711-a2, 671-3a
8.	CoC 92061	92061-1a, 92061-1b, 92061-2a, 92061-3a, 92061-3b, 92061-3c, 92061-4a, 92061-5b
9.	CoJ 46	46-1a2
10.	CoJ 64	64-2a, 64-3a

11.	CoS 687	687-1a, 687-2b$_2$, 687-2b$_2$-687-3b
12.	IND 84-406	406-1a, 406-2b
13.	IND 85-576	576-1a, 576-1b, 576-2a, 576-2b, 576-3a, 576-3b, 576-4b
14.	IJ 76-66	66-1b
15.	IK 76-66	76-66-1a
16.	NCO 310	310-1a$_2$
17.	Rhizosphere soil	SS1
18.		SS2
19.		SS3

The endophytic bacteria have been thought to be weekly virulent pathogens but have recently been discovered to have several beneficial effects on host plants, such as plant growth promotion and increased resistance against plant pathogens and parasites (Hallmann *et al.,* 1997). Systemic spread of endophtic bacteria has been demonstrated for *Erwinia* species, which was recovered from cotton roots, stems and unopened flowers, as well as mesocarp and endocarp of bolls (Misaghi and Donndelinger, 1990) and for *Pseudomonas aurofaciens*, which has been shown to invade aerial tissues of corn via internal movement from the roots.

All the 51 isolates were tested for their antagonism against sugarcane red rot pathogen *Colletotrichum falcatum* on potato dextrose agar. Pathogen mycelial growth was recorded on 5th and 7th day after planting. Zone of inhibition was recorded on 5th day. The result showed that among the 51 isolates, seven were effective in inhibiting mycelial growth of the pathogen found to be 687-2b$_1$, 46-1a$_2$, 312-2b, 71-1a, SS1, SS2, and SS3. The Standards biocontrol strains maintained in the laboratory Pf1 and FP7 showed inhibition zone of 10 and 11 mm respectively. Among the effective isolates the rhizosphere exhibited higher inhibition than the standard strains. In other strains the inhibition zone ranged from 5 mm to 8 mm (Table 3).

Antagonisms of the seven effective isolates were repeated on oatmeal agar and King's B media subsequently. All the isolates except 46-1a2 showed antagonism as like the standard strains. It was found that all the isolates exhibited almost similar inhibition zone. The isolate 687-2b1 recorded higher zone of 12 mm, which was higher than that of standard effective strains Table 4.

Table-3: Antagonism of sugarcane endophytes against *Colletotrichum falcatum*

Isolates	Colletorichum mycelial growth		Zone of inhibition (mm)
	5th day	7th day	
91-1b	28.0	40.0	-
91-4b	30.5	39.5	-
1148-1b	27.5	38.5	-
1148-2a	28.0	40.0	-
1148-2b	30.0	40.0	-
1148-3a	29.0	40.0	-
312-2b	23.0	30.5	7
419-1b	29.5	39.5	-
419-2a	28.0	39.0	-
419-2b	28.5	39.5	-
419-3a	28.5	39.5	-
419-3b	30.0	40.0	-
419-4a	28.0	40.0	-
6304-3a	28.0	39.5	-
71-1-1a	23.0	29.0	8
671-1a1	29.5	38.5	-
671-1a2	30.0	40.0	-
6711-3a	28.0	38.5	-
92061-1a	29.0	40.0	-
92061-1b	30.0	38.5	-
92061-2a	27.5	37.5	-
92061-3a	29.0	45.0	-
92061-3b	28.5	40.5	-
92061-3c	30.0	39.5	-
92061-4a	27.5	40.0	-
92061-5b	30.0	40.5	-
46-1a2	23.0	30.5	7
64-2a	29.0	39.0	-
64-3a	30.0	41.0	-
687-2b1	21.0	30.5	8
687-2b2	30.0	39.0	-
687-3b	30.0	41.0	-

84-406-1a	29.5	39.0	-
84-406-2b	28.5	38.5	-
85-576-1a	30.0	40.0	-
85-576-1b	28.5	38.5	-
85-576-2a	30.0	40.0	-
85-576-2b	30.0	40.0	-
85-876-3a	29.0	38.0	-
85-876-3b	30.5	38.5	-
85-876-4b	29.0	38.0	-
IJ 76-66-1a	28.5	38.5	-
IK 76-66-1b	30.0	40.0	-
310-1a2	30.0	40.0	-
SS1	22.0	28.0	13
SS2	22.5	28.5	12
SS3	22.5	28.5	13
Pf1	22.5	29.5	10
FP 7	20.0	28.5	11
Control	30.0	45.0	

Table-4: Antagonism of different efficient isolates against *Colletotrichum falcatum* in oatmeal agar

Isolates	Mycelial growth		Zone of inhibition (mm)
	5th day (mm)	7th day (mm)	
Control	30.0	44.0	-
312-2b	20.0	32.5	11
71-1-1a	22.0	30.0	8
46-1a2	28.5	40.5	-
687-2b1	23.0	32.0	12
SS1	21.0	30.0	10
SS2	22.0	29.0	9
SS3	22.0	31.5	10
PF1	21.5	31.5	10
Fp7	21.5	30.0	11

The present study revealed survival of useful bacterial antagonists in sugarcane stalk tissues. Only a limited number of sugarcane endophytes were found to be antagonistic to red rot pathogen, indicating that efficient endophytes could be identified from sugarcane stalks by screening large number of samples. The red rot pathogen *Colletotrichum falcatum* infects sugarcane stalks, which is economically important to manufacture sugar or jaggery. Loss to sugar industry caused by the disease over the last one century is well known and many epidemics of the diseases in different decades had created havoc in sugar industry. The pathogen also gained virulence in the recent years (Chona, 1980; Singh and Singh, 1989).

Among the 51 isolates, 39 were gram-positive rods; seven were gram-negative rods, one-gram positive coccobacilli. All the three isolates from rhizosphere were gram-negative rods. In general, the results indicated that the proportion of gram-negative bacterial endophytes was found to be significantly higher than gram-negative bacteria. The biochemical tests such as Gelatin liquefaction, levan formation, and growth at 4°C and 45°C and pH 5.7-citrate test and mannitol utilisation showed in Table 5.

Table 5: Biochemical characterisations of sugarcane endophytes antagonistic to *C. falcatum*

Efficient Isolates	Gelatin Liquefaction	Levan formation	Growth at 4°C	Growth at 45°C	Growth at pH 5.7	Citrate test	Mannitol test
312-2b	-	-	+	-	+	+	-
71-1-1a	+	-	-	+	-		
46-1a_2	+	-	-	+	-		
687-2b_1	+	-	-	+	-		
SS1	+	-	+	-	-		
SS2	+	-	+	-	-		
SS3	+	+	+	-	-		
Pf1	+	-	+	-	-		
FP7	+	-	+	-	-		

The strain 687-2b1, which has shown growth at 4°C, pH 5.7 and positive for citrate test, was grouped as *bacillus cereus*. While the strains 46-1a21, 3122b, 71-11a positive to gelatin liquefaction and growth at 45°C were grouped under *Pseudomonas aeruginosa*. All the three-rhizosphere strains, which were positive to gelatin liquefaction and growth at 45°C, were grouped as *Pseudomonas fluorescens*. Isolate SS3 also showed positive to levan formation. The results of the present study indicate that endophytic strains from sugarcane could be exploited for managing red for disease as biocontrol agent. For isolation of endophytes hybrid sugarcane varieties and clones of *Saccharum spontaneum* and *Erithanus species* were used. In addition of biological control, endophytic bacteria improved plant growth in different crops. Hurek *et al.* (1994) reported that inoculation of rice with diazotrophic endophyte *Azoareus* species BH72 significantly promoted plant growth. Since application methods have been standardised for the biocontrol bacteria in sugarcane, the identified endophytic strains could be mass multiplied following the standardised procedure for commerical exploitation.

References

Baker, C.A. and Henis, J.M.S. 1997. Commercial production and formulation of microbial biocontrol agents. In: *New directions in Biological Control: Alternatives for Suppressing Agricultural Pets and Diseases*. Edited by R.R. Baker and P.E. Dunn, A.R. Liss, New York. pp. **334-344.**

Barber, C.A. 1901. Sugarcane disease in Godawari and Ganjam Districts. Madras Dept. *Land Records and Agri. Bull.* **43: 181-194.**

Chona, B.L. 1980. Red rot of sugarcane and sugar industry a revivew. *Indian Phytopath.* **33: 191-206.**

Hallmann, J., Quadt-Hallmann, A., Mahaffee, W.F. and Kloepper, J.W. 1997 Dept. of Plant Pathology, Biological Control Institute, Alabama Agricultrual Expt. Station, Aubum, USA. *Can J. Microbiol.* **43: 895-914.**

Hurek, T., Reinhold Hurek, B., Van Montagu, M. and Kellemberger, E. 1994. Root colonisation and systemic spreading of *Azoarcus* sp., strain BH72 in grasses. *J. Bacteriol.* **176: 1913-1923.**

Misaghi, I.J. and Donndelinger, C.R. 1990. Endophytic bacteria in symptom-free cotton plants. *Phytopathology.* **80: 808-811.**

Rott, P., Bailey, R.A., Comstock, J.C., Croft, B.J. and Saumtally, A.S. 2000. *A Guide to Sugarcane Diseases.* CIRAD/ISSCT Montpellier, France.

Singh, K. Singh, R.P. 1989. Red rot. In *Diseases of Sugarcane. Major Disease.* Eds, Ricaud, C. Egan, B.T. Gillaspie, A.G. Jr, and Huges, C.G. Amsterdam, The Netherlands, Elsevers B.V. pp. **169-188.**

Viswanathan, R. and Samiyappan, R. 1999. Red rot disease in sugarcane: a major constraint for the Indian sugar Industry. *Sugarcane.* **5: 9-15.**

Viswanathan, R. and Samyiappan, R. 2000. Red rot disease in sugarcane: Challenges and prospectus. *Madras Agricul. J.* **87: 549-559.**

Viswanathan, R. and Samiyappan, R. 2002. Role of oxidative enzymes in the plant growth promoting rhizobacteria mediated induced systemic resistance in sugarcane against *Colletotichum falcatum. J. Plant Dis. Prof.* **109.**

□□□

16

Effect of *Nerium odorum* Soland. on the Periodicity of Estrous Cycle in Female Albino Rats

Dixit Mukesh, Prakash A.O.
SOS Pharmacy, Jiwaji University, Gwalior (M.P.)

Introduction

The administration of antifertility agents of hormonal virtues *viz.* estrogenic or antiestrogenic are known to act on the reproductive tract in female rats. There are some antifertility agents which do not disturb the periodicity of the estrous cycle being non-hormonal and their *modus operandi* lies with the toxic action at the level of development of blastocyst. Extract of *Nerium odorum* are known to impair fertility in female rats. Therefore, keeping in view the antifertility activity in the leaf extract of *Nerium odorum*, the present investigation has been undertaken to know whether the extract of *Nerium odorum* has any effect on the periodicity of estrous cycle in female rats.

Material and Methods

Animals and their maintenance

Virgin female rats of Wistar strain weighing 130±10 gm were collected from the animals colony of the department. These animals were maintained under uniform husbandry conditions of light and temperature (25°C-26°C) and were fed with "Gold Mohar" rat pelleted diet and water *ad libitum*.

Preparation of leaf extract

Weighed dried leaves of *Nerium odorum* were grinded powdered and extracted with 50 per cent ethanol using dumping method. The powder form was dumped with 50 per cent ethanol in air tight glass jar and after about 15 days, the upper clear supernatent was separated and evaporated to dryness under low temperature and reduced pressure in a rotatory film vacuum evaporator. The oily black brown semi solid mass so obtained was considered as extract which was kept in sealed bottle and stored in refrigerator till its use. A small amount of this extract was taken out at the time of need.

Preparation of standard doses and its administration

In earlier studies 25 mg/kg dose has shown significant antifertility activity and therefore for the present studies dose of 250 mg/kg body wt. was prepared by weighing 2.5 gm of dried extract which was macerated with gum acacia and finally suspended in mineral water upto 100 ml. 1.0 ml of this prepared extract was administered orally to a rat of 100 gm with the help of intragastric catheter.

Experimental protocol

The selected female rats were examined for different stages of estrous cycle as described by Prakash (1979 a).[1] Vaginal smear of each rat was examined daily at a definite time and the stages of estrous cycle were recorded (Fig.1). The female rats showing at least two regular cycles were selected for experimental purpose. The selected animals were distributed into different groups of control and experimental. A dose of 250 mg/kg body weight was administered orally to all three groups of experimental rats for 10, 20 and 30 days and the control group received vehicle only. Vaginal smear of each rat was examined daily at definite time during the experimental period. After the last treatment the administration of extract was discontinued and the record of various stages of estrous cycle as revealed through the vaginal smear was examined and analysed statistically using student 't'-test.

Result

The effect of 50 per cent ethanolic extract of *Nerium odorum* on the periodicity of estrous cycle in adult cyclic rats (Tables-1 and 2) revealed that in the control rats the diestrus stage covered almost half of the cycle followed by estrus of 30 per cent and then proestrus and metestrus of short durations. When the extract of *Nerium odorum* was administered for 10 days there was an increase in the duration of estrus phase

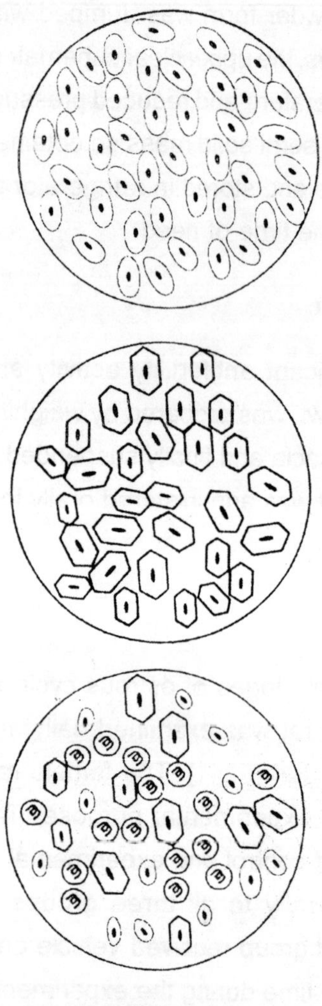

PROESTRUS

The round epithelial cells nucleated with dense cytoplasm.

ESTRUS

The hexagonal cornified cells located singly or in sheets.

METAESTRUS

The mixture of all type of cells, leucocytes, epithelial as well as cornified cells.

DIESTRUS

The leucocytes or leucocyte dominant cells.

Fig. 1: Various stages of estrus cycle in rats

significantly when the period of administration was increased to 20 and 30 days, there was further prolongation in the percent duration of estrus stage, consequently the percent duration of the other stage was reduced. The table also reveals that the administration of extract did not alter the percent duration of proestrus and metaestrus, however, shifting occurred from diestrus to estrus. It is also interesting to not that the period of 10 days of administration of extract could significantly induce the changes, which persisted even upto 30 days. Therefore, the effect of extract of *Nerium odorum* on the percent duration of estrous cycle clearly revealed that the effect was dose dependent irrespective to duration of treatment.

Table-1: Effect of 50% ethanolic extract of *Nerium odorum* Soland. on the periodicity if estrous cycle in adult rats (In terms of percent duration). 5 animals were used in each set. Values are mean ± SE

Group No.	Duration (Days)	Treatment	Percent duration of stages of estrous cycle				Restoration time for normalicy in cycle (days)
			Proestrus	Estrus	Metaestrus	Diestrus	
1	10	Control	12.0 ± 0.6	28.0 ± 1.0	14.0 ± 0.7	46.0 ± 2.1	6.2 ± 0.3
		Treated	12.0 ± 0.6	40.0 ± 2.0*	10.0 ± 0.05*	38.0 ± 1.9*	
2	20	Control	12.0 ± 0.6	29.0 ± 1.5	15.0 ± 0.7	44.0 ± 2.2	8.3 ± 0.5
		Treated	12.0 ± 0.6	45.0 ± 2.3*	13.0 ± 0.6	30.0 ± 1.5*	
3	30	Control	14.0 ± 0.7	30.0 ± 1.5	12.0 ± 0.6	44.0 ± 2.2	12.4 ± 0.7
		Treated	12.0 ± 0.6	48.0 ± 2.4*	10.0 ± 0.5	30.0 ± 1.58	

Statistical Analysis

* P value Vs. respective control <0.05

Discussion

During a reproductive cycel, the ovary undergoes cyclic changes and so as the uterus. Consequently the vaginal epithelium induced remarkable changes morphologically and histologically. The cyclic changes in the vagina under the influence of female sex steriods estrogen and progesterone are known as estrous cycles. Turner and Bagnara (1975)[2] have clearly shown that the duration of estrus cycle and also the conversion of one stage to another in a estrous cycle is hormonal dependent. The entire length of one estrous cycle in rats varies from 4-5 days and consists of proestrus, estrus, metaestrus and diestrus. The entire cycle is divided into follicular phase which contains proestrus and estrus whereas leucocytic phase contains metaestrus and diestrus.

Table 2: Effect of 50% ethanolic extract of *Nerium odourm* Soland. on the periodicity of estrous cycle in adult rats (In terms of duration in days). 5 animals were used in each set. Values are mean ± SE

Group No.	Duration (Days)	Treatment	Duration of stages of estrous cycle			
			Proestrus	Estrus	Metaestrus	Diestrus
1	10	Control	1.2 ± 0.22	3.2 ± 0.22	1.2 ± 0.22	4.4 ± 0.27
		Treated	1.2 ± 0.22	3.6 ± 0.44	1.2 ± 0.22	4.0 ± 0.35
2	20	Control	1.4 ± 0.27	6.2 ± 041	4.2 ± 0.41	4.0 ± 0.22
		Treated	2.0 ± 0.5	9.2 ± 0.22*	2.6 ± 0.27	6.2 ± 0.41
3	30	Control	6.2 ± 0.41	9.2 ± 0.22	2.6 ± 0.27	12.0 ± 0.35
		Treated	2.4 ± 0.44	15.2 ± 0.41*	3.2 ± 0.22	9.2 = 0.22

Statistical Analysis

* P value Vs. respective control <0.05

The conversion of follicular phase to leucocytic phase is governed by estrogen and progesterone of ovarian origin which are further influenced by the pituitary gonadotropins (Lerner, 1969).[3] Allen and Doisy (1923)[4] discovered the estrous cycle in female rats and have shown the significant role of esyrogen and progesterone in its inter or intra conversions. Thereafter, number of authors worked on the mechanism of estrogen and progesterone in relation to vaginal cornification. Estradiol-17β known as naturally occurring estrogen is also known to modify vaginal epithelial cells into cornified corn flake like cells (Turner and Bagnara, 1975 and Nalbandov 1970).[5] The intra conversions of follicular phase to leukocytic phase is the inter play of estrogen and progesterone which are secreted by the ovary. Progesterone is also known to act as an antiestrogen and thus it also helps to regulate the estrous cycle. Lerner (1969)[3] has studied the effect of steroidal estrogens on the periodicity of estrus cycle and has reported that administration of typical estrogen is known to induce vaginal cornification. However, when estrogenic agents are administered to the estrogen primed rats, the interruption of vaginal cornification occurs. Therefore, it is quite clear that a typical estrogen induces vaginal cornification and shows estrus stage whereas the progesterone induces diestrus stage. Regulation of estrogen and progesterone is carried out through the feed back mechanism.

Medicinal plants which have shown potent antifertility activity in laboratory animals have been studied for their effect on the estrous cycle as these have shown significant hormonal properties in immature rats. Phyto estrogens and other estrogenic herbal preparations are known to alter the periodicity of estrus cycle in rats. Ethanolic extract of direct berries of *Embelia ribes* when administered to cyclic rats increased the duration of estrus phase (Prakash *et al.,* 1979 a).[6] Similarly Kholkute *et al.* (1976)[7] have reported that benzene extract of *Hibiscus rosa-sinensis* induced a prolong diestrus stage in adult rats. Barethia (1995)[8] while working on the extract of *Apium graveolnes* have shown a typical characteristics effect on the estrous cycle. When the dose was administered for 10 days there was an increase in the follicular phase of the estrous cycle, however, when the extract administered for 20 and 30 days the same extract increased the leucocytic phase. This dual action of the extract has been explained on the estrogenic potentiality in the extract of *Apium graveolnes*. In the present investigation author has observed that the administration of 50 per cent ethanolic extract of *Nerium odorum* to adult cyclic rats significantly increased the periodicity of estrus phase.

Therefore, on the basis of present finding it is concluded that the 50 per cent ethanolic extract of *Nerium odorum* which have shown promising antifertility activity in rats induced prolongation in follicular phase in the estrous cycle in adult rats. The estrus stage was persisted as long as the extract was administered which clearly indicated a synergistic effect. Findings also suggested that the estrogenic influence on the estrous cycle appeared is mild as feed back was not initiated. Moreover, studies also suggested that the prolonged estrus stage of the cycle was returned to normal after discontinuation of the extract although in the long period of treatment took more time to restitute. Therefore, it can safety be said that estrogenic effect of extract appeared temporary and the animals might resume the normal estrous cycle after discontinuation of the treatment.

References

1. Prakash, A.O., Hibiscus rosa-sinensis Linn. Effect on estrous cycle in rats. *Probe*. (1979 a) **18, 79.**

2. Turner, D.C. and Bagnara, J.T., *General Endocrinology*. Fifth edition (1975). W.B. Sauders Company, Tokyo, Japan.

3. Lerner, L.J. The biology of non-steroidal antifertility agents. In Contraception. *The Chemical Control of Fertility*, (Ed. Daniel. Lednicer) (1969), p. **161**, Marcel Dekker Inc. N.Y.

4. Allen, E. and Doisy, E.A. An overian Hormone. Preliminary report on its localisation, extraction and partial purification, and action in test animals. *J. Am. Med. Assoc.* N.Y. (1923), **81, 819.**

5. Nalbandov, A.V. Reproduction in female mammals and birds: In A.V. Nalbandov (ed) *Reprod. Physiol.*, D.B. Taraporewala Sons and Co. Pvt. Ltd., Mumbai, India, 2 (1970), **124-127.**

6. Prakash, A.O. and Mathur, R. Studies on estrous cycle of albino rats Response to Embelia ribes Burn extracts. *Planta Medica*. (1979 a), **36, 134.**

7. Kholkute, S.D., Mudgal, V. and Deshpandey, P.J. Screening of indigenous medical plants for antifertility potentiality. *Planta Medica*. (1976), **29(2): 151.**

8. Barethia, R. Antifertility effect of *Apium graveolens* and *Crotalaria juncia* Linn. in female rats. Ph. D. Thesis (1995), Jiwaji University, Gwalior.

❑❑❑

17

Stress Amelioration by Ketoconazole in NaCl Stressed Cowpea

R. Somasundaram, P. Manivannan, R. Senthilkumar, R. Gopi, R. Sridharan, M. Gomathinayagam and R. Panneerselvam
Division of Stress Physiology, Department of Botany, Annamalai University, Annamalainagar–608 002 Tamil Nadu

Introduction

Soil salinity represents an increasing threat to agricultural production. High Sodium (Na^+) concentrations in soils are toxic to most higher plants. Osmotic stress, due to drought and salinity is the most serious problem that limits plant growth and crop productivity in agriculture (Boyer, 1982). In arid and semiarid regions, soil salinity is a common occurrence. The use of poor irrigation water and saltwater encroachment is also increasingly threatening agriculture in humid regions (Marschner, 1995).

Production of plants with tolerance to environmental stresses is one of the priorities in plant science research and certain chemicals have been demonstrated to regulate the expression of stress tolerance (Fletcher *et al.,* 2000, Senaratna *et al.,* 2000). Among the stress "Protectant" compounds triazoles, particularly paclobutrazol, have been very effective in inducing tolerance to a variety of stress (Fletcher and Hofstra, 1988). The triazole derivatives are fungitoxic and plant growth regulating properties and triadimefon (a trizaole compound) is reported to change the balance of important plant growth regulators including gibberellins, ABA and cytokinins (Fletcher and Hofstra, 1985).

Ketoconazole, a triazole compound widely used as human fungicide and it also exhibits the plant growth regulation properties like other triazole compounds.

This study aims to understand the effect of salinity on growth and metabolism of *Vigna unguiculata* under NaCl stress and also the ameliorative effect of ketoconozole on sodium chloride treated plants and the effect of ketoconazole alone.

Materials and Methods

The seeds of *Vigna ungiculata* (L.) Walp. var. vamban-6 (cowpea) were surface sterilised with 0.2 per cent HgCl. Two seeds were sown in each pot of 30 cm diameter and 30cm height containing 3 kg of soil mixture composed of red soil, sand and the farm manure (FYM) at 1:1:1 ratio. All the pots were watered to the field capacity with groundwater upto 35 days after sowing (DAS). Later plants were irrigated with groundwater marked as control and other treated with 100 mM NaCl, 100 mM NaCl + 20 mg^{L-1} ketoconazole and 20 mg^{L-1} ketoconazole alone respectively on 35 and 75 DAS. The experiment was laid out in a completely randomised design (CRD) with 3 replicates randomly on the 40th and 80th DAS and separated into root, stem and leaves and used for determining biochemical and antioxidant enzyme activities.

Biochemical constituents are extracted and estimated soluble proteins (Bradford, 1976), Total free aminoacids (Moore and Stein, 1948), Proline (Bates *et al.,* 1973), Glycine betaine contents (Grieve and Gratten, 1983), and antioxidant enzymes like Ascorbate peroxidase (Asade, 1987), and Catalase activities (Chandlee and Scandalios, 1984).

Results and Discussion

Sodium chloride treatment decreased the protein content in root, stem and leaves of the cowpea plants on 40th and 80th DAS when compared with control. Decreased protein content have been reported in a number of plants like *Atriplex halimus* (Bajji *et al.*, 1998) and sorghum (Azooz *et al.,* 2004). The combination of NaCl with ketoconazole increased the protein content in all parts of the plant when compared to NaCl stressed plants. However the increase in lower than control. Similar results were observed in the NaCl stressed tradimefon treated *Vigna unguiculata (*Gopi *et al.,* 1998). Ketoconazole treatment to the unstressed plants increased the protein content when compared to control. These earlier findings of enhanced protein content in raddish (Muthukumarasamy *et al.,* 2000) is correlated by the present observation. Sodium chloride treatment increased the amino acids, proline and glycine betaine content of all parts of the plants to a larger extent. Similar observations were made in soybean (Panneerselvam *et al.,* 1998), Sorghum (Azooz *et al.,* 2004) and *Arachis hypogaea* (Girija *et al.,* 2002). Ketoconazole treatment to the NaCl stressed plant markedly reduced the accumulation of amino acid, proline and glycine betaine content when compared with NaCl stressed plants however it was higher than that of control. Similar observations were made in NaCl treated *Vigna unguiculata* seedlings with triadimefon treatment (Saha and Gupta, 1993) Ketoconazole treated plants increased the aminoacid, proline and glycine betaine content when compared to control. It was reported in Radish (Muthukumarasamy *et al.,* 2000), Cucumber seedlings (Feng *et al.,* 2003) and wheat seedlings (Berova *et al.,* 2002) (Table 1).

Sodium chloride stress inhibit the ascorabate peroxidase (APX) activity in cowpea plants when compared to control. The higher level of endogenous ascorbate peroxidase is essential to maintain the antioxidant system that protect plants from oxidative damage due to biotic and abiotic stresses (Shigeoka *et al.,* 2002). Ketoconazole with NaCl treatment increased the ascorbate peroxidase activity when compared to NaCl stressed plants. Ketoconazole treatment alone also increased the ascrobate peroxidase activity when compared to control and other treatments. Paclobutrazol increased the level of APX activity in wheat cultivars (Kraus *et al.,* 1995; Berova *et al.,* 2002).

Catalase (CAT) activity has been reduced by the NaCl stress in all parts of the cowpea plants when compared with control. It was reported that salt stress decreased the catalase activity in green gram (Panda, 2001). Ketoconazole treatment to the NaCl stressed plants resulted in the increased catalase activity when compared to NaCl stressed plants. Ketoconazole treatment alone increased the catalase activity when compared to control. Similar results were observed in the H_2O_2 scavenging systems represented by APX and CAT are more important in imparting tolerance wheat varieties (Sairam *et al.,* 1998; Wille Kens *et al.,* 1997) (Table 2).

Ketoconazole treatment to the NaCl stressed plant increased the protein content might be the reason for decreased amino acids, proline and glycine betaine content when compared to the NaCl stressed plants. Catalase and APX activities increased when compared to NaCl stressed plants might be the reason for improving the tolerance to the plants by ketoconazole.

Table-1: Effect of NaCl, Ketoconazole and their combination of the protein, amino acids, proline and glycine betaine content in *Vigna unguiculata* (Values are the mean ± SD of 3 replicates)

	Days after sowing (DAS)	Root				Stem				Leaf			
		Control	100mM NaCl	100mM NaCl+ 20mg^{L-1} Ketoco-nazole	20mg^{L-1} Ketoco-nazole	Control	100mM NaCl	100mM NaCl+ 20mg^{L-1} Ketoco-nazole	20mg^{L-1} Ketoco-nazole	Control	100mM NaCl	100mM NaCl+ 20mg^{L-1}	20mg^{L-1} Ketoco-nazole Ketoco-nazole
Protein (mg g^{-1} dry wt.)	40	92.56 (±2.77)	69.04 (±1.72)	90.212 (±1.80)	111.28 (±2.78)	73.32 (±1.46)	55.52 (±1.38)	70.22 (±1.40)	87.88 (±1.75)	79.04 (±1.58)	69.28 (±1.38)	74.76 (±1.86)	92.96 (±2.32)
	80	113.88 (±3.41)	88.80 (±2.66)	110.56 (±3.03)	132.08 (±3.96)	92.56 (±2.77)	71.64 (±2.14)	89.84 (±2.69)	112.32 (±3.36)	104.52 (±3.13)	79.44 (±2.38)	97.93 (±2.93)	127.40 (±3.81)
Aminoacid (mg g^{-1} dry wt.)	40	1.699 (±0.041)	2.889 (±0.072)	2.741 (±0.068)	1.725 (±0.043)	1.249 (±0.031)	1.919 (±0.045)	1.822 (±0.045)	1.343 (±0.033)	1.788 (±0.044)	2.891 (±0.072)	2.656 (±0.066)	1.882 (±0.047)
	80	2.966 (±0.088)	4.469 (±0.134)	3.620 (±0.108)	3.106 (±0.093)	2.178 (±0.065)	3.848 (±0.115)	3.683 (±0.110)	2.249 (±0.067)	2.980 (±0.089)	5.369 (±0.161)	4.667 (±0.140)	2.285 (±0.068)
Proline (mg g^{-1} dry wt.)	40	0.489 (±0.012)	0.843 (±0.021)	0.569 (±0.011)	0.504 (±0.012)	0.492 (±0.012)	0.721 (±0.014)	0.593 (±0.011)	0.552 (±0.011)	0.428 (±0.010)	0.673 (±0.013)	0598 (±0.011)	0.498 (±0.009)
	80	0.992 (±0.029)	1.343 (±0.040)	1.089 (±0.032)	1.031 (±0.030)	0.987 (±0.029)	1.234 (±0.037)	1.087 (±0.032)	0.996 (±0.029)	0.928 (±0.027)	1.187 (±0.035)	0.984 (±0.029)	0.958 (±0.0280)
Glycine betaine (mg g^{-1} dry wt)	40	15.088 (±0.301)	22.512 (±0.450)	19.928 (±0.398)	18.403 (±0.368)	16.928 (±0.388)	22.616 (±0.452)	19.872 (±0.397)	18.608 (±0.378)	19.448 (±0.388)	24.296 (±0.485)	23.712 (±0.474)	20.024 (±0.400)
	80	26.554 (±0.796)	31.344 (±0.940)	29.968 (±0.899)	27.808 (±0.834)	25.392 (±0.761)	32.816 (±0.984)	29.544 (±0.886)	28.282 (±0.848)	43.488 (±1.304)	58.336 (±1.750)	49.807 (±1.494)	46.772 (±1.403)

Table-2: Effect of NaCl, Ketoconazole and their combination on the ascrobate peroxidase and catalase activity in *Vigna unguiculata* (Values are the mean ± SD of 3 replicates)

	Days after sowing (DAS)	Root				Stem				Leaf			
		Control	100mM NaCl	100mM NaCl+ 20mg^{L-1} Ketoco-nazole	20mg^{L-1} Ketoco-nazole	Control	100mM NaCl	100mM NaCl+ 20mg^{L-1} Ketoco-nazole	20mg^{L-1} Ketoco-nazole	Control	100mM NaCl	100mM NaCl+ 20mg^{L-1} Ketoco-nazole	20mg^{L-1} Ketoco-nazole
Ascorbate Peroxidase μg g^{-1} dry	40	0.382 (±0.009)	0.204 (±0.004)	0.280 (±0.007)	0.412 (±0.008)	0.602 (±0.012)	0.345 (±0.006)	0.572 (±0.011)	0.692 (±0.013)	0.678 (±0.013)	0.374 (±0.007)	0.621 (±0.015)	0.729 (±0.018)
weight	80	0.678 (±0.020)	0.322 (±0.009)	0.548 (±0.016)	0.864 (±0.025)	0.944 (±0.028)	0.629 (±0.018)	0.878 (±0.026)	0.977 (±0.029)	0.944 (±0.029)	0.715 (±0.021)	0.983 (±0.029)	0.998 (±0.029)
Catalase min^{-1} mg^{-1}	40	8.41 (±0.168)	5.88 (±0.117)	7.96 (±0.159)	10.24 (±0.204)	6.40 (±0.128)	4.64 (±0.116)	6.04 (±0.120)	8.80 (±0.233)	6.67 (±0.166)	4.92 (±0.123)	6.12 (±0.153)	7.14 (±0.178)
Protein	80	11.24 (±0.337)	8.64 (±0.259)	10.18 (±0.305)	13.36 (±0.400)	9.25 (±0.277)	7.56 (±0.226)	8.96 (±0.268)	11.24 (±0.337)	12.67 (±0.380)	9.71 (±0.291)	11.94 (±0.358)	14.54 (±0.436)

References

Asada, K. and M.T. Takahashi. 1987. Production and scavenging of active oxygen in photosynthesis. In Kely DJ. Osmond B, Arntzen CJ. (ed.). *Photoinhibition*, Amsterdam: Elsevier, **227-287.**

Azooz, M.M., M.A. Shaddad and A.A. Abdel—Latef. 2004. The accumulation and compartmentation of proline in relation to salt tolerance of three Sorghum, cultivars. *Indian J. Plant Physiol.*, **9(1): 1-8.**

Bajji, M., J. Kinet and S. Lutts. 1998. Salt Stress effects on roots and leaves of *Atriplex halimus* L. and their corresponding callus cultures. *Plant Sci.,* **137: 131-142.**

Bates, L.S., R.P. Waldern and I.D. Teare. 1973. Rapid determination of free proline for water-stress studies. *Plant and Soil.*, **39: 205-207.**

Berova, M.Z. Zlatev and N. Stoeva. 2002. Effect of palobutrazol on wheat seedlings under low temperature stress. *Bulg. J. Plant Physiol.*, **28(1-2): 75-84.**

Boyer, J.S. 1982. Plant productivity and environment. *Science.* **218: 443-448.**

Bradford, M.M., 1976. A rapid and sensitive method for the quantitation of microgram quantities of protein utilising the principle of protein dye binding. *Anal. Biochem.*, **72: 248-254.**

Chandlee, J.M. and J.G. Scandalios. 1984. Analysis of variants affecting the catalase development program in maize Scutellum. *Theor. Appl. Genet.*, **69: 71-77.**

Feng, Z., Guo and Feng. 2003. Amelioration of chilling stress by triadimefon in cucumber seedlings. *Plant Growth Regul.*, **30: 277-283.**

Fletcher, R.A. and G. Hofstra. 1998. Triazoles as potential plant protectants. In: Berg, D. and M. Plempel (eds), *Sterol Biosynthesis Inhibitors in Plant Protection*. Ellis Horwood Ltd., Cambridge, England. **321-331.**

Fletcher, R.A., Giley, A., Sankhla, N. and T.M. Davis. 2000. *Triazoles as Plant Growth Regulators and Stress Protectant Hort. Rev.*, John Wiley and Sons Inc., **24: 56-138.**

Flectcher, R.A., Hofstra, G. 1985. Triadimefon a plant multi protectant. *Plant Cell Physiol.*, **26: 775-780.**

Girija, C., B.N. Smith, P.M. Swamy. 2002. Interactive effects of Sodium chloride and Calcium chloride on the accumulation of proline and glycinebetaine is peanut (*Arachis hgpogaea* L.). *Environmental and Experimental Botany.* **47: 1-10.**

Gopi, R., B.M. Sujatha, L. Karikalan and R. Panneerselvam. 1998. Triadimefon induced variation in growth and metabolism in the NaCl Stressed *Vigna unguiculata* (L.) Seedling. *Geobios,* **25(4): 235-241.**

Grieve, C.M. and S.R. Gratten. 1983. Betaine status in wheat in relation to nitrogen stress abd transient salinity stress. *Plant and Soil.*, **85: 3-9.**

Kraus, T.E., B.D. McKerise and R.A. Fletcher. 1995. Paclobutrazol induced tolerance of wheat leaves to paraquat may involve increased antioxidant enzyme activity. *J. Plant Physiol.*, **145: 570-576.**

Maraschner, H. 1995. *Mineral Nutrition of Higher Plants.* (Academic Press: San Diego).

Moore, S. and W.H. Stein 1948. Photometric method for use in the Chromatography of amino acids. *J. Biol. Chem.*, **176: 367-388.**

Muthukumarasamy, M., S.D. Gupta and R. Panneerselvam. 2000. Influence of triadimefon on the metabolism of NaCl stressed radish. *Biol. Plant.*, **43: 67-72.**

Panda, S.K. 2001. Oxidative response of green gram seeds under salinity stress. *Indian J. Plant Physiol.*, **6: 438-440.**

Panneerselvam, R., M. Mutukumarasamy and S.N. Rajan 1998. Amelioration of NaCl stress by triadimefon in Soybean seedlings. *Biologia Plantarum*, **41(1): 133-137.**

Saha, K. and K. Gupta. 1993. Effect of LAB-150978-A plant growth retardant on sunflower and mungbean seedlings under salinity stress. *Indian J. Plant Physiol.*, **36(3): 151-154.**

Sairam, R.K., P.S. Deshmukh and D.C. Saxena. 1998. Role of antioxidant systems in wheat genotypes tolerance to water stress. *Biol. Plant.*, **41: 387-394.**

Senaratna, T., Touchell D., Buun E., and K. Dinon. 2000. Acetyl salicylic acid (Aspirin) and salicylic acid multiple stress tolerance in bean and tomato plants. *Plants Growth Reg.* **30: 157-161.**

Shigeoka, S.T. Ishikawa, M. Tamoi, Y. Miyagawa, T. Takeda, Y. Yabuta and Kazuya Yoshimura. 2002. Regulation and function of ascorbate peroxidase isoenzymes. *Journal of Experimental Botany.* **53(372): 1305-1319.**

Willeknes, H., S. Chamnogpol, M. Davey, M. Schraudner, C. Langebartels, M. Van Montagu, D. Inze and W. Van Camp. 1997. Catalase is a sink for H_2O_2 and is indispensable for stress defense in C_3 plants. *EMBO Journal.* **16: 4806-4816.**

❑❑❑

18

A New Echinostome Cercaria from An Aquatic Snail, *Vivipara bengalensis* (Annandale, 1921) of Manipur

Indrani Chongtham and R.K. Gambhir

Parasitology Lab., Department of Life Sciences, Manipur University, Canchipur, Imphal–795 003, Manipur, India

Introduction

A survey of collection of snails in Manipur for larval trematodes reveals the occurrence of a new echinostome cercaria, *C. mekolensis* n.sp., from an aquatic snail *Vivipara bengalensis*. About 70 snails (*Vivipara bengalensis*) were collected from different freshwater bodies of Manipur. Out of these three snails were found infected with the new echinostome cercaria. It belongs to the Echinatoides group of Sewell (1922). The present specimen does not agree with any of the known species of the Echinatoides group and therefore, a new specific name is proposed to accommodate it.

Materials and Methods

Snails were collected from different water bodies of Imphal districts of Manipur. They were examined for cercariae in good light. The cercariae obtained were studied live employing neutral red as intravital stain. The larvae were fixed in hot 70 per cent alcohol and permanent preparations were made by using alum acetocarmine for staining. Measurements were taken using ocular-micrometer. Sketches were made with the help of camera lucida.

Description: (Fig. 1)

The cercariae emerge out of the snail host in large numbers during noon hours when light was plenty. They swim actively with the help of lashing movement of their tail. While swimming the anterior end elongates and turns ventrally. The cercariae crawl for a very short distance with the help of sucker on the bottom of the container. The cercariae remain alive in tape water for a few hours after their emergence. They show positively phototropic behaviour.

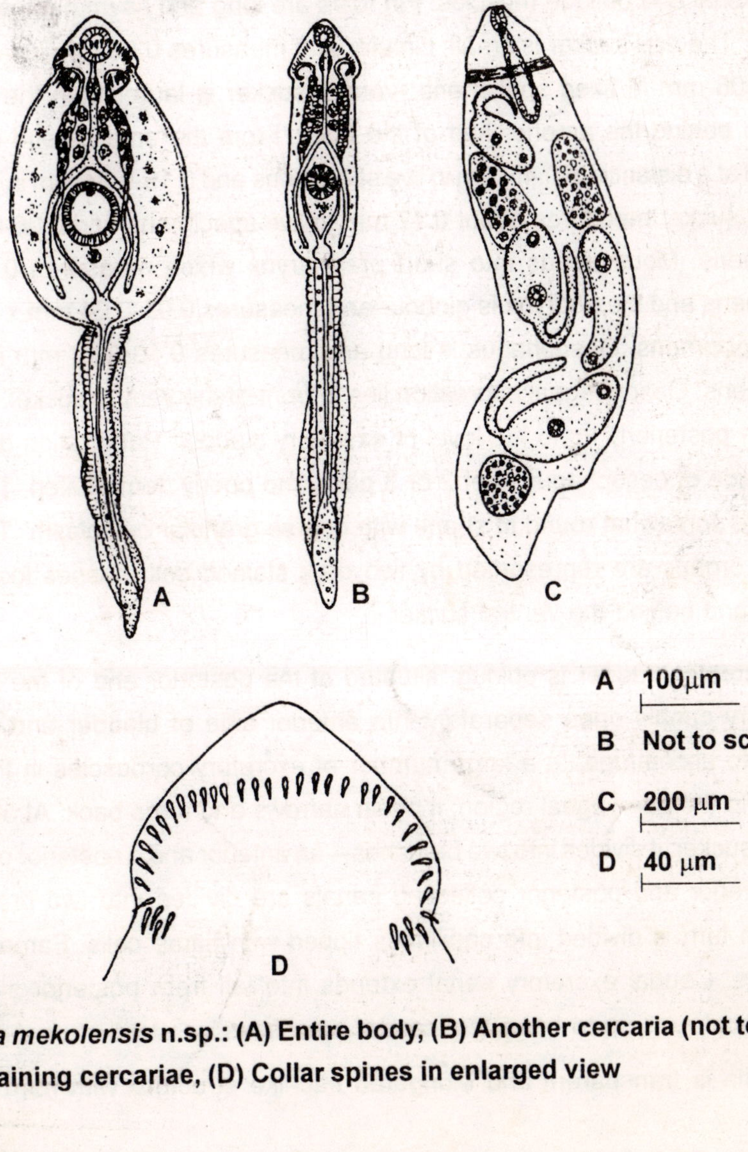

Fig. 1: *Cercaria mekolensis* n.sp.: (A) Entire body, (B) Another cercaria (not to scale), (C) Redia containing cercariae, (D) Collar spines in enlarged view

The body is elongate-oval with narrow anterior and broad posterior ends. The cuticle is thick and without spines (aspinose). Body measures 0.66 mm x 0.38 mm in live specimens and 0.26-0.45 mm x 0.10-0.27 mm in fixed specimens. A distinct muscular collar is present at the anterior end of the body. The collar spines are 38 in number arranged in a single row including 4 end-group spines (Fig. 1-D). Tail is aspionse with an invaginable caudal process at its tip. In live specimen the tail is almost as long as the body, measuring 0.63 mm x 0.08 mm and in fixed specimens it becomes longer than the body and measures 0.47-0.63 mm x 0.03-0.06 mm. It consists of circular longitudinal and oblique muscles. Fin folds are long and asymmetrical on two sides of the tail. The oral sucker is small, circular and measures 0.08 mm in live specimen and 0.03-0.06 mm in fixed specimens. Ventral sucker is larger than the oral sucker and located behind the anterior half of the body. From the anterior end of the body, it is located at a distance of 0.45 mm in live specimens and 0.15-0.24 mm in fixed specimens. Ventral sucker measures about 0.12 mm in live specimens and 0.06-0.09 mm in fixed specimens. Mouth leads into short prepharynx which measures 0.03 mm in fixed specimens and the pharynx is globour and measures 0.01-0.04 mm x 0.01-0.03 mm in fixed specimens. Oesophagus is long and measures 0.10-0.12 mm in length in fixed specimens. Oesophageal bifurcation lies in front of the ventral sucker. Intestinal caeca extends posteriorly up to the level of excretory bladder. Penetration gland situated on either side of oesophagus is of 2 or 3 pairs and poorly demarcated. The cystogenous cells are somewhat round in shape with coarse granular cytoplasm. The rudiments of genital organs are represented by two dark stained cell masses located at anterior border and behind the ventral sucker.

Excretory bladder is oblong, situated at the posterior end of the body. Two main excretory canals open separately into anterior side of bladder and runs forward. It dilates to accommodate a large number of excretory corpuscles in the oesophageal region. In the pharyngeal region, it again narrows and turns back. At about the level of ventral sucker, it divides into two branches—an anterior and a posterior collecting canals. The anterior and posterior collecting canals are divided into two branches, each of which in turn is divided into capillaries tipped with flame cells. Fame cells are 36 in numbers. Caudal excretory canal extends into tail from posterior side of excretory vesicle and runs up to base of a caudal processes.

Redia is transparent and elongated sac-like structure with narrow anterior and broad posterior ends. It measures 1.62-2.79 mm x 0.13-0.35 mm in live specimens and 1.25-1.71 mm x 0.35-0.43 mm in fixed specimens. Birth pore is located behind the collar region. Mouth is terminal and leads into a pharynx which measures 0.04 mm in live specimens and 0.04-0.06 mm x 0.04-0.05 mm in fixed specimens. Gut is short with yellowish contents and extends up to 0.29 mm in length. The marginal lappets are present in the hind region of the body. Each redia contains 1-4 mature cercariae, few in developing stage and 3-4 germ balls.

Discussion

Host-*Vivipara bengalensis* (Annandale, 1921)

Period of collection-05/04/2005 to 22/05/2005

Locality-Mekola (Imphal West District, Manipur, India)

No. of specimens examined-(70)

No. of specimens found infected-(3)

Percentage of infection-(4. 28%)

A review of the literature reveals that 12 species of Echinatoides group of echinostome cercariae have so far been reported from different snail hosts of India. Of these, *Cercaria unnaoensis III* and *Cercaria spinosa* described from *Vivipara bengalensis* come close to the present species. But the present species described from *Vivipara bengalensis* is "aspinose" in contrast to *Cercaria unnaoensis III* and *Cercaria spinosa* which are "spinose". Moreover the tail of the present species is longer (0.47-0.63 mm) in contrast to that of *Cercaria unnaoensis III* and *Cercaria spinosa* which are 0.30-0.50 mm and 0.40-0.52 mm respectively. The present species has 38 collar spines including 4 end-group of spines on either side whereas *Cercaria unnaoensis III* and *Carcaria spinosa* have 37 and 37-38 collar spines including 5 end-group spines in both (Table 1). The other species described so far from different snail host differ from the present one in the number and position of penetration gland, flame cell patterns and number of collar spines in general.

In view of the reason stated above, the present cercaria is therefore, regarded as a new species and is named as *Cercaria mekolensis* n.sp. from "Mekola" the locality of collection, Manipur.

Table-1: Chart showing comparison of morphometric measurements of two very closely related Indian species of cercaria from same host along with that of the new species from India

Sl. No.	Characteristics	*Cercaria* sp.		
		Cercaria unnaoensis III	*Cercaria spinosa*	Present new species
1	Host and locality	*Vivipara bengalensis* *Limenea auricularia* *Indoplanorbis exustus* Locality-Priyadarshini Heal, Unnao (U. P.)	*Bellamya bengalensis* Locality- Nawabganj, Unnao (U. P.)	*Vivipara bengalensis* Locality-Mekola, imphal West District, Manipur.
2	Body measurement/ Dimension (mm)	0.26-0.38 x 0.10-0.17	0.27-0.40 x 0.11-0.18	0.26-0.15 x 0.10-0.27
3	Tail measurement/ Dimension (mm)	0.30-0.50 x 0.03-0.05	0.40-0.52 x 0.03-0.05	0.47-0.63 x 0.03-0.06
4	No. of collar spines	37 including 5 end-groups of spines	37-38 including 5 end-groups of spines	38 including 4 end-groups of spines
5	Body covering/cuticle	Spinose	Spinose	Aspinose with thick cuticle
6	No. of penetration Gland	3 pairs situated on either side of oesophagus	3 pairs situated on either side of oesophagus	2-3 pairs situated on either side of oesophagus
7	Flame cell pattern	2[(3+3+3) + (3+3+3)] = 36	2[(3+3+3) + (3+3+3)] = 36	36
8	Description of Fin Fold	Two short fin folds present near tip of tail	Fin folds asymmetrical on two sides of tail	Fin folds are long and asymmertical on two sides of tail
9	Redia measurement/ Dimension (mm)	0.50-1.10 x 0.13-0.15	0.45-0.31 x 0.14-0.19	1.25-1.7 x 0.35-0.43

References

Dawes, Ben. 1968. *The Trematoda: With special reference to British and other European forms*. Cambridge University Press, Cambridge. p. **644.**

Mukherjee, R.P. 1992. *Larval Trematodes of India*. Part II. Zoological Survey of India, Kolkata. p. **98.**

Pandey, K.C., H.S. Singh and S.S. Lal. 1982. *Studies on Some Freshwater Cercaria*. A publication from the Department of Zoology, Meerut University, U.P. **1(2): 17-38.**

Rao, N.V. Subba. 1989. *Freshwater Molluscs of India*. Zoological Survey of India, Kolkata:. p. **289.**

Tonapi. G.T. 1980. *Freshwater Animals of India: An Ecological Approach*. Oxford and IBH Publishing Co., New Delhi, Mumbai and Kolkata. p. **341.**

❑❑❑

19

Impact of Synthetic Pyrethroid Residues on Aphid Resurgence and Plant Metabolites in *Gossypium hirsutum* L.

R. Ravindhran, Anne Xavier* and S. Ignacimuthu

Entomology Research Institute, Loyola College, Chennai–600 034

*Department of Botany, Holy Cross College, Tiruchirapalli–620 002

Introduction

Although significant advances have been made in the development of non-insecticide methods of pest management in cotton, insecticides are still indispensable. The use of synthetic organic insecticides has triggered the process by which insects turn into pests. Control of this pest depends on the application of broad-spectrum insecticides. The synthetic pyrethroids have attracted considerable attention in recent years as a potential replacement for naturally occurring pyrethrins and other types of synthetic insecticides against cotton bollworm *Heliothis armigera* (Sorathia and Chari, 1981; Thote *et al.,* 1988, Nisha George and Kalyanasundaram, 1994). The use of insecticides for protecting crops from pest infestation inevitably leads to the accumulation of residues in different parts of the crop plants. Pyrethroids are reported to be relatively stable for long duration (Tiwari, 2005). In cotton, repeated and frequent application of synthetic pyrethroids has resulted in resurgence of secondary pest *Aphis gossypii.* The present investigation was carried out to study the effect of synthetic pyrethroid residues on aphid resurgence and plant metabolites of cotton.

Materials and Methods

Field experiment: A micro plot experiment (plot size 10 m × 8 m) was conducted on three varieties (LRA 5166, MCU 5 and SAVITA) of cotton (*Gossypium hirsutum* L.). Experiments were done in a completely randomised design. Three replications were maintained. Insecticides *viz.*: Cypermethrin 75g a. i. ha^{-1}, Fenvelerate 75g a. i. ha^{-1} and Deltamethrin 15g a. i. ha^{-1} were sprayed successively on each variety of cotton at an interval of 15 days commencing the first spray at 45 days after sowing (DAS). Spray fluid used per hectare was 1000 liters. The control plants received only water as spray.

Assessment of pest population: Population count of *Aphis gossypii* was recorded in the early morning hours from 6 a.m. to 7 a.m. on 3 leaves each selected from top, middle and bottom regions of the plant on 10 randomly selected plants. Pests were counted one-day before and 7th day after third spray. The resurgence ratio was calculated using Henderson Tilton formula (Jayaraj and Regupathy, 1986).

Biochemical parameter: Biochemical content (total sugar, amino acids, protein, phenol and gossypol) of leaf samples and leaf exudate were separately analysed after third spray using standard methods (Sadasivam, and Manickam, 2000; Mahadevan and Sridhar, 1982). Residues were quantified in the leaves on 0, 1, 3, 5, 7 and 10th day after third spray as per the method of Kuttalam (1990). The results were compared using Duncan Multiple Range Test (DMRT) (Duncan, 1995).

Results and Discussion

Phrethroid residues and aphid resurgence: Gas chromatographic analysis of pyethroid residues on the leaves of cotton is presented in Table 1. During the initial period (24 hours after third spray) the disipation rate of all the pyrethroids was comparatively faster with the advancement of days. Among the pyrethroids, the rate of dissipation of deltamethrin was high and it was followed by cypermethrin and fenvelerate. However the persistence of pyrethroid residues was found to be approximately 17 to 25 per cent on the seventh day in all the varieties. No remarkable variations were observed between varieties. Parmar (1988) observed that the rate of dissipation of synthetic pyrethroid was slow when compared to organophosphorus and carbamate insecticides.

Foliar application of synthetic pyrethroids resulted in the resurgence of aphids in all the varieties under study (Table 2). The resurgence ratio was highest in deltamethrin treated leaves and it was lowest in fenvelerate treated leaves. A significant increase in *aphid resurgence* was obtained in the variety SAVITA followed by LRA 5166 and MCU 5. There may be some stimulation when aphids are confined to aged deposits of fenvelerate seven days after application (Jackson and Wilkins, 1985). Bartlett (1968) found that compounds with a relatively long persistence of insecticide caused aphid resurgence once they has reached the sub lethal level. In our study, high aphid resurgence was observed in pyrethroid treated leaves of all the varieties. Persistence of pyrethroids was associated with aphid resuragence.

Table-1: Persistence (ppm/g) and dissipation (%) of pyrethroid residues in the leaves of cotton

Days clapsed	Cypermethrin	Fenvelerate	Deltamethrin
		Mean (ppm) + S. D. (%)	
0	3.065 + 0.02	3.2263 + 0.01	0.698 + 0.02
1	1.451 + 0.015 (52.7)	1.512 + 0.01 (53.6)	0.381 + 0.02 (45.4)
3	0.765 + 0.02 (75.0)	1.115 + 0.02 (65.8)	0.145 + 0.02 (79.2)
7	0.501 + 0.01 (83.7)	0.792 + 0.015 (75.7)	0.091 + 0.015 (87.0)
10	0.231 + 0.03 (92. 5)	0.441 + 0.01 (86.5)	0.0

Values in parentheses indicate the percentage of dissipation of pyrethroid residues.

Biochemical Parameters: Table 2 presents the biochemical contents of pyrethroid treated cotton leaves. Analysis of variance showed that deltamethrin treatment had induced significant accumulation of total sugar (35.23mg/g), amino acids (6.03mg/g), and protein, but had reduced total phenol content (18.46mg/g). Similar results were obtained in the cypermethrin and fenvelerate treatments. Among the three varieties, SAVITA recorded high total sugar, amino acids and protein. Phenol content was low (Table 3).

Gossypol, a sesquiterpene, is a major defence chemical against cotton insects. In the present study, a significant increase in gossypol content was observed in the

Table-2: Effect of foliar application of pyrethroids on aphid resurgence and biochemical constituents of three varieties of Cotton (Treatment mean)

Leaf samples							Leaf exudates			
T-mean	Aphid Resurgence	Total sugar (mg/g)	Amino Acids (mg/g)	Protein (mg/g)	Phenol (mg/g)	Gossypol (mg/g)	T-mean	Total sugar (mg/g)	Amino Acids (mg/g)	Phenol (mg/g)
T_1		27.05b	3.47c	11.82b	26.01a	2.71a	T_1	1.66b	0.46b	0.47a
T_2	0.65b	35.23a	4.30b	17.99a	20.35b	1.79bc	T_2	6.23a	2.78a	0.30a
T_3	0.67b	30.32b	4.18b	14.92ab	20.41b	1.98b	T_3	5.89a	1.89a	0.34a
T_4	1.03a	37.33a	6.03a	16.99a	18.46b	1.57c	T_4	8.31a	2.48a	0.37a
C.D. (0.05)	0.14	6.86	0.71	4.35	2.68	0.33	C.D. (0.05)	3.54	1.12	0.19

T_1–Control, T_2–Cypermethrin, T_3–Fenvelerate, T_4–Deltamethrin, T-mean–Treatment mean, C.D.–Critical Difference

Means followed by a common letters under each column are not significantly different at the 5% level by Duncan's Multiple Range Test (DMRT)

Table-3: Effect of foliar application of pyrethroids on aphid resurgence and biochemical constituents of three varieties of cotton (means of three varieties)

Leaf samples							Leaf exudates			
V-mean	Aphid Resurgence	Total sugar (mg/g)	Amino Acids (mg/g)	Protein (mg/g)	Phenol (mg/g)	Gossypol (mg/g)	V-mean	Total sugar (mg/g)	Amino Acids (mg/g)	Phenol (mg/g)
V_1	0.81b	30.62b	3.41c	14.49ab	20.67b	1.77b	V_1	5.90a	1.45b	0.52a
V_2	0.61c	28.70b	4.06b	13.58b	23.45a	2.32a	V_2	3.97a	1.46b	0.28b
V_3	1.24a	38.13a	6.03a	18.21a	19.79b	1.94b	V_3	6.70a	2.80a	0.31b
C. D. (0.05)	0.14	5.95	0.62	3.77	2.32	0.28	C. D. (0.05)	3.06	0.97	0.15

V_1-LRA 5166, V_2-MCU5, V_3-SAVITA, C.D.-Critical Difference.

Means followed by a common letters under each column are not significantly different at the 5% level by Dunan's Multiple Range Test (DMRT).

untreated check. Among the pyrethroids, deltamethrin treatment significantly reduced gossypol content (1.57mg/g). SAVITA showed remarkable low level of gossypol content.

Foliar application of pyrethroids had influenced greater exudation of total sugar, amino acid and lower level of total phenol content. Leaf exudates obtained from deltamethrin treated leaves showed high level of total sugar (8.31 mg/g) and it was on par with the cypermethrin (6.23 mg/g) and fenvelerate (5.89mg/g) treated leaves. However total phenol estimated from leaf exudate found to be low in all the treatments.

In the present study, persistence of pytrethroid residues effectively increased the total sugar, amino acid, and protein and reduced the level of total phenol and leaf gossypol content. These changes might have attracted more aphids which in turn led to high aphid resurgence. The altered biochemical status of the host plant following insecticide application was reported to cause resurgence (Tamil Selvan *et al.*, 1990; Joseph Dominick and Mohanasundaram, 1992; Sadasivam and Manickam, 2000).

From the present study, it could be concluded that frequent application of pyrethroids alters the physiology of host, which in turn might possibly render the cotton plant vulnerable to sucking pest *Aphis gossypii.*

References

Bartlett, B. 1968. Outbreaks of two spotted spidermites and cotton aphids following pesticide treatment. 1. Pest stimulation Vs natural enemy destruction as the cause of outbreaks. *J. Econ. Entomol.* **61: 501-506.**

Duncan, D.B. 1995. Multiple Range and Multiple F-Tests. *Biometrics*, **11: 1.**

Jackson, A.E.M. and Willkins, R.M. 1985. The effect of sub-le*hal dosages of the synthetic pyrethroid Fenvalerate on the reproductive rate of the aphid Myzus persicae. *Pesticide Sci.* **16: 364-368.**

Jayaraj S. and Regupathy, A. 1987. Studies on the resurgence of sucking pest of crops in Tamil Nadu. In: *Resurgence of Sucking Pests. Proc. Natl. Symp.* (Ed) S. Jayaraj, Tamil Nadu Agricultural University, Coimbatore, **225-240.**

Joseph Domnick and Mohanasundaram, M. 1992. Effect of insecticides on the biochemical nature of the host plant and its relation to resurgence of the whitefly, *Bemisia tabaci* on cotton. *Pestology.* **16(1): 7-10.**

Kuttalam, S. 1990. Occupational exposure of field workers on the insecticides applied in cotton ecosystem. Ph. D. Thesis submitted to and approved by Tamil Nadu Agricultural University, Coimbatore.

Mahadevan, A. and Sridhar, R. 1982. *Methods in Physiological Plant Pathology*. (Sivakami Publications, Chennai, India): p. **328.**

Nisha George and Kalyana Sundaram, M. 1994. Chemistry of synthetic pyrethroid insecticides-some recent advances. **53: 933-945.**

Parmar , B.S. 1988. Evaluation of protection schedules on agricultural crops and post harvest products. In *Progress in Agrochemical Research*. I.A.R.I. Publishers, New Delhi, p. **22.**

Sadasivam, S. and Manickam, A. 2000. *Biochemical Methods for Agricultural Sciences*. Wiley Eastern Ltd., Chennai, India, pp. **37-201.**

Sorathia, B.K. and Chari, M.S. 1981. Evaluation of synthetic pyrethroids fenvelerate for the control of bollworm on hybrid 4 cotton. *Pesticides*, **15(1): 14-17.**

Tamilselvan, C., Sundaraj, R., Regu, K. and David, B.V. 1990. Influence of deltamethrin on the bio-chemical parameters of cotton and biology of the whitefly *Bemisia tabaci* (Gennadius). *Pestology* **14: 17-19.**

Tiwari, S.N. 2005. Residual activity organophosphorus and pyrethroid insecticides against *Rhyzopetha dominica* (F) (Coleopetra, Bostrychidae) in stored wheat. *Pestology* **29(7): 7-14.**

Thote, R.L., Takey, Y.M. and Nimbalkar, S.A. 1988. Control of bollworms on cotton with synthetic pyrethroids alone and in alteration with conventional insecticides, *Pestology*. **12(7): 16-20.**

❑❑❑

20

On a New Trematode of the Genus Encyclometra Baylis and Cannon, 1924 (Trematoda: Encyclomentriinae) from the Stomach of a Snake in Manipur

I. Gyaneswori and R.K. Gambhir
Department of Life Sciences, Parasitology Laboratory,
Manipur University, Canchipur Imphal–795 003

Introduction

During a survey of trematodes of reptiles, 3 snakes were dissected of which two snakes were found to be parasitised by trematodas. The trematodes were collected from stomach of the snake. *Tropidonotus piscator*. On further detail study the trematodas were found to belong to the genus *Encyclometra* (Baylis and Cannon, 1924). The present specimen comes closer to the *E. japonica* (Yoshida and Ozaki, 1929), yet differ in many characteristics and described as a new species.

Material and Methods

The collected trematodas were fixed in AFA and pressed between glass slides and preserved in 70% alcohol having 5% glycerine. The preserved specimen is stained in Bulloy's acetoalum carmine, dehydrated in alcohol grades, clear in xylene and mounted in DPX. Measurements were taken in millimeter, diagrams were drawn with the aid of camera-lucida and measurements were taken with occular micrometer.

Description

Body fusiform, spinose with pointed posterior end and measures 3.09-6.89 mm in length and 1.44-1.77 mm in width. Oral sucker circular 0.45-0.62 x 0.52-0.67 mm. Prepharynx short muscular pharynx well developed 0.20-0.44 x 0.40-0.43 mm. Oesophagus short, ventral sucker or acetabulum slightly larger than oral sucker at anterior half of the body and measures 0.61-0.88 x 0.53-1.02 mm. Intestinal caeca simple, sub-equal, left-caccum extending upto the hind region and right caecum slightly at anterior to it. (Fig. 1: *Encyclometra imphalensis* n. sp.)

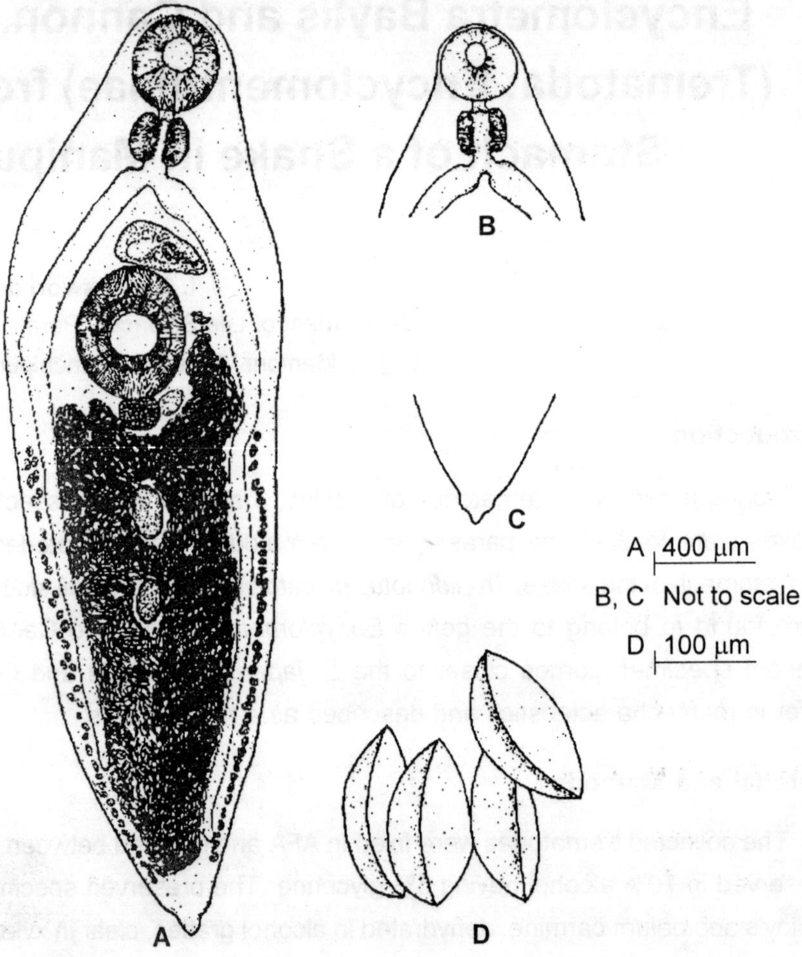

Fig. 1: *Encyclometra imphalensis* n.sp. (A) Entire worm, (B) Anterior end (not to scale), (C) Posterior end (not to scale), (D) Eggs

Testis tandem and post-equatorial or in middle third of the body. Anterior testis slightly smaller than posterior testis and measures 0.23-0.39 x 0.22-0.31 mm and posterior testis measures 0.23-0.44 x 0.20-0.36 mm. Ovary is oval, just behind the ventral sucker and measures 0.12-0.20 x 0.20-0.30 mm. Recepticulum seminis is oval situated close to ovary. Vitellaria numerous and extending along the intestinal caeca covering nearly the whole length of the hind body. Excretory bladder not visible. Eggs are oval, operculated and pointed at both ends measuring 0.07-0.14 x 0.02-0.50 mm.

Discussion

Baylis and Cannon (1924) erected the genus *Encyclometra* with *caudata* as its type species which is recovered from the intenstine of a snake. Later on Yeh (1953) reviewed this genus and validated the following three species only *viz. E. columbrianum* (Reed, 1819, Dollfus 1929); *E. japoncia* (Yoshida and Ozaki, 1929) and *E. asymetrica* (Wallace, 1936). Again, Srivastava and Ghosh, 1968, described *E. bargara* from an Indian banded krait, *Bargarus fasciatus* at Kolkata, which was later synonymised as *E. japonica*. In 1982, Nirupama Agarwal and K.C. Pandey reported the species *E. japonica* (Yoshida and Ozaki, 1929) from Lucknow.

In the present specimen the body is longer, perpharynx well developed, ventral sucker larger than the oral sucker posterior testis is larger than anterior one, eggs are also larger in size than that of *E. Japonica* and the adult body have a pointed posterior end. By considering the characteristic that differs from its nearst relative *E. japonica*, a new species is proposed to accommodate this species giving the name *E. imphalensis* n.sp from the locality of the host.

Host—*Tropidonotus piscator*
Habitat—Stomach
Locality—Chinga makha, Imphal, Manipur
Date of collection—10th May 2005.

References

Nirupama Agarwal and K.C. Pandey, 1982. Remarks on some known Trematodas of snakes, *Tropidonotus piscator. Reading in Zoology* (1982), **1(2): 2-7.**

Yamaguti, S., 1985. Systema Helminthum, vol I. *Digenetic Trematodas of Vertebrates*, Part I, p. **979**, New York Intersciences.

Ben Dawes, 1968. The Trematoda with special reference to British and other European forms. The Syndies of the Cambridge University Press, p. **491.**

❑❑❑

21

Estimation of Physiological Parameters and Their Association in Influencing Biological and Grain Yields in Rice (*Oryza sativa* L.)

P.R. Rao and S. Ravichandran

Directorate of Rice Research, Rajendranagar, Hyderabad–500 030

Introduction

"Rice is life", was the theme of United Nations Organisation (UNO) for the year 2004, as the just concluded year 2004 was celebrated as "International Year of Rice (IYR)". This shows the importance of rice, the important cereal crop over all other commodities. Hence, it is utmost important to understand the physiological factors that influences biological yield (dry matter) and economical yield (grains). Realising this importance, a statistical empirical model was fitted by making use of various physiological and growth parameters such as plant height, tiller number, leaf area, root weight, leaf weight, stem weight and chlorophyll content recorded at 30, 60, and 90 days after planting (DAP). And at maturity, data on panicle number per m^2, grain number per m^2, grain number per panicle, total dry matter and harvest index was collected. To study these characters, and experiment was conducted with five new plant types (Khush,) 1995) from IRRI, four high yielding varieties (HYV) and one hybrid. These new plant types (NPT) has 8-9 tillers as compared to the HYVs, which normally have 15-20 tillers. Apart from this, the NPTs possess higher chlorophyll content with 200-250 grains per panicle and efficient root system than the traditional high yielding variety plant types. For brining uniformity, while estimating economical and biological yields, one hybrid *viz.* KRH$_2$ was included in the experiment. For estimating biological yield

(BY) and economical yield (EY), data on the above mentioned parameters were obtained. Following Draper and Smith (1998), step-wise regression models are developed for estimating biological and economic yields by utilising data on yield influencing independent parameters. These models would help in estimating the yield at different points of maturity with the availability of data.

Material and Methods

A field experiment was carried out during kharif and rabi (2003-2004) at Directorate of Rice Research, Rajendranagar, Hyderabad. The soil was alkaline in reaction (pH 8.2), low in available nitrogen (217 kg/ha), and medium in available phosphorus (25.52 kg/ha) and potassium (217.46 kg/ha). The experiment was conducted in a Randomised Block Design (RBD) with four replications. The main treatment consisted of five new plant types from tropical Japonica background developed at IRRI, Philippines; four high yielding varieties *viz.*, IR 64, Jaya, NDR 359, Rasi and one hybrid KRH_2 from Karnataka, India. Fertiliser dose was applied @ 120 : 60 : 40 (NPK kg/ha), nitrogen was applied in 3 split doses initially after 10-15 days after planting thereafter at active tillering and panicle initiation stage. Spacing followed was 20 × 10 cm. Chlorophyll was measured using SPAD 502, Minolta. The leaf area was measured by delta T image and data was recorded for dry matter analysis at 30, 60 and 90 days after planting and at maturity yield and yield components were recorded. Following Draper and Smith (1998), regression models were developed for various sets of data say 30 DAP, 60 DAP, 90 DAP and at maturity. By making use of data on 30, 60 and 90 DAP, estimation of biological yield as dependent variable and factors influencing biological yield was carried out using step-wise regression procedure. Simple and multiple correlation coefficients were also worked out. For estimating EY, only data collected at the time of maturity was utilised. Estimation of EY was carried out by fitting a multiple regression model by keeping grain yield as dependent variable and the other factors such as panicle number, grain number per m^2, grain number per panicle. Total dry matter, 1000 grain weight as harvest index were treated as independent variables. Simple and multiple correlation coefficients among variables were worked out.

Results and Discussion

Correlation coefficients of biological and economical yields with other independent variables were worked out for data on 30 DAP and is presented in Table 1. Table 1 reveals that the correlation coefficients due to biological yield with stem weight, leaf weight and leaf area are highly significant whereas the correlation coefficients due to plant height, tiller number and root weight and chlorophyll are significant. Regression model was fitted for estimating biological yield at 30 DAP and the variation explained by the model is not significant. Hence regression model fitted for data on 60 and 90 DAP. A multiple regression model was fitted to data on 60 DAP for estimating BY and is estimated to be:

$$\mathbf{BY_{est}\ (60\ DAP) = -0.14 * V_2 - 0.01 * V_3 - 0.002 * V_4 - 0.003 * V_5 + 1.108 * V_6 + 0.99\ V_7 + 0.24 * V_8} \quad \mathbf{(1)}$$

Table-1: Correlation Matrix of biological yield with related variables at 30 DAP

Variable	V_1	V_2	V_3	V_4	V_5	V_6	V_7	V_8
V_1	1							
V_2	0.27	1						
V_3	0.20	–0.07	1					
V_4	0.43	0.08	0.20	1				
V_5	0.35	–0.02	–0.01	0.24	1			
V_6	0.86	0.42	0.22	0.52	0.29	1		
V_7	0.94	0.17	0.16	0.35	0.35	0.71	1	
V_8	0.13	0.04	–0.01	–0.18	–0.04	0.06	0.15	1

V_1: Biological Yield	V_2: Plant Height	V_3: Tiller Number	V_4: LAI
V_5: Root Weight	V_6: Leaf Weight	V_7: Stem Weight	V_8: Chlorophyll

Correlation coefficients with BY and other factors were worked out and are presented in Table 2. Table 2 shows that again the maximum correlation coefficients are due to stem weight, leaf height and leaf area. It is important to note that the contribution of root weight to yield has considerably reduced, which shows that this factor does not influence yield at 60 days. This model explains 90% of the variation present in the data set. Again a model was fitted for the various growth parameters recorded after 90 DAP and the biological yield as estimated is:

$$\mathbf{BY_{est}\ (90\ DAP) = 1.24 * V_2 + 0.70 * V_3 - 0.001 * V_4 + 0.10 * V_5 + 0.34 * V_6 + 1.15 * V_7 - 0.62 * V_8} \quad \mathbf{(2)}$$

Table-2: Correlation Matrix of biological yield with related variables at 60 DAP

Variable	V_1	V_2	V_3	V_4	V_5	V_6	V_7	V_8
V_1	1							
V_2	0.56	1						
V_3	0.19	–0.02	1					
V_4	0.67	0.42	0.18	1				
V_5	0.09	–0.22	0.10	–0.10	1			
V_6	0.80	0.51	0.27	0.92	0.04	1		
V_7	0.96	0.51	0.13	0.46	0.09	0.59	1	
V_8	0.15	0.10	0.36	–0.10	0.20	–0.03	0.21	1

V_1: Biological Yield	V_2: Plant Height	V_3: Tiller Number	V_4: LAI
V_5: Root Weight	V_6: Leaf Weight	V_7: Stem Weight	V_8: Chlorophyll

Simple correlation coefficients presented in Table 3 reveals that at 90 DAP, again stem weight contributes maximum to BY followed by leaf weight and leaf area. It is to be noted that the biological yield is a combination of stem weight, leaf weight and leaf area at all the three stages of plant growth. However, other factors like plant height, tiller number, root weight and chlorophyll do not contribute significantly to biological yield in all the three stages of plant growth. Hence, we may able to estimate variation in yield with the availability of information on stem weight, leaf weight and leaf area. Hence, by making use of all the above information, a multiple regression model for estimating biological yield was fitted and the fitted models is:

$$\mathbf{BY_{est} = 0.04 * leaf\ area + 0.01 * leaf\ weight + 1.73 * stem\ weight} \quad \mathbf{(3)}$$

Model (3) explains 98% of variation present in the data set.

Table-3: Correlation Matrix of biological yield with related variables at 90 DAP

Variable	V_1	V_2	V_3	V_4	V_5	V_6	V_7	V_8
V_1	1							
V_2	0.23	1						
V_3	0.11	–0.16	1					
V_4	0.62	0.47	–0.11	1				
V_5	0.19	–0.17	–0.10	–0.03	1			
V_6	0.62	0.45	–0.20	0.92	0.06	1		
V_7	0.80	0.40	–0.16	0.76	0.22	0.79	1	
V_8	0.23	0.34	–0.09	0.39	0.14	0.39	0.36	1

V_1: Biological Yield	V_2: Plant Height	V_3: Tiller Number	V_4: LAI
V_5: Root Weight	V_6: Leaf Weight	V_7: Stem Weight	V_8: Chlorophyll

Similarly, correlation coefficients due to economic yield with panicle number, grain number/m^2, grain number/panicle, harvest index and biological yield were worked out using SAS (1990) software package and are presented in Table 4. It is seen from Table 4 that the maximum contribution of EY is based on grain number/m^2 followed by BY and grain number/panicle. Hence, EY is estimated to be:

$$EY_{est} = -1.41 * V_2 + 11.52 * V_3 - 6.62 * V_4 + 0.62 * V_5 - 10.75 * V_6 + 16.33 * V_7 \quad (4)$$

Table 4: Correlation matrix of Economical yield with related variables at maturity

Variable	V_1	V_2	V_3	V_4	V_5	V_6	V_7
V_1	1						
V_2	0.18	1					
V_3	0.86	0.35	1				
V_4	0.64	–0.52	0.60	1			
V_5	0.77	0.13	0.58	0.40	1		
V_6	–0.15	–0.41	–0.63	–0.20	0.01	1	
V_7	0.61	0.13	0.50	0.62	–0.10	–0.33	1

V_1: Grain Yield	V_2: Panicle number	V_3: Grain No./m^2	V_4: Grain No./panicle
V_5: Total Dry Matter	V_6: 1000 Grain Weight	V_7: Harvest Index	

Multiple correlation coefficient is highly significant and it explains a high percentage of variation present in the data set. However, a regression model of EY with only three factors such as grain number/m^2, grain number per panicle and total dry matter recorded a maturity could be fitted and would be utilised for estimating EY as these factors contribute maximum to EY or grain yield. Hence, a regression model was fitted for estimating EY based on the above three factors and is:

$$EY_{est} = 9.75 * \text{grain number/m}^2 + 0.96 * \text{grain number/panicle} + 0.24 * \text{TDM} \quad (5)$$

Hence, for estimation of biological yield, data recorded at 90 DAP on parameters such as stem weight, leaf weight and leaf area is sufficient. For this purpose, model (3) can be effectively utilised. For estimation of economical yield, model (5) can be utilised. So, for estimation of economical yield, information on grain number per m^2, grain number per panicle and total dry matter recorded at the time of maturity is enough. It can also be concluded that information on stem weight, leaf weight, leaf area, grain no/m^2 and grain number/panicle are required for estimation of both biological and economical yield in various cultivars. By having information on the above parameters, we can judge the superiority of elite genotypes among group of cultivars. Generally, scientists collected information on various growth parameters for estimating the economical and biological yields. From the present study, it can be suggested that information on many parameters are not required for estimation of yields. Since collection of information on various parameters involve a lot of time and money, it is recommended from the present study, it is essential to collect information on the above mentioned parameters only for estimation of yields. Statistical modelling of economic and biological yield of the estimated with the help of available information is discussed in the paper. Models developed explain high percentage of variation presenting the data. These models would help the scientists in estimating potential yield of various rice varieties in the event of available information. This would help the experiments to make effect in increasing the yield by providing required input at various stages of crop growth. These models would also help in ascertaining the yield of particular crop well before harvest. Performance of these models could be improved upon by making use more number of explanatory parameters contributing to yield.

References

Khush, G.S., 1995. Breaking the yield frontier of Rice. *Geo Journal*. 35, pp. **329-332.**

Draper and Smith, 1998. *Applied Regression Analysis*, 3rd edition, John Wiley and Sons, New York.

SAS, 1990. *SAS Users' Guide*, Version 6.12, SAS Institute Incorporation. U.S.A.

❑❑❑

22

A Preliminary Taxonomic Survey of Host-Range of *Dendropthoe falcata* (L.F.) Ettingsh. In the Forest Trees of Shakthi- Auroville, Vanur Taluk, Villupuram District, Tamil Nadu

D. Kadamban and J. Anuradha
Department of Botany, K.M. Centre for P.G. Studies, Pondicherry–605 008

Introduction

Fungi, bacteria, various and nematodes are the organisms, which certainly cause damage to plants, especially to economically valuable plants. It is surprising to know that the parasitic flowering plants are also considered as pathogens. 277 genera and 4100 species of flowering plants are reported to be parasitic on host plants cultivated by humans out of which only 25 genera cause negative impact and are thus considered pathogens (Nickrent and Musselman, 2004).

Loranthus longiflorus Desr., now known as *Dendrophthoe falcata* (L. F.) Ettingsh. belongs to the family Loranthaceae of the order Santalaes. This order contains 158 genera and ca. 2240 species and among them three families deserve special mention in causing damage to economically important plants *viz*. Loranthaceae, Santalaceae and Viscaceae. Loranthaceae, a large family found mainly in tropics, is composed of 74 genera and ca. 900 species.

Loranthaceae contains a majority of stem parasites that are commonly known as 'Mistletoes'. The Anglo-saxon word Mistelt derives from the old German word "mist" for dung and "tan" meaning twing. This is an apropos name given, that birds defecate the seeds of Mistletoes on to tree branches. Most Mistletoes are woody shrubs often with brittle stems and leaves (Calder and Bernhardt, 1983).

Lushington (1902) has found *Loranthus* L. as one among the five genera of this family to be common in the forests of South India. He devised a classification for Loranthaceae based upon parts of plants which are always available (stem, leaves etc.) and can be examined at the time of inspection.

All the members of the genus *Loranthus* are semi parasites and grow on the branches of the host plants by means of haustoria or suckers where the parasitic roots join the host. In addition to the nourishment obtained from the host plants, the green leaves also meet partially their requirements of food.

Many species of *Loranthus* are found in India. Hooker (1896) and Lushington (1902) have described not less than 59 species. The Bombay Dept. of Agriculture, Mumbai (1931) listed 28 hosts of the parasite, *L. longiflorous*, of which 13 are fruit trees, *Mangifera indica* being frequently and seriously damaged. Brand (1938) reported that *L. longiflorus* var. *falcata* is the main species attacking at least 110 tree species, including teak, *Acacia, Albizia* and *Melia* in Chennai. Cooke (1908) has recorded about 12 species from Mumbai and 110 tree species from Chennai attacking many important timber species. Locy (1936) reported the common occurrence of *Loranthus longiflorus* on *Mangifera indica, Dalbergia latifolia, Aegle marmelos* and *Terminalia catappa* in and around Patna. Partridge (1911) has observed only two species of *Loranthus longioflorus* and *Loranthus scurrula* from the forests of Hyderabad. Prof M. Sayeeduddin and Salam (1935) of the Osmania University have worked on the angiospermic parasites of Hyderabad and have noticed the parasite *Loranthus* on 20 of the important garden and forest plants of Hyderabad state. *Achras sapota* and *Euginea jambos* are the two important garden fruit tress that are severely damaged. *Loranthus longiflorus* was very common on *Bassia latifolia* and *Diospyros melanoxylon*. Troup (1921) reported the occurrence of this parasite on *Shorea robusta* and *Pinus longifolia* in Orissa. However, *Dendrophthoe falcata* (L.f.) Ettingsh shows quite a good number of host ranges, particularly of angiosperms.

Materials and Methods

The present study was conducted to enumerate the host range of *Dendrophthoe falcata* (L.f.) Ettingsh causing severe damage to the forest trees of Shakti-Auroville.

Shakti, is one of the many communities in Auroville, located at 11° 59' 15' N 79° 50' 19' E at an altitude of 35 m, in Vanur Taluk, Villupuram district, before Kuilapalayam village. The area is about 35 acres of man made forest and the plant diversity is about 400, including many indigenous and exotic herbs, shrubs, climbers and trees. More than 180 Tropical Dry Evergreen arborescent species and 125 exotic species were planted, during the last 23 years. *Khaya senegalensis* of Meliaceae is the oldest tree which was planted 23 years ago. The present attempt in compiling a list of host plants cultivated and ingested by *Dendrophthoe falcata* (L.f.) Ettingsh, will act as ready companion for agriculturists, agronomists and horticulturists as it will render all possible assistance and provide ample information about host-parasite relation.

A typical maritime tropical climate with a dissymmetric rainfall regime prevails in the study area. The weather is generally humid and hot for most part of the year. The south-west monsoon contributes an average of 356.6 mm of annual rain whereas the north–east monsoon contributes 317.8 mm rainfall. The minimum temperature was 18°C in January and the maximum was 41°C in May. The relative humidity averages to 69 per cent.

The relationship between *Dendrophthoe falcata* (L. f.) Ettingsh with its host was studied among the arborescent species growing at Shakti-Auroville. Plants infested with the parasites have been identified using regional and State Floras (Matthew, 1983; Gamble, 1925). The taxa are arranged alphabetically with their families by adding correct nomenclature, citations and information regarding flowering, fruiting periods and description of host plants etc.

Systematic Enumeration

The hosts are arranged alphabetically with their families:

(1) *Acacia auriculiformis* A.cunn. Ex Benth. **Mimosaceae**

A handsome evergreen tree, unarmed, to 10m with pendulous branches, Flowers: Yellow Flowering—December to February; Fruiting—February to April.

(2) *Acacia holoserica* A. Gunn. ex G. Don **Mimosaceae**

Tree reaching an height of 8 m with numerous branches beginning from the base, giving it a multistemmed appearance with phyllodes. Flowers small, bright yellow in spikes.

(3) *Aegle marmelos* (L.) Corr. Serr. **Rutaceae**

A deciduous, armed tree to 8m bearing ovoid berry (woody). Flowering – March; Fruiting – January to December.

(4) *Albizia lebbeak* (L.) Benth. **Mimosaceae**

A deciduous tree to 15m, with a crown of equal spread, Leaf Pinnate. Petals greenish, pod flat. Flowering – February to April; Fruiting – April.

(5) *Albizia guachepele* (kunth) Dugand. **Mimosaceae**

A large tree growing more than 15m high, girth over 2.5m, bark flaking in rectangular patches, brown to greenish – green, young growth rusty pubescent. Leaflets pubescent on both sides but more densely pubescent beneath. Calyx and corolla rusty pubescent, calyx olive green, corolla lobes brownish, stamens white fading to yellow, slightly fragment. Pod–dark brown, velvety, twisted, dehiscing along the adaxial side. Seed white, compressed.

(6) *Annona reticulata L.* **Annonaceae**

A deciduous tree to 8m. Alternate, simple leaf; Petals-greenish; fruit – Berry; Flowering and Fruiting – January to December.

(7) *Anthocephalus cadamba* (Roxb.) Miq. **Rublaceae**

A medium to large sized deciduous tree with clean cylindrical stem with horizontal branches, bark dark grey with longitudinal fissures peeling off in thin scales, leaves elliptic; flowers yellow or orange in globose heads.

(8) *Artocarpus heterophyllus* Lam. **Moraceae**

Tree to 10m. Perianth thick – fleshy, fruit edible; Flowering – January to April, Fruiting – May to September.

(9) *Calliandra haematocephala* (L.) Druce **Mimosaceae**

Tree to 8m, Leaf – Pinnate, Petals purplish, Pod oblong; Flowering – July to December; Fruiting – December.

(10) *Cassia fistula* L. **Caesaliniaceae**

Deciduous tree to 8m. Petals yellow, Pod-oblong; Flowering – March to May; Fruiting- May.

(11) *Cassia siamea* Lam. **Caesalpiniaceae**

Tree to 10m. Petals yellow, Pod – strap-shaped; Flowering and Fruiting – January to December.

(12) *Chloroxylon swietenia* DC. **Rutaceae**

Deciduous unarmed tree, Paripinnate leaf, petals – greenish and clawed. Seed oblong apically winged; Flowering- March to April; Fruiting – January to December.

(13) *Coccoloba uvifera* L. **Polygonaceae**

Evergreen shrub with leaves often viewed in red. The fragrant white flowers are very small and borne on spikes. Female trees bear reddish fruits pear shaped and hang down in grape like clusters and attract birds and children.

(14) *Cordia sabestina* L. **Boraginaceae**

Evergreen large shrub to a height of 5.8m with simple leaves. Flowers in terminal racemes with scarlet colours; Flowering during January to March.

(15) *Cordia dichotoma* forst. f. **Boraginaceae**

Small deciduous tree, flowers white appear in large corymbose cymes.

(16) *Couroupita guianensis* Aublet **Lecythidaceae**

Deciduous tree to 20m, Petals–rose/red. Androphore large. Flowering in successive waves.

(17) *Ficus amplissima* Smith **Moraceae**

Tree to 18m, petiole glandular at the apex below. Figs monoecious.

(18) *Ficus benghalensis* var. *krishnae* **Moraceae**

Small tree, upto 10m high, with aerial roots wiry, Leaves alternate and form a cup on the dorsal surface at base 15-20 cm long, rounded at apex, entire, cuneate at base, often planted in gardens and in pots for the unusual cup shaped leaves.

(19) *Ficus religosa* L. **Moraceae**

Tree – 8 to 20m, leaves broadly ovate, petiole glandular at the apex below.

(20) *Filicium decipiens* (Wight and Arn.) Thwaites **Sapindaceae**

Evergreen shola tree to 15m. Leaves imparipinnate, petals – greenish, Drupe- dark blue at ripening. Flowering – October to January; Fruiting- April.

(21) *Grevillea robusta* A. Cunn, ex. R. Br. **Proteaceae**

Evergreen introduced tree to 35m, Leaf – bipinnatified having silvery texture on abxial side, flowers – orange, follicle – oblique, Flowering – March to May Fruiting – June. Shade tree in coffee plantations.

(22) *Guettarda speciosa* L. **Rubiaceae**

Copiously branched tree to 8m, corolla – white fragrant, Flowering – March to June.

(23) *Ixora pavetta* Andrews **Rubiaceae**

Slender shrub to 8m, Corolla – white, drupe, Flowering – March to April; Fruiting-January to February.

(24) *Labramia bojeri* A. Dc. **Sapotaceae**

A tree 8-15m tall; leaves at the apices of thick branchlets, thick, obovate-oblong, upto about 25 cm long, 9 cm wide, rounded or emarginate at the apex, cuneate at the base, glandular tomentose beneath; Flowers fasciculate beneath terminal leaves on leafless branches; calyx glandular; corolla lobes 6, lanceolate with two lateral lanceolate appendages as long as the lobes; stamens 6; ovary 10 – locular; Fruit ovoid.

(25) *Lannea coromandelica* (Houtt.) Merr. **Anacardiaceae**

Tree to 12m. Leaf imparipinnate, Drupe obovoid; Fruiting – April.

(26) *Lantana camara* L. **Verbenaceae**

Armed, straggling shrub to 3m. Corolla red-yellow, Drupe dark blue. Flowering and Fruiting – January to December.

(27) *Malphigia glabra* L. **Malphigiaceae**

Small evergreen shrub to tree with 2-3m. tall. Leaves opposite, inflorescence sessile in axillary cymes. Pinkish to reddish in colour.

(28) *Manilkara littoralis* (Kurz) Dubard **Sapotaceae**

A large evergreen tree, upto 36m in height and over 4m in girth found as Endemic to Andaman and Nicobar Islands. Bark blackish brown, thin, furrowed in old trees, leaves elliptic or obovate, crowded towards the ends of branchlets; Flower solitary, axillary greenish white, fragrant; berry depressed globose.

(29) *Manilkara zapota* (L.) P. Royen **Sapotaceae**

Tree to 15m. Corolla white, berry globose, seeds oblong, compressed. Flowering – February to April; Fruiting – May.

(30) *Melia azedarach* L. **Meliaceae**

Tree to 12m. Petals-lilac, staminal tube dark purple, Drupe oblong. Fruit ripening yellow by August. Flowering – January to March.

(31) *Peltophorum pterocarpum* (DC.) Backer ex Heyne. **Caesalpiniaceae**

Densely foliaceous tree to 12m. Leaf-pinnate, Petals golden yellow, pod stipitate, ornamentally planted. Flowering – March to May; Fruiting – April.

(32) *Persea americana* Miller **Lauraceae**

Medium to large tree 9-20m in height. Leaves are often pubescent and reddish when young, becoming dark green and leathery when mature. Flowers are yellowish green and borne on inflorescence and the fruit is a berry which at maturity is purple or reddish.

(33) *Pongamia pinnata* (L.) Pierre **Fabaceae**

Deciduous tree to 15m. Leaf – imparipinnate, corolla-pinkish-white. Pod-obliquely oblong. Flowering – February to June; Fruiting – April to October.

(34) *Psidium guajava* L. **Myrtaceae**

Tree to 5m. Leaf decussate, petals-white, Berry globose.

(35) *Punica granatum* L. **Punicaceae**

Tree to 6m, Leaf decussate, petals blood red. Flowering – March to May; Fruiting – May.

(36) *Samanea saman* (Jacq.) Merr. **Mimosaceae**

Tree to 15m. Leaf alternate, pinnate, petals brightly coloured. Flowering – March to June; Fruiting- May.

(37) *Syzygium malaccense* (L.) Merr. Ex Perry **Myrtaceae**

Tree with a height of 20m, branched near base and with broadly ovoid canopy. Leaves opposite and red when young. Inflorescence exclusively on defoliate twig parts, short and dense with dark red petals with numerous stamens.

(38) *Swietenia mahagoni* (L.) Jacq. **Meliaceae**

Tree to 25m, petals-cream coloured, capsule oblong. Flowering – April Fruiting –January to December.

(39) *Tabebuia argentea* (Bureau et K. Schum.) Britt. **Bignoiaceae**

Trees upto 10m with bright yellow flowers with palmately divided leaves. Flowers appear during leafless time-summer. (February-April). The tree is often damaged by strong wind or uprooted in cyclonic weather due to its shallow root system.

(40) *Tabebuia rosea* (Bertol.) DC. **Bignoniaceae**

Densely foliaceous tree to 15m. Leaf digitate, corolla white/pinkish. Flowering-January to December.

(41) *Tectona grandis* L. f. **Verbenaceae**

Deciduous tree to 15m, Corolla yellowish, Drupe globose, flowering–June to September; Fruiting-July.

(42) *Triplaris surinamensis* Cham. **Polygonaceae**

Tree, to 3m or more; leaves oblong to elliptic, to 15 in. long, acuminate; flowers in great panicles, calyx wings 2 in. long, 2-3 times as long as tube. Female flowers creamy – white, becoming red in age.

(43) *Thespesia populnea* (L.) Sol. Ex. Corr. Serr. **Malvaceae**

Tree to 8m, Leaf cordate, petals-yellow, capsule globose. Flowering–opening yellow by 10 A.M. closing pink in the evening, Fruiting-January to December.

(44) *Ziziphus mauritiana* Lam. **Rhamanaceae**

Tree 5-10m, stipular thorns erect/recurved leaf-orbicular-rotund, petals greenish, drupe-oblong-globose. Flowering–July to November, Fruiting-October to December.

Results and Discussion

The relationship between *Dendrophtoe falcata* (L.f.) Ettingsh and its host was studied among the arborescent species growing at Shakit-Auroville. The parasite is non-host specific and found growing on 44 tree species among 180 Tropical Dry Evergreen arborescent species and 125 exotic species. Among the infested trees, 23 of them are indigenous and 21 are exotic. Luxuriant growth of the parasite is observed on *Samanea saman, Pongamia pinnata, Grevillea robusta*, and *Acacia auriculiformis*, whereas the parasite is not well established in the hosts of *Cassia fistula, Chloroxylon swietenia, Ficus benghalensis* var. *krishnae* and *Cordia sabestina*. The parasites are well represented on six species of Mimosaceae, four species of Moraceae and in Caesalpiniaceae and Sapotaceae with three each. Compared with the earlier reports (Fischer, 1926; Sambandam, 1966; Ravindranth and Narasimha Roa, 1959; Singh, 1959; Ghosh, 1968, 1969; Ghosh *et al.*, 2002). *Acacia holoserica, Albizia guachepele, Callinadra haematocephala, Coccoloba uvifera, Filicium decipiens, Labramia bojeri, Malpighia glabra, Manilkara littoralis, Persea americana, Syzygium malaccense and Triplaris surinamensis* are the new hosts recorded from the study area.

From the study, it is concluded that the parasite *Dendrophthoe falcata* is not a host specific parasite. Its preferential growth on certain host is quite significant which is evident from its luxuriant growth.

References

Bombay Dept. Agri. Leaflet 4; 1931, Report on Neilgherry Loranthaceous Parasitical Plants. *Loranthus*.

Brand, A.R. 1938; Working Plan for Nilambur Forest Division, Madras (1938-39–1952-53).

Calder, D.M. and P. Bernhardt. 1983. *The Biology of Mistletoes*. Academic Press, New York.

Cooke, T. 1908. *The Flora of the Presidency of Bombay*; Vol. II, **545-551.**

Fischer, C.E.C. 1926. *Loranthaceae* of Southern India and their host plants. *Rec. Bot. Surv. India* **11: 159-195.**

Gamble, J.S. 1925. *Flora of the Presidency of Madras*; Vol. VII; **1247-1254.**

Ghosh, R.B. 1968. *Loranthaceous* parasites of the Indian Botanic Garden, Calcutta. *Sci. and Cult.* **34: 426.**

Ghosh, R.B. 1969. On a newly recorded host species of *Dendrophtoe falcata* (L.F.) Ettingsh. *J. Bomb. Nat. Hist. Soc.* **67(2): 354.**

Ghosh, R.B., Das, D. and V.K. Maji, 2002. A preliminary census of host plants of *Dendrophthoe falcata* (L.f.) Ettinghs, in the tropical garden of India. *J. Econ. Taxon. Bot.*, **26: 517-523.**

Hooker, J.D. 1890. *The Flora of British India*; Vol. V; **203-223.**

Locy, R.C. 1936. Some more unrecorded host plant *Loranthus longiflorus*. *Curr. Sci.* **5: 875-876.**

Lushington, A.W.; Identification of *Loranthaceae* by their Leaves; *Ind. For.*; Vol XXVIII; No. 2; Feb. 1902; **58-66.**

Matthew, K.M. 1983. *Flora of the Tamil Nadu Carnatic*. Vol. 3.

Nickrent, D.L. and L.J. Musselman. 2004. Introduction to parasitic flowering plants. *The Plant Health Instructor*. DOI: 10. 1094/PHI-I-2004-0330-01.

Partridge, E.A. 1911. *Forest Flora of H.E.H. the Nizam's Dominion*.

Ravindra Nath, V. and V.L. Narasimha Rao, 1959. New host of *Dendrophthoe falcata* (L.f.) Ettingsh. *J. Indian Bot. Soc.*, **38, 204.**

Sambandam, C.N. 1966: Some new combination of *Loranthus longiflorus* Desr, and host species. *Annamalai Univ. Agric. Mag.* **63-64.**

Sayeeduddin, M. and M.A. Salam. 1935. A somewhat combinations parasite. *Curr. Sci.* **4: 160-162.**

Singh, B. 1959. More unrecorded hosts of *Dendrophthoe falcata* (L.f.) Ettingsh. *Hort. Adv.* **3: 131-134.**

Troup, R.S.; *Silivoculture of Indian Trees*; 1921 Ed; Vol. 1; **93**, and Vol. III: **799.**

❑❑❑

23

Lentic Fishery: Fishery of a Tropical Wetland (Beel) in Assam

Devashish Kar

Department of Life Science, Assam (Central) University, Silchar–11, India and President, Conservation Forum, Silchar

Introduction

Fishes mean different things to different people. To many, fish is best known as an item of food. To a fisherman, fish is a source of earning livelihood. To an amateur, it is an article of hobby or sport. To an ichthyologist, a true fish is a cold-blooded vertebrate having gills for respiration and paired and unpaired fins for locomotion and balancing.

The occurrence of fish had been known to the Indians since three millennium BC (Hora, 1956). Remains of fish, with indications of their use as food, have been obtained from excavations at Mohenjodaro and Harappa, which dates back to 2500-1500 BC. King Somesvara (1127 AD) was the first to record the common sport fishes of India. Bloch's (1785) work was the first modern record on Indian fishes. This was followed by Lacepede's book (1798-1803). Hamilton (1822) described 269 ichthyospecies from the Ganges and its tributaries. Day's work (1878, 1889) eclipses the taxonomy and distribution of almost all the fish groups of India. Recent contributions on the fish and fish groups of India, in general, have been made by Hora (1920, 1924, 1930, 1935, 1940, 1945, 1956), followed and extended by Misra (1959), Menon (1951, 1974), Jayaram (1981), Jhingran (1993). On the other hand, significant works on the lentic and lotic water fishes of North-East (NE) India, in particular, have been done by Dey (1973, 1981), Dey and Kar (1989 a, b, c; 1990); Sen (1982), Lahon (1983), Kar (1990, 1999, 2000 a, b; 2002, 2003 a, b; 2004, 2005 a, b, c); Kar and Dey (1982 a, b; 1986;

1987, 1990 a, b, c, 1991, 1993, 1996; 2000, 2002), Kar *et al.*, (1995, 1996, 1998, 2002 a, b, c; 2003, 2004, 2005); Kar and Barbhuiya (2000 a, b; 2004); Dhar *et al.* (2004); Sen (1982), Lahon (1983), Goswami (1985), Nath (1987), Yadava (1987), Nath (1989), Yadava *et al.* (1989).

Sampling and data collection were done for two years (Jan 2002-Dec 2003). Fish indentification and the Biodiversity of fishes have been studied after Menon (1974, 1999), and Jayaram (1981, 1999). Data on annual fish yield have been extrapolated from daily catch statistics recorded at all the landing stations of the Beel (FAO, 1974) while the trend, population dynamics and cyclic variations have been constructed by 12 months' moving average method (Coxton and Cowden, 1950). The fish catching devices have been studied after Brandt (1980) and Dey (1981). The physico-chemical characteristics of water and soil have been analysed after APHA (1995) and Jackson (1973) respectively. Other limnological parameters have been studied after Welch (1948).

Lentic Potential of Assam

The simple aspect of ichthyology is the logical unity of theoretical and practical knowledge of fish to produce efficient and rational methods of raising more fish. Contrary to such background, the naturally-formed lakes and wetlands constitute great potential of fishery resources in India. Of the 0.72 x 10^6 ha lake coverage in India, the North-Eastern (NE) region of the country, particularly, Assam, portrays significant potentialities of such water bodies for productive fisheries. Besides lotic territories, the region is gifted with myriads of different kinds of wetlands, which could be broadly of three categories, *viz.*, (a) perennial ones (locally called 'Beel'), (b) seasonal ones (locally called 'Haor'), and (c) river-formed oxbow types (locally called 'Anua') (Kar, 2002). All these constitute 81 per cent of the total lentic area (>0.12 x 10^6 ha) in Assam (Kar, 2005 c).

In Assam, there are 1392 numbers of such wetlands having a total of 22,896 number of fisheries of different categories. Out of this, the number of registered wetlands is only 394 (30.38%) covering an area of 70,000 ha; of which, 19,000 ha is in good condition; 15,000 ha is in semi-derelict condition and 35,000 ha is in derelict condition (Anon, 1980). The remaining wetlands are mostly unregistered and under the control of both public and private sectors (Kar, 2003 a).

In this content, it is interesting to note that the district of Cachar, Karimganj and Hailakandi covers the highest number of such unregistered wetlands in Assam. But, inspite of the vast resources, the existing fish production rate in Assam is <50 per cent of the total requirement based on a minimal consumption of 5 kg/person/year. Absence of adequate ichtyological data is one of the reasons why fish culture is not encouraged in these vast water bodies although the fish prone water area constituted by such wetlands, otherwise, justifies the tremendous potentialities of the Beel fisheries of the region (Kar, 2000 a). The importance of these Beels is also judged from the fact that the Beels directly act as prospective spawning grounds of large growing riverine fishes, particularly, the coveted Indian major carps (IMC). It is with this basic aim that a detailed pilot study had been conducted on the Limnology and fishery of Sone Beel, the biggest wetland in Assam in order to portray various aspects of its limnology and fishery (Kar. 1990, 2005 a). The following in an account of the fishery of Sone Beel.

Fishery, Fish Yield, Fisherfolk and Fishing Efforts of Sone Beel Wetland

The Sone Beel

The 'Sone Beel', situated in Karimganj district (92°24' 50" E—24°44' 30"N), is the biggest (Δ 3458.12 ha at Full Storage Level) tectonic (Mio-pliocene) Beel in Assam. It is unique in having continuous inflowing and outflowing channels. The Beel has a long shoreline of 35.4 km, a shore development of 1.69, a gross of 101, 54 x 10^6 m^3/sec and a mean depth of 0.29 m. The area of the Beel diminishes to 409.37 ha at the Dead Storage Level (DSL). The continuous inlet and outlet chancels of the Beel recorded maximum discharges of 33.91 and 87.07 m^3/sec respectively. The Beel water is circumneutral with pH varying from 6.0 to 7.9. Dissolved Oxygen (DO) ranging from 2.6 to 5.9 mg/lit, Free carbon di-oxide (FCO_2) fluctuating from 0.9 to 14.5 mg/lit and Total Alkalinity (TA) lying between 25.0 to 76.0 mg/lit (Dey and Kar 1987; Dey and Kar, 1989 e; Kar, 2005 e).

Ichthyospecies of the Beel and their zoogeography

In Sone Beel, 70 valid ichthyospecies belonging to 49 genera under 24 families and 11 orders have been recorded (Dey and Kar 1990; Kar, 2005 b). Of these, 59 species (*Chitala chitala, Notopterus notopterus, Amblypharyngodon mola, Asidoparia morar. Barilius bendelisis, Catla catla, Cirrhinus mrigala, C. reba, Cyprimus carpio, Danio devario, Esomus danricus, Labeo bata, L. ealbasu, L. gonius, L. nandina, L.*

rohita Puntius chola, P. conchonius, P. sarana sarana, P. tieto, Rasbora doniconius, R. elanga, Salmostama basaila, Botia dario, Lepidocephalus guntea, Acanthocobitis botia, Sompleptes gongota, Mystus cavasius, M. corsula, M. tengata, M. vititatus, Aorichthys, seenghala, Rita rita, Ompok bimaculatus, Wallago attu, Ailia coila, Pseudeutropius atherinoides, Chupisoma garua, Eutropiichthys vacha, Silonia siliondia, Pangasius pangasius, Gagata nangra, Glyptothorax telchitta, Clarius batrachus, Heteropneustes fossilis, Chaca chaca, Channa gachua, C. marilius, C. punctatus, C. striatus, Badis badis, Nandus nandus, Anabas testudineus, Colisa fasciatus, C. lalius, Macrognathus aculeatus, M. panclus, Mastocembelus armatus and *Tetyradon cutcutia*) belong to the Primary Fresheater Group (Myers, 1949); while the remaining 11 species (*Pisodonophis boro, Gudusia chapra, Hilsa ilisha, Xenentodon cancila, Aplocheilus panchax, Amphipnous cuchia, Chanda baculis, C. nana, Rhinomugil coresula, Siçamugil cascasia* and *Glossogobius giuris*) belong to the Peripheral Class (Nichols, 1928; Darlington, 1957). In other words, the primary class formed 84.28 per cent while the Peripheral class made 15.72 per cent of the total fish composition. Significantly, the cyprinids contained 35.59 per cent of the Primary forms. On the basis of distributional pattern, it is more appropriate to categories families Nototeridae, Nandidae, Anabantide and Mastocembelidae as Primary forms (Mirza, 1975) rather than as Secondary types (Jayaram, 1974) although these fishes have better salinity tolerance (Sen, 1982). On the basis of Indian and Extra-Indian territorial distribution (Motwani *et al.,* 1962), the fish denizens of Sone Beel include 20 'Widely distributed species' of little significance and eight species of 'Northern India'. Ecomorphologically (Dey, 1973), *G. telchitta* is the only true 'Rheophilic species' occurring in Sone Beel. The 'Semi-torrential group' contained five species with weak and minimal body adaptations. A critical analysis of the migratory propensity of the fishes of Sone Beel revealed that *A. mola, A. testudineus, C. baculis, C. nama, C. punctata, C. striata, G. charpa, H. fossilis, L. guntea, M. aculeatus, M. pancalus, M. armatus, A. seenghala, M. vittatus, N. nandus, O. bimaculatus, P. chola, P. conchonius, S. bacaila, W. attu* and *X. cancila* are the 'resident forms' with perennial species *P. chola*, constituting, on average, 24.60 per cent of the total fish yield of the Beel. Among the 'casual visitors', the IMCs enter the Beel for breeding, while *C. garua, E. danrica, E. vacha, M. corsula, M. tengara, R. corsula, R. rita* and *S. gongota are* of 'seasonal occurrence'. Lastly, the 'Plainwater group' includes 57 species of the Beel with least corporal modification and insignificant migratory habit (Kar and Dey, 1986; 2005 e).

Commercial Production and Composition

The annual average fish yield from Sone Beel has been recorded as 335.18 metric tones (mt), of which, IMC constituted 0.73 per cent, other carps and minnows 29.93 per cent, clupeids 8.31 per cent, featherbacks 0.27 per cent, loaches 1.68 per cent, bagrids 7.94 per cent, siluroids 3.72 per cent, schilbeids 0.88 per cent, pangasiids 0.00112 per cent, sisorids 0.015 per cent, live cat fish 1.92 per cent, chacids 0.025 per cent, needle fish 0.18 per cent, killi fish 0.03 per cent, murrels 7.53 per cent, glass fish 13.91 per cent, perches 7.45 per cent, mullets 0.045 per cent, gobies 6.31 per cent, eels 7.43 per cent, and puffer fish 0.94 per cent (Kar and Dey, 1987; Kar, 2005 d).

Average per hectare fish yield from Sone has been estimated as 96.8 kg/ha obtained by Vaas and Schurman (1949) in some tropical freshwaters of Java. The poor fish yield could be assigned due to siltation of the Beel (Welcomme, 1979) exposing the fish in the shallower areas to shortage of living space through an eventual diminution of the water spread area. However, the Beel soil with pH 5.0 to 5.9, organic carbon (OC) 0.25 to 1.74 per cent, available phosphorus 0.15 to 1.93 mg/100 g; and, water with pH 6.0 to 7.9, DO 2.4 to 5.9 mg/lit are not considered very suitable for rich fish production (Banerjea, 1967; Michael, 1969).

IMC fishery

In Sone Beel, the IMC consists of all the five species (*Catla catla, Cirrhinus mrigala, Labeo calbasu, L. gonius* and *L. rohita*) and its fishery is found not be a detached and isolated one as the crop is dependent on the stocks in the adjoining rivers (Kar, 1990; Dey and Kar, 1990: Kar and Dey, 2000). The landing of IMC from Sone has been recorded to be 2571.52 kg with a per hectare yield of 0.74 kg on the average (Kar, 2005 e). The species-wise average monthly relative yield (MRY) have been 0.017, 0.130, 0.26, 0.076, 0.36 per cent respectively. Concomitantly, the species-wise annual relative yield (ARY) successively have been 0.015, 0.02, 0.21, 0.39, and 0.23 per cent during the first year; and, 0.01, 0.13, 0.16, 0.07 and 0.24 per cent during the second year. The IMCs entering Sone Beel varied in size from 19.0 to 66.8 cm for *C. catla;* 12.9 to 65.0 cm for *C. mrigala*; 15.0 to 54.3 cm for *L. calbasu*; 6.8 to 50.0 cm for *L. gonius*; and 12.0 to 78.5 cm for *L. rohita.* As these fishes prefer a lotic territory to lentic one, in Sone Beel too, they tend to occur in regions which are more fluviatile in nature. Indeed, direction, coupled with ingress of water through the perennial channels could be the influential factors for recruitment of these coveted species during monsoon. Likewise, temperature is not found to have any direct impact on the movement of the IMCs. Their occurrence in Sone Beel almost throughout the year, without much impact being caused by temperature, lends support to this view. Field investigations further revealed that, with the onset of rains, shoals of IMC enter the Beel through the outlet when they are caught in appreciable number without much discrimination in size. Such non-judicious and irrational harvest of the IMC including their juveniles, cause imbalance in the IMC fishery of the Beel culminating in overall poor contribution of this coveted group in the total fish yield of the Beel (Kar, 2005 d, e).

Hilsa

Concomitant to IMC, the Hilsa (locally called 'ilish') also forms an important fishery of Sone Beel and its crop is also related to its stock in the adjoining rivers. In Sone Beel (Kar and Dey, 1982 a, b; Dey and Kar, 1989 a; Yadava *et al.,* 1989), there is a single run of hilsa with the fish entering the Beel in shoals on the advent of monsoon unlike in other lentic (Jones and Sujansingani, 1951) and lotic (Pillay, 1958) bodies of India. The total landing of hilsa from Sone Beel has been recorded as 163.5 kg during the first year and 220.7 kg during the following year; the trend thereby showed average per hectare yield of 0.055 kg during the investigating period. Concomitantly, the MRY recorded during the first session have been 0.04 (April), 0.23 (July), 0.11 (August), 0.01 (September) and 0.00403 per cent (October); while, during the following session, the MRY values have been 0.07 (May), 0.09 (June), 0.06 (July), 0.04 (August), and 0.37 per cent (September). The ARY have been 0.04 and 0.07 per cent during the first and second sessions respectively. The hilsa entering the Beel varied in size from 30-60 cm and from 23.0-53.4 cm during first and second sessions respectively. Unusual turbidity of water may not always be considered as a criterion for increased rate of natural recruitment of hilsa, because Sone Beel exhibited a good population of the fish within a turbidity of range of 20.5 to 83.6 TU (Dey and Kar, 1987). Migratory behaviour of shad does not seem to be directly influenced by temperature. However, their occurrence in Sone Beel only during April to October indicated that the fish prefer to live in Sone Beel during the hot season when its water is warm. After hilsa enter the Beel through the outlet, they are caught in large number without any discrimination in size. Subsequently, intensive fishing in the Beel does not spare much of the stragglers to perform the return journey.

Concomitantly to Sone Beel, not only hilsa, but also its advance fry stages have been recorded for the first time by Kar and Dey (2002) in Chatla Haor seasonal floodplain wetland in Cachar district of Assam revealing a new vista about the biology and breeding behaviour of the fish.

Conservation of IMC and Hilsa

Protection of both gravid and juvenile fish is the best course for the conservation of IMC and hilsa fisheries. However, the conditions prevalent in Sone Beel exhibit many difficulties for the successful achievement of this objective. The enormity of the Beel coupled with intense fishing throughout the year make it very difficult to restrict fishing and fish capture for any period or to observe a 'closed season' for some months. This would cause immense hardship to the fisherfolk in the absence of alternative source of income. Notwithstanding the above limitation, fishing restrictions during the monsoon season along the inlet and outlet belts would probably afford an opportunity for the conservation of these fishes in Sone Beel (Kar, 2002; 2003 b; 2005 d, e). The Epizootic Ulcerative Fish Disease Syndrome (EUS) has been posing a serious threat to fish sustenance and conservation (Kar, 1999, 2005 e; Car *et al.*, 1995, 2003).

Fishing Centres and Fish Landing Stations

Fishing Centres (FC)

Sone Beel, not being a very deep wetland, the entire waterspread area is used for intensive fishing throughout the year both on domestic (littoral) and commercial (limnetic) basis. While such a trend is more discernible during the Full Storage Level (FSL) of the Beel, eight distinct 'Fishing Centres', locally called 'Bundhs', are leased out principally to the 'Kaibarta' fishing community (Dey and Kar, 1989 b) at Dead Storage Level (DSL) of the Beel every year, mostly during the period from November to April. Such fishing zones, which lie at deeper portions of the Beel at DSL, are kept encircled with bamboo mat (height 2-8m) and fishes are captured and marketed regularly as warranted throughout the period of operation. The names of such 'bundhs' with their fishing areas (FA), months of operation (MO), watere level (WL) and fish yield (FY) at the centres, as recorded (Kar, 2005 e), are gives below:

Bundh	FA (ha)	MO	WL (m)
Belala	1.12	Aug-Apr	0.10-4.00
Chirakhaora	48.50	Nov-Apr	0.10-2.00
Nunail	6.50	Sep-Nov	2.00-4.00
Putighat	2.00	Nov-Apr	0.10-2.00
Kurerkata	15.05	Dec-Mar	0.25-1.50
Kanagajar	50.00	Oct-Dec	1.50-3.00
Batoitila	105.90	Nov-Apr	1.10-2.00
Vashankuri	8.46	Dec-Mar	0.25-1.50

Fish Landing Stations

There are three principal fish landing stations (LS) situated along the shoreline of Sone Beel (Fig. 4). These are Debodwar LS, kalibaribazar LS and Phakhoagram L.S. The position of LS are shifted with the change in the water level of the Beel. An account of these LS along with landing area (LA), leasing amount (la), month operation (MO) and quantity of fish landed (QL) in each (Kar, 2005 e), is summarised hereunder:

LS	LA (m^2)	la (Rs)	MO	QL (mt)
Debodwar	700	1451.00	May-Sep	
	2800	6672.00	Oct-Dec	159.90
	1900	3500.00	Jan-Apr	
Kalibaribazar	400	15937.00	May-Sep	
	468	16500.00	Oct-Now	258.60
	1491	20000.00	Nov-Apr	
Phakhoagram	182	Nil (Railway Plot)	whole year	217.10

Fisherfolk and Fishing Efforts

Fisherfolk

The activities of the fishermen and their socio-economic condition have been influencing the development and utilisation of the fishery resources of the wetlands in a significant way. The aspect gained importance in view of the inferior social status and overfishing of the Beels by the unprogressive but simple, poverty-stricken people. In Sone Beel, fisherfolk belonging to four principal communities (Dey and Kar, 1989 b) inhabit 39 villages around the Beel (Kar, 2000 b; 2003 b; 2005 e). An account of these communities is tabulated below:

Fishermen community	Relative abundance (%) out of		Fishing area (ha/fisherman)
	Total fishermen	Total inhabitants	
Kaibarta	68.14	47.67	0.32
Patni	14.66	12.24	1.51
Maimal	12.29	8.59	1.81
Namasudra	4.91	3.43	4.53
		30.07 (Non-fishermen)	

Moreover, on the basis of fishing intensity, the fisherfolk of Sone Beel have been further classified, after FAO (1962), into three cateogries, *viz.*, Professional, Part-time and Occasional (Kar, 1990), an account of which, with relevant information (Kar, 2005 e), is tabulated below:

Fisherman category	Relative abundance (%) amongest				Total
	Kaibarta	Patni	Maimal	Namasudra	
Professional	55.04	6.91	5.54	2.51	70.00
Part-time	8.00	3.00	2.60	1.40	15.00
Occasional	4.50	4.75	4.75	1.00	15.00
Total	67.54	14.66	12.89	4.91	100.00

Fishing efforts and crafts

Fishing efforts

24 different kinds of fishing gears and methods under 7 categories (FAO, 1963; Brandt, 1980) have been found in operation in Sone Beel (Kar, 1990; 2005 e; Kar and Dey, 1991, 1993, 1996; Kar *et al.*, 1998). An account of these gears along with their CPGH (Catch/Person/Gear/Hour: Dey, 1981) of some of the principal gears is tabulated below:

Category	Gear (local name)	Size (m)	Meash size (mm)
I. Hook and Line gears	1. Lar barshi	L.400-600	-
	2. Kupa barshi	L.450-500	-
	3. Tanga barshi	L.350-450	-
II. Traps	4. Dori	H.0.45-1.35 W.0.30-0.60	Bamboo slits 1 mm apart
	5. Gui	H.0.90-200 W.0.75-1.00	Bamboo slits 10 mm apart
	6. Paran	H.0.20-0.40 W.0.10-0.14	Bamboo slits 7 mm apart
	7. Khati bundh	H.2.00-3.00 L/unit 4.50	Dense bamboo mat
III. Trawls	8. Chhatjal	L.3.50-4.50 W.in front 2.00-3.00	Upper part: 4.0 Lower part: 1.0
	9. Pelain	L.2.50-3.00 W.in front 1.00-2.00	1.0-5.0
IV. Scooping gear	10. Dheki jal	L..800-12.00 W.in front 7.0-9.00	Wider part: 50.0 Narrower part: 6.0
	11. Darar jal	L.4.50-5.00 W.4.00-5.00	5.0
	12. Kuchrung jal	L.1.25-1.50 Dia: 1.00-1.15 Cir: 3.00-3.50	5.0
V. Entangling gears	13. Patan jal:		
	Big variety	L.300-400 H.4.00.4.50	46.0
	Medium variety	L.300-390 H.3.00-3.50	19.0
	Small variety	L.300-400 H.1.50-2.00	15.0

	14. Haran jal	L.4.50-6.50 W. of widest portion 3.50-4.50	Upper part: 8.0 Lowet part: 6.0
VI. Encirecling gears	15. Maha jal	L.700-750 H.3.75-6.00	Episurface: 5.0 Underwater: 1.0
	16. Dal jal	L.80-90 H.2.00-5.00	1.0
	17. Chat jal	L.110-120 H.2.50-3.50	1.0-8.0
	18. Ghuran jal	L.30.00-40.50 H.1.50-2.50	1.0-6.0
	19. Jhaki jal	L.4.50-5.40 Cir. 11.25-14.00 Dia. 4.00-4.50	6.0
	20. Tana jal	L.45.00-67.50 H.3.60-4.50	46.0
	21. Rekh jal	L.5.00-5.50 Dia. 4.50-5.50 Cir. 11.50-15.00	35.0-38.0
VII. Miscellaneous	22. Arar jal	L.3.50-4.50 H.2.50-2.75	4.0-5.0
	23. Fal jal	L.100-2000 H.3.00-4.00	1.0
	24. Impoundment:		
	With bailing vessel ('Kuin')	L..0.50-5.00 W.0.20-0.45	-
	Without bailing Vessel ('Debli')	-	-
	'Dhagamera'	-	-

Crafts and their salient features

Craft Crew	Size (m)	Construction	Propulsion
Plank-built boat		Wooden planks nailed together to form a rigid frame	
Flat bottom:			
Large: 20-25	L.25.0-35.0 W.8.0-10.0 H.1.5-2.0		Manual
Small: 3-6	L.10.0-20.0 W.2.0-4.0 H.0.75-1.0		Manual
Round bottom: 3-6	L.10.0-20.0 W.2.0-4.0 H.0.75-1.0		Manual
Dug-out boat: 1-3 (Canoe)	L.5.0-10.0 W.1.0-2.0 H.0.5-.75	Hollowing-out single log of wood	Manual

Concluding Remarks

Beels in Assam are to be considered as gifts of nature, a treasure worthy to adore. Such potential fishery beds are to be properly utilised to the fullest extent through tangible schemes to result in prolific harvest of fish for the region. And, an overall improvement and jusicious harnessing of these water bodies will be imperative and have no alternative, if the fish production of the NE region in general and Assam in particular needs to be augmented in real perspective.

References

Anonymous. 1980. *Statistical Handbook of Assam*, Directorate of Economics and Statistics, Govt. of Assam.

APHA. 1995. *Standard Methods for the Examination of Water and Wastewater*. American Public Health Association (USA). p. **1193.**

Banerjea, S.M. 1967. Water quality and soil conditions of fish ponds in some States of India in relation to fish production. *Indian J. Fish.* **14: 115-144.**

Bloch, M.E. 1985. *Naturgeschichte der auslandischen Fische*. Berlin: 3 Vols.

Brandt, A.V. 1980. *Fish Catching Methods of the World*. Fishing News Books, Ltd., Surrey, UK. p. **418.**

Coxton, F.E. and Cowden, D.J. 1950. *Applied General Statistics*. Sir Isaac Pitman and Sons (London). p. **944.**

Darlington, P.J. 1957. *Zoogeography: The Geographical Distribution of Animals*. John Wiley and Sons, New York (USA). p. **675.**

Day, F. 1878. *The Fishes of India*, being a Natural History of the Fishes known to inbahit the Seas and Freshwaters of India, Burma and Ceylon, 195 pls, Text and Atlas in 4 parts, W_M Dawson and Sons Ltd. (London). p. **778.**

Day, F. 1989. *The Fauna of British India, including Ceylon and Burma: Fishes*, Vols. I and II. pp. **548** and **509.**

Dey, S.C. 1973. Studies on the Distribution and Taxonomy of the Ichthyfauna of the hill streams of Kamrup-Khasi-Garo Regions of Assam with special reference to the functional morphology of some rheophillic Fishes, D.Sc. Thesis, University of Kolkata (India). p. **299.**

Dey, S.C. 1981. *Studies on the hydrobiological conditions of some commercially important lakes (Beels) of Kamrup district of Assam and their bearing on fish production*, 18 Figs., 8 pls. Final Technical Report, North-Eastern Council, Govt. of India (Shillong). p. **177.**

Dey, S.C. and Kar, D. 1987. Physico-chemical Complexes of Water and Soil in Sone, an Ichthyologically potential tectonic lake of Assam. *J. Assam Sci. Soc.* **30: 1-11.**

Dey, S.C. and Kar, D. 1989 a. An account of *Hilsa ilisha* (Hamilton) of Lake Sone in the Karimganj district of Assam. *Bangladesh J. Zoology*. **17: 69-73.**

Dey, S.C. and Kar, D. 1989 b. Fishermen of Lake Sone in Assam: Their socio-economic status. *Science and Culture*. **55: 395-398.**

Dey, S.C. and Kar, D. 1989 c. Aquatic Macrophytes of Lake Sone in Assam. *Environment and Ecology*. **7: 253-254.**

Dey,. S.C. and Kar, D. 1990. Fish yield in Sone, A tectonic lake of Assam. *Matsya*. **15-16: 39-43**.

Dhar, N.; Goswami, M.M.; and Kar, D. 2004. Cooperative Movement: Need for the hour in Capture Fishery sector. In: *Cooperative Movement in Assam. Problems and Prospects* (Eds. P.J. Goswami, and P.R. Acharjee). G.C. College Employees' Credit and Thrift Cooperative Society Ltd. (Silchar). pp. **113-116.**

FAO. 1962. *Rapport au Government de la Re 'publique du Nigher sur la situation et' evolution de la pe^che au Niger: base 'sur le travail de M. Jaques M. A. Daget', Rapp. FAO/PEAT,* **1525:** p. **27.**

FAO. 1963. *Modern Fishing Gears of the World.* Proc. 2nd. World Fishing Gear Congress (London).

FAO. 1974. Design of Fisheries Statistical Surveys: Inland Waters. Food and Agricultural Organisation. *Fisheries Teach. Pap.* **133: 1-122.**

Goswami, M.M. 1985. Limnological Investigations of a tectonic lake of Assam, India with their bearing on fish production. Ph.D. Thesis, University of Gauhati, Assam. p. **395.**

Hamilton, B. 1822. *Account of the Fishes found in the River Ganges and its tributaries.* Archibald Constable and Co. Edinburgh *UK). p. **405.**

Hora, S.L. 1920. Revision of the Indian Homalopteridae and of the genus *Psilorhynchus* (Cyprinidac) *Rec. Indian Mus.* **19: 195-215.**

Hora, S.L. 1924. Fishes of Siju Cave, Garo Hills, Assam. *Rec. Indian Mus.* **26: 27-32.**

Hora, S.L. 1930. Ecology, Bionomics and Evolution of torrential fauna with special reference to the organs of attachement. *Phil. Trans. Roy. Soc. London* (B), **218: 171-282.**

Hora, S.L. 1945. Symposium on the factors influencing breedings of carps. *Proc. Nat. Inst. Sci. India.* **11(3).**

Hora, S.L. 1956. Fish painting of 3rd Millennium BC from Nal (Baluchistan) and their geographic significance. *Mem. Indian Mus.* **14: 78-84.**

Jackson, M.L. 1973. *Soil Chemical Analysis.* Prentice Hall of India Pvt. Ltd. New Delhi. p. **498.**

Jayaram, K.C. 1974. Ecology and Distribution of Freshwater Fishes, Amphibians and Reptiles. In: *Ecology and Biogeography in India* (Ed. M.S. Main), W. Junk Publishers, The Hague (The Netherlands). p. **475.**

Jayaram, K.C. 1999. *The Freshwater Fishes of Indian Region*, Narendra Publishing House, Delhi. p. **551.**

Jhingran, V.G. 1997. *Fish and Fisheries of India.* Hindustan Publishing Corporation, Delhi. p. **727.**

Jones, S. and Sujansingani, K.H. 1951. The hilsa fishery of the Chilka Lacke. *J. Bombay Nat. Hist. Soc.*, **50: 264-280.**

Kar, D. 1990. Limnology and Fisheries of Lake Sone in the Cachar district of Assam (India), *Matsya.* **15-16: 209-213.**

Kar, D. 1999 b. *Microbiological and Environmental Studies in relation to Fish Disease in India.* Gordon Research Conference, Connecticut, USA.

Kar, D. 2000 a. Present status of Fish Biodiversity in South Assam and Tripura. In: *Fish Biodiversity of North-East India* (Eds. A.G. Ponniah and U.K. Sarkar), NBFGR-NATP Publication No. 2. National Bureau of Fish Genetic Resources, Lucknow. pp. **80-82.**

Kar, D. 2000 b. Socio-economic Development of the Fisherwomen through Aquaculture with emphasis on Integrated Farming in the villages around Chatla Haor Wetland in Silchar, Assam. Seminar presented on DBT-sponsored Awareness Workshop on Biotechnology-based Programmes for Women and Rural Development, NEHU, Shillong 1. pp. **13-14.**

Kar, D. 2002. Present status of Biodiversity of Fishes of Barak valley region of Assam with a note on their Management and Conservation, pp. **3-10.** Proc. UGC- Sponsored State-level Seminar on Biodiversity of Assam, Session Charaman's Lecture, 30 Jan 2000 (Eds. M.K. Bhattacharjee, M. Dattachoudhury and P.B. Mazumdar), Karimganj College, Assam University (Assam). pp. **3-10.**

Kar, D. 2003 a. Fishes of Barak drainage, Mizoram and Tripura. In: *Environment, Pollution and Management* (Eds. A. Kumar, C. Bohra and L.K. Singh), APH Publishing Corporation, New Delhi. pp. **203-211.**

Kar, D. 2003 b. Peoples' Perspective of Fish Conservation in the Water bodies of South Assam, Mizoram and Tripura. In: *Participatory Approach for Fish Biodiversity Conservation in North-East India* (Eds. P.C. Mahanta, and L.K. Tyagi), National Bureau of Fish Genetic Resources (ICAR), Lucknow. pp. **325-328.**

Kar, D. 2004. A Glimpse into the Fish Bioresources of North-East India with a note on their Management, Conservation and Biotechnological Potential. Proc. (Invited Lecture) DBT-sponsored National Symposium on Biodiversity Conservation and Sustainable Utilisation of Environmental Resources: Tripura University, 10-11 Jan., 2004. p. **9.**

Kar, D. 2005 a. Fish Genetic Resources and habitat Diversity of the Barak drainage. In: *Aquatic Ecosystems, Conservation, Restoration and Management* (Eds. T.V. Ramachandra, N. Ahalya and C. Rajsekara Murthy), Capital Publishing Company, Bangalore, pp. **68-76.**

Kar, D. 2005 b. Fish Fauna of River Barak of Mizoram and of Tripura with a note on Conservation, *J. Freshwater Biol.* **16.**

Kar, D. 2005 c. Fish Diversity in the Major Rivers in Southern Assam, Mizoram and Tripura. In: *GIS/Spatial Analyses in Fishery and Aquatic Sciences* (Eds. T. Nishida, P.J. Kailola and C. E. Hollingworth) Proc. 2nd International Symposium on GIS and Spatial Analysis in Fishers and Aquatic Sciences, 2-6 Sep., 2002, University of Sussex, Brighton (UK) and Fisheries and Aquatic GIS Research Group, Japan. pp. **679-691.**

Kar, D. 2005 d. Sustainability issues of Inland Fish Biodiversity and Fisheries in Barak drainage (Assam), in Mizoram and in Tripura, Proc. International Symposium on Improved Sustainability of Fish Production systems and Appropriate Technologies for utilisation, Cochin University of Science and Technology, School of Industrial Fish and Fisheries, **1.**

Kar, D. 2005 e. *Fundamentals of Limnology and Aquaculture Biotechnology*. Foundation Books (Cambridge University Press), Hyderabad. p. **454** (in Press).

Kar, D. and Dey, S.C. 1982 a. *Hilsa ilisha* (Hamilton) from lake Sone in Cachar, Assam. Proc. Indian Sci. Congr. **69:** p. **77.**

Kar, D. and Dey, S.C. 1982 b. An account of Hilsa ilisha (Hamilton) of Sone Beel (Cachar district, Assam, India). Proc. All-India Sem. Ichthyol. **2:** p.**3.**

Kar, D. and Dey, S.C. 1986. An account of the ichthyospecies of lake Sone in the Barak valley of Assam (India), Proc. All-India Sem. Icthyol. **6:** p. **29.**

Kar, D. and Dey, S.C. 1987. An account of the Fish and Fisheries of lake Sone in the Barak valley of Assam (India). Proc. Workshop Development of Beel Fisheries in Assam, **1:** p. **13.**

Kar, D. and Dey, S.C. 1991. Gill nets in Lake Sone of Assam with their Economics and Impact on Fishery. *J. Applied Zool. Res.* **2: 76-79.**

Kar, D. and Dey, S.C. 1993. Variegated Encircling gears in lake Sone of Assam. *J. Apll. Zool. Res.* **4: 171-175.**

Kar, D. and Dey, S.C. 1996. Scooping Gears of Lake Sone in Assam. *J. Applied Zool. Res.* **7: 65-68.**

Kar, D. and Dey, S.C. 2000. Yield and Conservation of Indian major carps of Lake Sone in Assam. *Environment and Ecology*. **18: 1036-1038.**

Kar, D. and Dey, S.C. 2002. On the occurrence of Advanced fry of *Hilsa (Tenualosa) ilsha* (Hamilton-Buchanan) in Chatla Haor Seasonal wetlansd of Assam. Proc. Zool. Soc. Calcutta. **55: 15-19.**

Kar, D.; Roy, A.; Dey, S.C. Menon, A.G.K.; and Kar S. 1995. Epizootic Ulcerative Syndrome in Fishes of India. World Congress of In Vitro Biology, *In Vitro*, **31: 7.**

Kar, D.; Dey, S.C.; Kar, S.; Michacl, R.G.; and Gadgil, M. 1996. Ichthyoecology, Management and Conservation Fish Resources of Lake Sone in Assan (India). *Tiger Paper* (FAO, UN). XXIII: **27-32.**

Kar, D.; Dey, S.C.; Kar, S.; and Remachandra, T.V. 1998. Trawls of lake Sone in Assam. *J. Applied Zool. Res.* **10: 170-172.**

Kar, D.; Laskar, B.A.; Madal, M.; Lalsiamliana; and Nath, D. 2002 a. Fish Genetic Diversity and habitat parameters in Barak draniage, Mizoram and tripura. *Indian J. Environment and Ecoplanning.* **6: 473-480.**

Kar, D.; Laskar, B.A. Nath, D.; Mandal, M.; and Lalsiamliana. 2002 b. Tor progenius (McClelland) under threat in River jatinga, Assam. *Science and Culture*. **68: 211.**

Kar, D.; Laskar, B.A.; and Nath, D. 2002 c. Tor sp. (Masheer fish) in river Mat in Mizoram. *Aquacult*. **3: 229-234.**

Kar, D.; Dey, S.C.; and Datta, N.C. 2003. *Welfare Biology in the New Millennium*. Allied Publishers Pvt.Ltd., Bangalore. p. **97.**

Kar, D.; Roy, A.; and Dey, S.C. 2004. An overview of Fish Genetic Diversity of North-East India. In: *Proc. National Workshop on Rational use of Water Resources of Aquaculture*. (Eds. S. K. Garg. and K. L. Jain), 18-19 March, 2004, CCS Haryana Agricultural University, Hisar, **1: 164-171.**

Kar, D.; Laskar, B.A.; Nath, D.; and Barbhuiya, M.H. 2005. Length-weight relationship of *Neolissochilus hexagonolepis* (McClelland) and *Garra lissorhynchus* (McClelland) collected from jatinga river in Assam. *Indian J. Fish* (in press).

Kar, D. and Barbhuiya, M.H. 2000 a. Length-weight Relationship and Condition Factor in *Gudusia chapra* (Ham-Buch) and *Botia dario* (Ham-Buch) from Chatla Haor (Flood plain wetland) in Cachar district of Assam. *Environment and Ecology*. **18: 227-229.**

Kar, D. and Barbhuyia, M.H. 2000 b. Ichtyodiversity of Chatla Haor: A Floodplain wetland in Barak valley region of Assam. In: *Advances in Zoology and Environmental Degradation and Biodiversity* (Eds. B. N. Pandey and B. K. Singh), Daya Publishing House, New Delhi. pp. **3-6.**

Kar, D. and Barbhuiya, M.H. 2004. Abundance and Diversity of Zooplankton in Chatla haor, a floodplain wetland in Cachar district of Assam. *Environment and Ecology*. **22: 247-248.**

Kar, D. and Barbhuiya, M.H. 2005. Length-weight relationship and Relative Condition Factor in *Hilsa ilisha* (Hamilton) of Barak drainage in Assam. *Indian J. Environment and Ecoplanning*. **10.**

Lacepe' de, S. 1798-1803. *Historie Naturelle des Possions*, Vol. I-V.

Lahon, B. 1983. Limnology and Fishers of some commercial Beels of Assam, India. Ph.D. Thesis, University of Gauhati. p. **349.**

Menon, A.G.K. 1951. The role of the Eastern Ghats in the distribution of the Malayan fauna and flora to Peninsular India. *Proc. Nat. Inst. Sci., India*. **17: 475-497.**

Menon, A.G.K. 1974. A Checklist of the Fishes of the Himalayan and the Indo-genetic Plains. Inland Fish. *Soc. India, Barrackpore*: p. **136.**

Menon A.G.K. 1999. Checklist: Freshwater Fishes of India. Occasional Paper No. 175. Zoological Survey of India, Kolkata. p. **136.**

Mechael, R.G. 1969. Seasonal trends in Physico-chemical factors and Plankton of freshwater fish pond their role in fish culture. *Hydrobiologia*, **33.**

Mirza, M.R. 1975. Freshwater Fishes and Zoogeography of Pakistan. *Pakistan J. Zool.* **2: 249-250.**

Misra, K.S. 1959. An aid to the identification of Commercial Fishes of India and Pakistan. *Rec. Indian Mus.* **57: 1-320.**

Motwani, M.P.; Jayaram, K.C.; and Sehgal, K.L. 1962. Fish and Fishers of Brahmaputra River System, Assam, I. Fish Fauna with observation on their zoogeographical significance. *Trop. Ecol.* **3: 17-43.**

Myers, G.S. 1949. Salt tolerance of freshwater Fish Groups in relation to geographical problems. *Bijdr tol de Dierk*, **28: 315-322.**

Nath, S. 1987. The Fish and Fisheries of the Barak river system in Assam, India with special reference to the Ecomorphology and distribution of its ichthyofauna. Ph.D. Thesis, University of Gauhati, Assam. p. **259.**

Nath, P. 1989. Systematics, distribution and ecology of ichthyofauna of Arunachal Pradesh with particular reference to the Limnology of Itanagar Lake. Ph.D. Thesis, University of Gauhati, Assam. p. **362.**

Nichols, J.T. 1928. Fishes of the White Nile (with Table of World's Freshwater Fish Faunane). *American Mus. Novitates* No. 319.

Pillay, T.V.R. 1958. Biology of hilsa, *Hilsa ilisha* from river hooghly. *Indian J. Fish.*, **5: 201-257.**

Schneider, J.G. 1801. *Systema Ichthyologics.*

Sen, N. 1982. Studies on the Systematics, Distribution and Ecology of the Iehthyfauna of Meghalaya and their bearing on the Fish and Fisheries of the State. Ph.D. Thesis, University of Gauhati, Assam. p. **576.**

Vaas, K.F. and Schurman, J.J. 1949. On the Ecology and Fisheries of some Japanese Freshwaters. Common Gen. *Agric. Res. Sta. Buitenzorg Java*, **97: 1-60.**

Welcomme, R.L. 1979. *Fisheries Ecology of the Floodplain*, Longmans, London p. **317.**

Yadava, Y.S. 1987. Limnology and Productivity of Dhir Beel of Assam. Ph.D. Thesis of University of Gauhati, Assam: p. **467.**

Yadava, Y.S.; Goswami, M.M.; Kar, D.; and Choudhury, M. 1989. Present status of Hilsa resources in Assam. In: *Fish Genetics in India* (eds. P. Das and A.G. Jhingran), Today and Tomorrow's Printers and Publishers, New Delhi. pp. **252-256.**

❑❑❑

24

An Account of the Indigenous Plants Used by the Meiteis of Manipur as Food

A. Victor Singh, P. Kumar Singh and A. Sophia Chanu
Department of Life Sciences, Ethnobotany and Plant Physiology Lab.
Manipur University, Canchipur, Imphal, Manipur

Introduction

The "Meiteis" is the indigenous major community of Manipur. It is one of the seven N. E. States of India situated at the border area with Myanmar as its immediate neighbour having 93.2°E and 94.47°E longitude, 28.50°N latitude and on altitude of 781 to 2020m. from the sea level. Imphal valley, the capital is 790m. altitude comprises 8 per cent of the total area of Manipur with an area of 1834 sq. km. Out of its total average of 22,327 sq. km. 92 per cent comprises of hills. Populations in hills and valley comprises of 36 per cent and 64 per cent respectively. Hence, Meiteis who generally are valley dwellers comprises 13,61,521 lakh out of a total of 23,88,634 (A/C 2001 Census). The State in divided into 9 districts (Imphal West, Imphal East, Thoubal, Bishenpur, Senapatu, Ukhrul, Tamenglong, Chandel, and Churachandpur). History reveals of them as the original dwellers of the valley, even before the English Dominion. There are other minorities *viz*. Muslims (riverbed dwellers), Nepalies (hill dwellers) and tribals (hill dwellers).

Different communities used different dialects but adopted Meiteilon (Manipuri) used by Meiteis as the common language of the State. Valley being surrounded by hills and mountain ranges is safe from extreme wind, snow and rain. And hills with various climatic factors at different elevations influence numerous indigenous and endemic floras.

The term "food" is applied to edible plants which store up reserve food in roots, stems, leaves and fruits that are eaten either cooked or raw. Wild plants play an important role in the life of Meiteis since time immemorial as they spend on them as food stuff in day-to-day life. The meitei does also have a rich herbal lore so far as the useful plants are concerned. This is probably due to their intimate connection with their surrounding forest and plants. Meiteis generally love spices, bitter test, hot test (*Capsicum frutescens*) and Salty test but lesser of Sweet test.

Methodology

Repeated visit to a maximum number of villgaes of all the 9 districts were made to meet Headman and elderly people who knew about the wild plants used as food.

First hand information from the local people was collected. Efforts were made to see the plants in wild and to collect plant specimens with their flowers (in case of Angiosperms) and with sori (in case of ferns). Plant materials used as vegetables were collected with their reproductive parts, herbarium sheets were prepared and later identifed referring to (B.S.I.) herbaria Shillong. Local names were collected as far as possible. Plant specimens collected during the trips as well as afterwards were preserved in herbarium sheets.

The Meiteis of the villagers hardly use oil and avoid fried vegetables. they rarely used it during fish and meat cooking. Underground plants are also roasted in fire and make special dishes.

Photographs of some useful plant materials were taken.

Discussion

The collected plants mentioned on the next pages are mostly wild with a few cultivated at present. They believed that the plants are of multiple uses as the plants also show certain medicinal properties over and above their utility as food. No record has so far been made on the account and is evidently the first record of the indigenous plants used by the Meitei Community as food.

It is expected that the present study will bring into focus how the Meiteis enjoy on various indigenous foodplants over and above the common one's showing a unique dietary habit. Further this account will give some indication about the availability of a multiple variety of plants in the state. The work is expected to encourage other researchers in Ethno botanical research of the Meiteis.

List of Food Vegetables used by *Meitei* Communities of Manipur

Sl. No.	Plant sp. with Family	Local Name	Parts used
1	*Abelmoschus esculentus* (L.) Monech. MALVACEAE	Bhelandri	Fruit and leaf
2	*Achyranthes aspera*. L. AMARANTHACEAE	Khujumpere	Young arial shoots
3	*Adhatoda vasica* Nees. ACANTHACEAE	Nongmangkha	Leaf
4	*Aegle marmelos* (L.) Corr. RUTACEAE	Heirikhagok	Fruit
5	*Agaricus campestris* Linn. AGARICACEAE	Chenggum	Whole plant
6	*Alangium chinensis* (Lour) Rchder. ALANGIACEAE	Kokal	Leaf
7	*Allium cepa*. L. LILIACEAE	Tilhou	Centric leaves
8	*A. hookerii*. L. LILIACEAE	Maroinapakpi	Leaf and root
9	*A. porrum*. L. LILIACEAE	Maroi nakupi	Leaf
10	*A. sativum*. LILIACEAE	Chanam	Root, leaf and modified leaf
11	*Alocasia cullata*. Schott. ARACEAE	Palukabi	Rhizome
12	*A. indica* (Roxb.) Schott. ARACEAE	Pan	Rhizome
13	*A. macrorrhiza*. ARACEAE.	Honggu	Rhizome
14	*Alpinia galanga*. Wild. ZI NGIBERACEAE	Kanghu	Leaf
15	*A. nigra*. (Gaertn.) Burtt. ZINGIBERACEAE	Pullei	Rhizome
16	*Alternanthera sessilies* (DC) R. Br. AMARANTHACEAE	Phakchet	Tender shoot
17	*Amaranthus tricolor* L. AMARANTHACEAE	Chengkruk	Young shoot
18	*A. spinosus* Linn. AMARANTHACEAE	Chengkruk tingkhangpanbi	Young shoot
19	*Amorphophallus campanulatus*. Bl. ARACEAE	Haopal	Young shoot
20	*Anisomeles indica* (Linn.) O. Kurtze. LABIATAE	Thoiding Angouba	Young shoot
21	*Ananas comosus* (L.) Merr. BROMELIACEAE	Kihom	Fruit
22	*Annona squamosa*. Linn. ANNONACEAE	Sitaphal	Fruit
23	*Antidesma acidum* Retz. EUPHORBIACEAE	Chingyensil	Leaf
24	*Aphanamixis polystachya*. Wall. MELIACEAE	Heriangkhoi	Fruit

25	*Aralia armata*. Seen. ARALIACEAE	Naosek-manbi	Leaf
26	*Areca catechu*. L. PALMAE	Kwa	Seed
27	*Arisaema* sp. ARACEAE	Lincheisu	Rhizome
28	*Artocarpus heterophyllus* Lank. MORACEAE	Theibong	Fruit
29	*A. lakoocha*. Roxb. MORACEAE	Harikothong	Fruit
30	*Arundinella callosa*. Murao. POACEAE	Laiwa	Young shoot
31	*Auricularia auricularia*. AURICULARIACEAE	Uchina	Fructification
32	*Averrhoa carambola*. L. OXALIDACEAE	Heinajom or Heinoujom	Fruit
33	*Azadirachta indica*. MELIACEAE	Neem	Leaf
34	*Bambusa arundinacea*. Willd. GRAMINEAE	Saneibi	Young shoot
35	*Basella alba* L. BASELACEAE	Urok-sumban	Young shoot
36	*Benincasa hispida*. (Thunb.) Cogn. CUCURBITACEAE	Torbot	Fruit
37	*Brassica campestris*. L. CRUCIFERAE	Hanggam	Leaf, seed
38	*B. oleracea var. botrytis*. CRUCIFERAE	Kobithamchet Manbi	Inflorescence
39	*B. oleracea var. capitata*. (L.) Alef. CRUCIFERAE	Kobiful	Bud
40	*B. rapa*. CRUCIFERAE	Kobimula	Modified stem
41	*Cajanus cajan*. Spreng LEGUMINOSAE	Mairongbi	Seed
42	*Calamus tenuis* Roxb. PALMAE	Heiri	Fruit
43	*Cannabis sativa*. L.. CANNABINACEAE	Ganja	Inflorescence
44	*Cassia hirsuta* Linn. CAESALPINIACEAE	Thaonam	Young shoot
45	*Capsella bursa-pastoris* (Linn) Moench CRUCIFERAE	Chantruk	Young shoot
46	*Carica papaya*. L. CARICACEAE	Awathabi	Fruit
47	*Celtis australis* L. URTICACEAE	Heikreng	Fruit
48	*Celosia argentea*. L. AMARANTHACEAE	Haolei angouba	Leaf
49	*Centella asiatica* (L) Urban. UMBELLIFERAE	Peruk	Leaf
50	*Chenopodium album* L. CHENOPODIACEAE	Monsaobi	Young shoot
51	*Cinnamomum tamala*. Nees and Ebern. LAURACEAE	Tezpat	Leaf
52	*C. zeylanicum*. Breyn. LAURACEAE	Usingsa	Bark
53	*Cissus adnata*. Roxb. VITACEAE	Kongouyen	Leaf

54	*Citrullus lanatus*. Schrat CUCURBITACEAE	Tarbuja	Fruit
55	*Cissus javaca*, DC. VITACEAE	Kongouyen	Leaf
56	*Citrus hystrix* L. RUTACEAE	Heiribob	Fruit
57	*C. medila* L. RUTACEAE	Heijang	Fruit
58	*C. maxima* (Burm. f.) Merr, RUTACEAE	Nobab	Fruit
59	*C.reticulata*. Blaneo. RUTACEAE	Komla	Fruit
60	*C. sinensis* (Linn.) Osbeck. RUTACEAE	Mausmi/Heithum	Fruit
61	*Clerodendrum serratum* (L) Moon. VERBENACEAE	Moirangkhanam	Young leaf
62	*C. indicum* (Unn.) Kurtze VERBENACEAE	Kuthap	Leaf
63	*Clitocybe multiceps* Peck FUNGUS	Tektekpal	Whole plant
64	*Cocos nucifera*. L. PALMAE	Yubi	Fruit
65	*Colocasia gigantean*. Hork. ARACEAE	Yendem	Whole plant
66	*Cariandrum sativum* L. APIACEAE	Phadigom	Leaf and petiole
67	*Crotolaria juncea* L. LEGUMINOSAE	U-hawamation	Young shoot
68	*Cucumis sativus*. Linn. CUCURBITACEAE	Thabi	Fruit
69	*Curcuma angustifolia* Roxb ZINGIBERACEAE	Yaipan	Flower
70	*C. longa*. Valeton. ZINGIBERACEAE	Yaingang	Modified stem and leaf
71	*Cucuribita maxima*. CUCURBITACEAE	Mairel Asangba	Fruit
72	*C. pepo*. CUCURBITACEAE	Mairel	Fruit
73	*Cycas pectinata*. Griff. CYCADACEAE	Yengdang	Young leaf
74	*Cyperus esculentus*. Linn. CYPERACEAE	Kaothum	Tuber
75	*C. rotundus*. CYPERACEAE	Sembang Kaothum	Leaf
76	*Desmodium microphyllum*. PAPILIONACEAE	Yensil nakupi	Leaf
77	*Dicrocephala latifolia* DC. ASTARACEAE	Laukok	Leaf
78	*Dillenia indica*. L. DILLENIACEAE	Heigri	Fruit
79	*Dioscorea alata*. DISCOREACEAE	Haa	Modified root
80	*Drymeria cordata*. Willd. CARYOPHYLLACEAE	Tandan Paibi	Leaf
81	*Elaeocarpus serratus* Linn. ELAEOCARPACEAE	Chorphon	Fruit
82	*Eryngium foetidum*. L. UMBELLIFERAE	Awa-phadigom	Leaf
83	*Euphorbia hirta* Linn. EUPEHORBIACEAE	Pakhangleiton	Leaf
84	*Euphoria longana* Lamk. SAPINIDACEAE	Nonganghei	Fruit

85	*Euryale ferox*. Salisb. NYMAPHEACEAE	Thangjing	Fruit
86	*Fagopyrum esculentum* Monech POLYGONIACEAE	Wakhei yendem	Tender shoot
87	*Ficus bengalensis*. L. MORACEAE	Khonganagbot	Young leaf
88	*Ficus glomerata* Roxb. MORACEAE	Heibung	Fruit
89	*F. hispida*. L. f. MORACEAE	Ashiheibong	Fruit
90	*E. palmata* Forsk MORACEAE	Heiba	Leaf
91	*Ficus semicordata* Forsk MORACEAE	Heirit	Fruit
92	*Flacourtia jangomas*. Raeusch. FLACOURTIACEAE	Heritroi	Fruit
93	*Garchinia pedunculata*. Roxb GUTTIFERCEAE	Heibung	Fruit
94	*Goniothalamus sesquipedalis*. Hook f. Thoms ANNONACEAE	Leikham	Young leaf
95	*Glycine max*. Meriall. LEGUMINOSEAE	Nunghawai	Seed
96	*Gynura cusimbus*. D. Don. ASTERACEAE	Terapaibi	Young shoot
97	*Hedyotis auriculata* (Linn.) RUBIACEAE	Langban Koukha	Tender shoot
98	*Hedychium coronarium*. Koening ZINGIBERACEAE	Takhellei	Modified stem
99	*H. elipticum*. ZINGIBERACEAE	Loklei	Modified stem
100	*Helianthus annuus*. L. ASTERACEAE	Numitlei	Seed
101	*Hibiscus cannabinus* Linn. MALVACEAE	Sougri	Leaf
102	*H. sabdariffa*. Linn. MALVACEAE	Silo-Sougri	Leaf
103	*Houttuynia cordata*. Thunb. SAURURACEAE	Tokningkhok	Leaf and stem
104	*Ipomea aquatica*. Forsk. CONVOLVULACEAE	Kolamni	Leaf and stem
105	*I. Batatas*: (Linn.) Lam. CONVOLVULACEAE	Mangra	Root
106	*Juglans regia* L. JUGLANDACEAE	Heijugak	Fruit
107	*Justiocia adhatoda*. L. ACANTHACEAE	Nongamakha Asinba	Leaf
108	*Lentinellus cochleatus* (Fungi)	Uyen	Whole plant (fructification)
109	*Leucaena glauca*. Benth. CONVOLVULACEAE	Chiginglei tujomba	Tender lead, fruit, seed
110	*Litchi chinensis*. (Geartn.) Sonn. SAPINDACEAE	Linchu	Fruit
111	*Lagenaria siceraria*. (Mot.) Standl. CUCURBITACEAE	Khongdrum	Fruit

112	*Lasiophaera gigantea*. (Batsch. ex. fr.) FUNGI	Phougak Chengum	Fruitification
113	*Lactuca sativa*. Linn. ASTERACEAE	Salad/Hanggam Angouba	Leaf
114	*Lemanea australis* Atkins. PHAEOPHYCEAE	Nungsham	Whole plant
115	*Leucas aspera*. Spreng. LABIATAE	Mayanglambung/ Mayangkimbum	Young shoot
116	*Luffa acutangula*. CUCURBITACEAE	Sebotchuthek-naiba	Fruit
117	*L. cylindrical*. CUCURBITACEAE	Sebot	Fruit
118	*Lysimachia monnieria*. SCROPHULARIACEAE	Leibak kundo	Upper portion of the plant
119	*Mangifera indica*. L. ANACARDIACEAE	Heinou	Fruit
120	*Manihot esculanta* (Bantz.) EUPHORBIACEAE	U-manggra	Modified root
121	*Marsilea minuta* Linn. MARSILIACEAE	Ising yensang or Ising tensil	Leaf
122	*Mentha arvensis*, L. LABIATAE	Nungsihidak	Leaf and stem
123	*M. viridis*. L. LABIATAE	Nungsihidak	Shoot
124	*Microcos paniculata*, L. TILIACEAE	Heitup	Fruit
125	*Mimosa pudica*. L. LEGUMINOSAE	Kangphal Ikaithabi	Young shoot
126	*Mimusop elengi*. L. SAPOTACEAE	Bokul	Fruit
127	*Momordica charantia*. L. CUCURBITACEAE	Karot Akhabi	Fruit
128	*Moringa oelifera* Lam. MORINGACEAE	Shajna	Young shoot
129	*Morus alba*. Poir. MORACEAE	Kabrangchak	Fruit
130	*Musa paradisiaca*. L. MUSACEAE	Laphu	Fruit, petiole
131	*Neptunia olenacea* Lour. MIMOSACEAE	Ising ikaithabi	Tender shoot
132	*Nelumbo nucifera*. Gaetn. NYMPHEACEAE	Thambal	Seed, Rhizome
133	*Nicotiana tabacum*. L. SOLANACEAE	Heedakmana	Leaf
134	*Nymphae alba*. Linn. NYMPHAEACEAE	Tharo-angouba	Petiole, tuber
135	*N. stellata*. Willd. NYMPHAEACEAE	Thariktha	Inflorescence
136	*N. nouchali* Bum. F. NYMPHAEACEAE	Tharo-Angouba	Tuber
137	*Nyctanthes arbortristis* L. OLEACEAE	Singgarei	Leaf
138	*Ocimum cannum* L. LABIATAE	Mayangba or Mayangton	Young shoot Inflorescence

139	*Oenanthe javanica*. (BL). DC. UMBELLIFERAE	Komprek	Aerial portion
140	*Olea europea* L. OLEACEAE	Chorphon	Fruit
141	*Paederia foetida* L. RUBIACEAE	Oinam	Tender leaf, shoot
142	*Parkia roxburghil* L. LEGUMINOSAE	Yongchak	Inflorescence, fruit
143	*Polygonum barbatum* Linn. POLYGONACEAE	Yelang.	Young shoot
144	*Portulaca oleracea*. L. PORTULACACEAE	Leipak Kundo	Leaf and stem
145	*Phaseolus lunatus* Linn. PAPILIONACEAE	Kalandri	Seed
146	*Phogacanthus thyrsiflorus* Nees. ACANTHACEAE	Nongmangkha angangba	Inflorescence and leaf
147	*Phaseolus vulgaris*. LEGUMINOSAE	Rajma	Seed
148	*Phyllanthus emblica*. L. EUPHORBIACEAE	Heikru	Fruit
149	*P. niruri*. Hook. EUPHORBIACEAE	Heikruman, or Chakpa-heikrus	Young shoot
150	*Phoenix sylvestris*. Roxb. PALMAE	Thangtup	Seed
151	*Pistia* sp. (Araceae). EUPHORBIACEAE	Kangjao	Modified Petiole leaf
152	*Piper betle*. Blume. PIPERACEAE	Panamana or Kwa mana	Leaf
153	*Pisum sativum*. Linn. PAPILIONACEAE	Hauwaitharak	Fruit
154	*Plantago major*. L. PLANTAGINACEAE	Yempat	Leaf and petiole
155	*Plumeria acuminata*. Ait. APOCYNACEAE	Khagi-Leihao	Young leaf
156	*Pogostemon purpurascens*, Dalz. LABIATAE	Tekta	Young shoot
157	*Podabrella microcarpa*. Bark. PODABREACEAE	Narin chenggum	Whole plant
158	*Polygonum chinensis*. Linn. POLYGONACEAE	Angom Yensil	Leaf
159	*P. hydropiper*. Linn. POLYGONACEAE	Lihar	Shoot
160	*P. plebejum*, R. Br. POLYGIONACEAE	Tharakmana	Leaf, young shoot
161	*P. posumbu*. Buch. POLYGONACEAE	Phakpai	Leaf
162	*Primula* sp. PRIMALACEAE	Kengoi	Tender shoot
163	*Prunus armeniaca*. L. ROSACEAE	Malhei	Fruit
164	*P. cerasoides*. D. Don. ROSACEAE	Chumberi	Fruit
165	*Psidium guayava*. L. MYRTACEAE	Pungthon	Fruit
166	*Psophocarpus tetragonolobus*. DC. LEGUMINISAE	Tengnumanbi	Fruit and root
167	*Punica granatum*. L. PUNICACEAE	Kamphoi	Fruit

168	*Pyrus communis*. L. ROSACEAE	Naspati	Fruit
169	*Quercus lamellosa*. Sm. FAGACEAE	Uyung	Seed
170	*Raphanus sativus* Linn. CRUCIFERAE	Hanggam mula	Modified stem
171	*Rhus succedanea* L. ANACARDIACEAE	Heimang	Fruit
172	*Rhus javanica*. L. ANACARDIACEAE	Heining	Fruit
173	*Rotala baccifera* L. LYTHRACEAE	Ising Kundo	Leaf
174	*Rubus niveus*. Thunb. ROASCEAE	Heijampet	Fruit
175	*Rorippa indica* L. BRASSICACEAE	Uchihanggam	Leaf
176	*Saccharum officinarum* L. POACEAE	Chu	Stem
177	*S. arundinaceum*. Retx. POACEAE	Singhut	Inflorescence
178	*Sagittaria sagittifolia*. L. ALISMATACEAE	Koukha	Rhizome
179	*Schizophyllum commune*. Nutt. SCHIZOPHYLLACEAE	Kanglayen	Fruitification
180	*Scirpus micronatus*. L.	Kaubak	Root
181	*Schum edule*. Sw. CUCURBITACEAE	Dash-Kush	Fruit
182	*Sesamum indicum*. L. PEDALLIACEAE	Thoding angouba	Seed
183	*Sesbania sesban*. Merr. LEGUMINOSAE	Chuchurangmei	Young shoot
184	*Smilax lancifolia*. Roxb. SMILACEAE	Kwambanbi	Young shoot
185	*Solanum xanthocarpum*, SOLANACEAE	Leipungkhanga	Fruit
186	*S. melongena*. L. SOLANACEAE	Pangal khamen	Fruit
187	*Spilanthes cleraceae*. Murr. COMPOSITAE	Tengnoupal	Fruit
188	*Stellaria media*. (L.) Vill. CARYOPHYLLACEAE	Yerum-keirum	Leaf
189	*Spinacia oleraceae*. L. CHENOPODIACEAE	Palangsak	Young shoot
190	*Syzgyium jambos*. (Linn). Alston. MYRTACEAE	Jam	Fruit
191	*Tamarindus indica*. LEGUMINOSAE	Mange	Fruit
192	*Tectona grandis*. L. VERBENACEAE	Chingsu	Young leaf and Young shoot
193	*Terminalia citrine*. Roxb. COMBRATACEAE	Manahi	Fruit
194	*Trigonella foenum-graecum* L. LOGUMINOSAE	Mithi	Young shoot
195	*Trapa natans*. L. TRAPACEAE	Heikak	Fruit
196	*Ustilago* sp. FUNGI	Kambong	Fungal spores
197	*Vicia faba*. Linn. PAPILIONACEAE	Hawaimubi	Fruit
198	*Vitis vinifera*. L.VITACEAE	Angoor	Fruit
199	*Vigna mungo*. FABACEAE	Sagolhawai	Seed
200	*Viola serpens*. Wall ex. Roxb. VIOLACEAE	Huikhong	Young seed

201	*Vulvariella esculenta*. (Mass) Singh. FUNGI	Charuyen	Whole part
202	*Wendlandia glabrata*. DC. RUBIACEAE	Feija	Young shoot
203	*Zanthoxyllum acanthopodium*. DC. RUBIACEAE	Mukthrubi	Young shoot and leaf
204	*Zingiber officinale*. Rose. ZINGIBERACEAE	Sing	Rhizome
205	*Zizania caduciflora*. (Turez). Hand Mazz POACEAE	Ising Kambong	Fungal spore

Discussion

The collected plants mentioned on the above pages are mostly wild with a few cultivated at present. They believed that the plants are of multiple uses as the plants also show certain medicinal properties over and above their utility as food. No record has so far been made on the account and is evidently the first record of the indigenous plants used by the Meitei Community as food.

It is expected that the present study will bring into focus how the Meiteis enjoy on various indigenous food plants over and above the common one's showing a unique dietary habit. Further this account will give some indication about the availability of a multiple variety of plants in the state. The work is expected to encourage other researchers in Ethno botanical research of the Meiteis.

References

Devi L.D.: *Folklore medicines of Ethno botanical Importance in Manipur*, Vol-I (1990).

Ghosh G.K.: *Herbs of Manipur*, Vol-I and Vol-II (1997a 1997b).

Loudon J.C.: *Encyclopedia of Plants*, Part-II (1973).

❑❑❑

25

Analysis of Fluoride in the Groundwater of Akola District—A Case Study

S.B. Thakare[1], Dr. A.V. Parwate[2], Dr. M. Rao[3]

1. Lecturer, College of Engineering and Technology, Babhulgaon (Akola) M.S.
2. Principal, A.V.C. College of Engineering, Mayiladuthurai (Nagai) T.N.
3. Principal, P.R. Engineering College, Vallam (Thanjavur) T.N.

Introduction

Fluoride has posed a great threat to human in our country. First-ever case of fluorosis was detected by the farmers of Andhra Pradesh State, among cattle during early 1930s. Shorett *et al.* published the first report on the endemic fluorisis in India since then edemic fluorosis has been identified in the 17 states of Indian union and is known to prevalent for the last six decades without receiving much attention to tackle it.

A national survey of drinking water in India indicated that over 43.5 million people living in 1.42 lakh habitations spread over 200 districts are at health risk due to water quality problems and fluoride is identified as one of the major contaminants. Fluoride is not found in the nature in free state. It occurs primarily in form of minerals like.

(1) Cryolite (Na_2AlF_6),
(2) Fluorite (Fluorspar) CaF_2
(3) Fluorpatite $Ca_5F(PO_4)_3$,
(4) Topaz Al_2SiO_4 $(OH, F)_2$
(5) Sellaite (MGF_2),
(6) Villiamite (NaF).

It is also a minor consituents many clay minerals and mica. Fluoride is released to the groundwater through weathering of primary silicates and associated accessory minerals. It is also partly are deposited in soil profile as carbonate concretions (Kankars) in soluble form of CaF_2 under certain climatic and hydrogeological conditions. Other sources of fluoride in groundwater are the insecticides, herbicides and phosphate fertilisers used in agricultural activities, Fluoride salts commonly in steel, aluminum, brick and tile industries may also contribute to groundwater due to industrial spillages. Fluoride in the combined state constitutes 0.078 per cent of the earth's crust.

Climate and Hydrogeology

Akola district is geographically situated at about 20° 44' North latitude to 21° North latitude and 77° 44' East longitude to 78° East longitude. The district is about 281.90 meters above mean sea level and area of the district is 10.575 sq. Km with 11.60 lakhs in habitants. Entire district comprises of rocks like basalt, granite, jambha rock and black stone. The rivers flowing through the district are Katepurna, Uma, Purna, Shahanur, Vidrupa, Pathar, Nirguna and Man. Hot and dry is the normal limate. Rainfall is during June to September and being received by South-west and North-east monsoon. The average rainfall is about 847 mm. Soil colour ranges from black to red. Depth of water table ranges from 4mt. to 100 mt. below ground level.

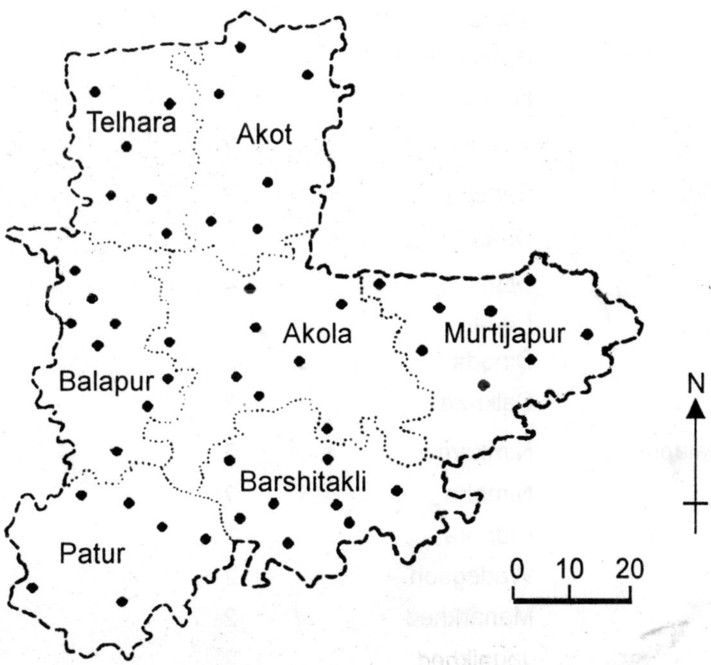

Map of Akola District

Material and Methods

Considering the present stock of the situation in groundwater quality and possible degradation of the same by environmental pollution, 100 samples of water from open wells and bore wells dispersed over the entire district have been analysed for quality surveillance. The water table was at its lowest ebb due to extraction from the aquifer to fulfil various daily demands during non-rainy season. So the samples are collected during the month of February 2005 to April 2004. Fluoride was analysed by Spands method as per the standard methods by uv spectrophotometer in the laboratory.

Table-1: Analysis of fluoride in groundwater

Sr. No.	Talukka	Village	No. of sources tested	Average fluoride concentration mg/l.
1	Akola	Babhulgaon	2	2.5
		Ghusar	2	2.80
		Shovar	2	1.80
		Kasli	2	0.95
		Yeota	2	0.60
		Bakharabad	2	2.80
		Dapura	2	1.30
		Yelwan	2	1.00
2	Akot	Fattepur	2	2.65
		Dharel	2	0.75
		Akot	2	1.90
		Panaj	2	0.36
		Dinoda	2	0.95
		Salkhed	2	1.60
3	Balapur	Nimkarda	2	1.40
		Nimbha	2	1.85
		Ridhora	2	1.70
		Wadegaon	2	1.65
		Manarkhed	2	0.52
		Jogalkhed	2	0.83
		Balapur	2	1.60

		Paras	2	0.80
		Barlenga	2	1.50
4	Barshitakali	Pinjar	2	2.10
		Chelka	2	1.70
		Walpi	2	1.65
		Titwa	2	1.50
		Zodga	2	0.90
		Shindked	2	1.30
		Dhaba	2	0.70
		Kinkhed	2	0.80
5	Murtizapur	Jamathi	2	0.90
		Mungshi	2	0.85
		Hirpur	2	1.70
		Hatgaon	2	1.58
		Anabhora	2	1.10
		Kalsara	2	0.70
		Jambha	2	0.50
		Kurum	2	1.30
6	Patur	Sawargaon	2	0.50
		Khanapur	2	0.40
		Shirla	2	0.80
		Patur	2	1.40
		Chondi	2	0.65
		Hingana	2	3.00
7	Telhara	Bamberi	2	0.60
		Belched	2	1.25
		Narsipur	2	1.50
		Saundala	2	1.75
		Telhara	2	0.80
		Adgaon (kh)	2	1.50

The study area selected for fluoride analysis in the ground water includes 51 villages from all the 07 Talukka of Akola district. It is observed from the analysis giver. In Table 1 that 29 villages have fluoride ion concentration above the permissible limit of 1 mg/lt. The maximum average fluoride ion concentration was found in 18 villages above 1.5 mg/lt, and the fluoride ion concentration below 0.5 mg/lt is found about 02 villages *i.e.* at Panaj in Akot taluka and at Khanapur in Patur talukka. It was found that there is a direct correlation between alkalinity and Fluoride ion concentration.

Conclusion

These findings suggest that the distribution of fluoride ions in the groundwater of Akola district is growing up. The concentration of fluoride in groundwater has direct relation to the contact time of water with the fluoride minerals in rocks.

An important measure to control the concentration of fluoride in groundwater is to adopt techniques to augment the groundwater storage through rainwater harvesting. Artificial recharge is the process by which the groundwater reservoir is augmented at a rate exceeding that under natural conditions. Hydrogeological surveys can help in identifying appropriate sites for construction of artifical recharge structures like percolation tanks, nala bunds, recharge shafts which can go a long way in diverting the surplus surface monsoon run-off to groundwater storage. Water conservation techniques like sub-surface dams and roof top rainwater harvesting have special significance in augmenting the groundwater storage which help in improving the quality of groundwater.

In area were the hydrogeological situation does not favour adoption of any technique for artificial recharge of groundwater and high fluoride occurs in both shallow and deeper aquifers, the only alternative is to either supply safe drinking water from surface/ groundwater source at some distance from the village or adopt technique of treatment of water.

To tackle to fluoride problem, the following approach may be adopted:

- **Indentification and delineation of areas affected**— Reconnaissance survey of habitations in fluorosis suspected areas and collection of site-specific data on health and water resources and quality aspects through standardised procedure using appropriately designed questionnaire and format may be completed to identify and delineate the area affected.

  In depth analysis of water sources, health status of delineated areas and evaluation of technology:

- Completion of chemical analysis of water sources along with community water supply sources.
- Engineering analysis of water quality data so collected
- Pre-intervention health assessment.
- Interaction with community and state level relevant agencies for planning of site-specific fluorosis control measures.
- Critical evaluation of identified defraudation technologies through a scientific ranking methodology and economical analysis of defraudation options *vis-à-vis* supply from alternate sources for delineated habitations to identify specific best option.

Preparation of Action Plan and implementation of solutions:

- Organising community awareness programmes on the health impact of fluorosis and the need for supplying fluoride free water.
- Demonstration and training programmes for public health engineers, district health agencies and local operators.
- Administering the medical and nutritional measures simultaneously making the provision of safe water supply through domestic defluoridation as interim measure.
- Design of water supply system utilising alternate source of preparation of process package including designs, drawings and specifications for defluoridation plant as best option.

Post-intervention evaluation:

- Post intervention health assessment on the similar lines to the pre-intervention health assessment.
- Periodic monitoring of measures taken and improvement in health.
- Post installation evaluation of the system adopted and recommendations for modification, if necessary, and site-specific 0 and M provision for sustained performance.
- Treatability studies for optimising the process control.

References

APHA 1995. *Standard methods for the examination of water and wastewater* (19th edn). American Public Health Association.

Azbar, N. and Turkman, A. 2000. Defluoridation in drinking waters. *Water Sci. Tech.*, **42: 848-858.**

Bhargava, D.S. and Killedar, D.J. 1991. Batch studies of water defluoridation using fishbone charcoal, *J. WPCF.* **63: 848-858.**

CPCB. 1994-95. *Laboratory analytical techniques. Series: Lats/9/1994-95.* Central Pollution Control Board, New Delhi.

CPHEEO. 1991. *Manual on water supply treatment.* Ministry of Urban Development, New Delhi.

Khabade, S.A. and Mule, M.B. 2003. Studies on Physico-chemical Parameters of Pundi Water Reservoir From Tasgaon Tahsil, *Indian J. of Environmental Protection*, **23: 1003-1007.**

Susheela, A.K. 1999. Fluorosis Management Programme in India, *Current Science*, **77: 1250-1256.**

❑❑❑

26

Phytoplanktonology of Pond at Deoghar

1a. Phytoplankton standing crop in relation to abiotic spectrum

Ashok Kumar and N.K. Singh
Dept. of Botany, Deoghar College, Deoghar

Introduction

Phytoplankton constitute the main food fraction for most of the aquatic fauna. Seasonal fluctuation in phytoplankton community is completely under the grip of physico-chemical regime of water bodies. In India, several investigators studied the limnobiology of lentic as well as lotic water bodies particularly of phytoplankton community (Zutschi and Khan, 1977; Mishra and Yadav, 1978; Saha and Singh, 1980; Prasad and Suxena, 1980; Singh *et al.,* 1983; Reddy and Venketswarlu, 1986; Bharti and Krishnamurty, 1991; Singh, 1993; Bose and Gorai, 1993). So far as the phytoplanktonology of ponds at Deogher (Santhal Pargana) is concerned, only fragmentary reports have accumulated in the recent past. Therefore, present investigation has been aimed at studying in phytoplankton standing crop variation in relation to physico-chemical characteristics of pond water (Nandan Pahar Pond) at Deoghar.

Materials and Methods

Samples of the surface water of Nandan Pahar Pond were collected at monthly intervals from June 1995 to May 1997. The water temperature was recorded by mercury centigrade thermometer. pH was determined by a lobibond pH comparater using indicators of different pH range. Transparency of the pond water was measured by sacchi disc method described by Welch (1948). Dissolved oxygen content, free CO_2, nitrate-nitrogen, phosphate-phosphorus, silicate-silica, chloride and total solids of the pond water determined by Winkler's iodometric mathod, by titration method, by phenoldisulphonic acid method, by ammonium molybdate stannous chloride method, by heteropolyblue method, by titrimetric method and by evaporation of the unfiltered sample respectively as described method by Trivedi and Goel (1986). Carbonate and bicarbonate alkalinity were estimated by titrimetric method as described by Welch (1948). For the phytoplankton standing crop 100 litres of surface water as filtered through bolting silk net (no-20xxx) having 173 meshes per linear inch and samples were preserved in 4 per cent formalin and Lugol's solution. The counting of phytoplankton was done by the method of Lackey (1938) of modified by Edmondson (1974).

Results and Discussion

Results (Text fig.–1) indicate that phytoplankton standing crop constituted peaks twice in a year *i.e.* one in May and other in November but the peak of greater magnitude was observed in May. Similar findings were also reported by Singh *et al.* (1983). During rainy season, the phytoplankton density declined considerably which may be possibly be due to flooding of rain water into the pond from suburbs coupled with unfavourable spectrum of environmental conditions. It, however, increased gradually in the later part of the year in and November constituted a secondary peak of somewhat lesser magnitude. So far as the percentage composition of different groups of phytoplankton flora is concerned, Chlorophytes with 39.64% outnumbered the other groups followed by Cyanophytes (35.18%) and Bacillariophytes (25.18%). Chlorophytes were recorded maximum during summer while the Cyanophytes and Bacillariophytes during last part of the rainy season.

The water temperature ranged between 16-35.5°C during the course of investigation with an average value of 25.125°C similar range of water temperature variation was also recorded by Bose and Gorai (1993) while they were studying seasonal fluctuation of plankton in relation to physico-chemical parameters of fresh water tank of Dhanabad, India. The maximum and minimum values were recorded in May and January. The water temperature exhibited negative correlation with dissolved oxygen, transparency and pH but positive correlation with free CO_2, nitrate-nitrogen, phosphate-phosphorus and chloride content of the pond water. The quantity of the dissolved oxygen decreased with the rise of mercury of the thermometer and *vice-versa* (Bose and Gorai 1993, Singh 1993) thereby showing negative correlation (Table-1). Actually at higher temperature the bacterial activity increased which led to increase in the nutrient status leading to positive correlation. Phytoplankton standing crop exhibited negative correlation with water temperature but it was not significant. The present finding is also in agreement with Jana (1979). This may be possibly due to bacterial decomposition at higher temperature and due to cloudy and unfavourable condition during rainy season which directly or indirectly influenced the correlation. The maximum density of phytoplankton was recorded in May when the water temperature was also quite high. Many investigators correlated higher phytoplankton production in such conditions (Seenayya 1971, Rana *et al.,* 1991: 1995). In the present investigation standing crop values of phytoplankters started increasing from November onwards with a slight decline in December and culminated in May. This was the indication that winter and summer temperature was probably suitable for the growth of phytoplankters. Padhi (1995) also recorded the dominance of Chlorophytes followed by Cyanophytes and Bacillariophytes. (Text fig.-1).

Seasonal Fluctuation Percentage Composition and Yearly Average of Phytoplankton (Nandan Pahar Pond)

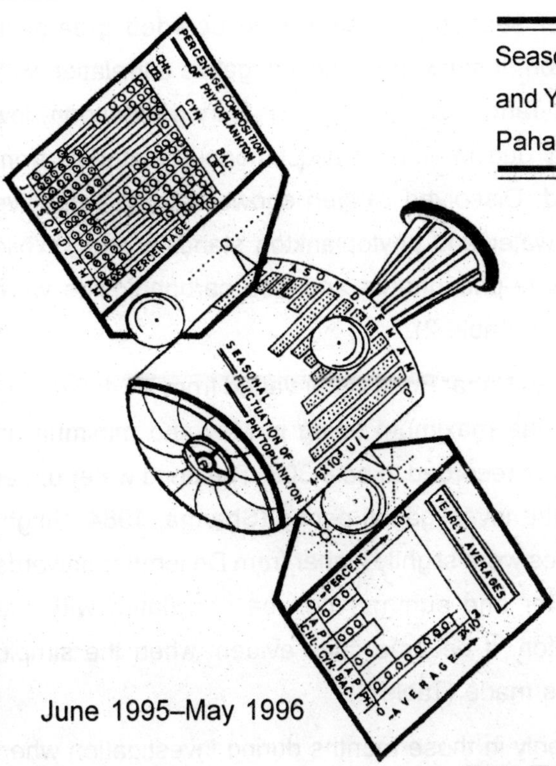

June 1995–May 1996

Text Fig. 1

Penetration of light in an aquatic system is dependent upon turbidity. Turbidity and the transparency in an aquatic system are inversely correlated. In the present investigation transparency in pond water fluctuated between 40-71.5 cm. with an average of 58.45 cm. (Table-1). Lower values of transparency were recorded during rainy season which may be possibly due to entry of allochtonous muddy water during monsoon which decreased the transparency of water. The present observation was also in conformity with that of Singh (1962), Singh *et al.* (1983), Singh (1993). During the period of low transparency, phytoplankter density declined to minimum and also the number of species decreased considerably. It showed strongly significant positive correlation (Table-2) with phytoplankton standing crop (Saha *et al.,* 1985; Singh, 1993). Transparency was found to be positively related with carbonate alkalinity, hydrogen ion concentration and dissolved oxygen whereas negatively correlated with nutrients of the water body (Table-2).

The dissolved oxygen content varied from 6.8-9.6 ppm with an average of 7.862 ppm (Table-1). The maximum value was estimated when the water temperature was minimum. On the whole during winter season, its value was recorded greater in comparison to summer and rainy season. It showed strongly negative correlation with water temperature (Saha *et al.,* 1985; Tanti and Saha, 1993). During monsoon, low oxygen content estimated was possibly due to entry of allochthonous materials from the suburbs with high oxygen demand. Dissolved oxygen showed strongly positive correlation with transparency of pond water and phytoplankton standing crop. While silicate-silica, nitrate-nitrogen, phosphate-phosphorus and free carbondioxide were negatively related with dissolved oxygen (Table-2).

Hydrogen ion concentration in Nandan Pahar Pond water varied from 7.6-8.9 during the course of two years investigation, the maximum being in May and minimum in October (Table-1). The average value was recorded to be 8.308. The pond water under study showed alkaline nature thought the investigation period (Sharma, 1984; Singh, 1993). Hydrogen ion concentration values were slightly higher from December onwards to May. Its higher values during winter and summer can be correlated with the comparative abundance of phytoplankton. It became quite evident when the simple correlation and regression analysis was made (Table-2).

Free carbondioxide was recorded only in those months during investigation when the carbonate alkalinity was found absent. It was reported during monsoon season

Table-1

Physico-chemical characteristics of Nandan Pahar Pond Water

Months	Atmos. Tem. (°C)	Water Tem. (°C)	Trans (Cm)	DO (ppm)	pH	Free CO_2 (ppm)	CO_3 (ppm)	HCO_3 (ppm)	NO_3-N (ug/l)	PO_4-P (Mg/l)	SiO_3-Si (Mg/l)	Total Solid (ppm)	Chloride (ppm)
(1995)													
June	32.0	29.0	65.0	7.8	8.4	-	4.0	158.0	533	0.091	11.39	800	61.4
July	30.0	29.0	47.0	7.6	8.0	2.5	-	130.0	538	0.169	13.21	835	59.3
August	30.0	28.0	44.3	7.0	7.8	4.5	-	133.0	603	0.234	13.41	760	60.0
September	30.0	29.0	44.7	7.0	7.8	4.8	-	133.0	600	0.234	13.41	750	58.3
October	32.0	30.0	47.3	6.8	7.6	6.4	-	140.0	597	0.078	12.50	640	50.0
November	25.0	22.0	58.0	8.5	8.4	-	2.0	158.0	547	0.021	12.34	700	56.0
December	23.0	20.0	60.3	8.8	8.6	-	10.0	140.0	551	0.035	12.00	697	55.0
(1996)													
January	20.0	18.0	65.0	9.5	8.7	-	11.0	140.0	543	0.039	11.37	671	56.0
February	20.0	19.0	66.0	8.0	8.5	-	12.0	137.0	549	0.024	11.00	650	59.4
March	24.0	22.0	71.0	7.8	8.8	-	11.0	140.0	533	0.020	10.00	673	63.2
April	30.0	25.0	70.0	7.5	8.7	-	11.0	142.0	539	0.019	9.37	679	67.0
May	34.0	30.5	71.5	7.2	8.9	-	13.0	135.0	530	0.016	9.00	675	70.0
June	36.0	27.0	55.0	7.6	8.4	3.0	3.0	160.0	541	0.091	11.00	690	65.3
July	30.0	29.0	46.0	7.5	8.0	-	-	127.0	572	0.124	13.00	697	66.2
August	30.0	28.0	44.3	7.0	7.8	4.0	-	130.0	600	0.233	13.45	700	65.0
September	30.0	29.0	40.0	7.0	7.8	4.0	-	129.0	609	0.123	13.00	690	63.0
October	31.0	30.0	44.0	7.0	7.6	6.0	-	133.0	597	0.078	13.15	691	64.5
November	28.0	25.0	56.4	7.5	7.9	4.0	-	126.0	567	0.032	12.11	725	56.0
December	26.0	20.0	60.0	8.0	8.4	-	3.0	155.0	544	0.041	12.00	690	48.3
(1997)													
January	17.0	16.0	67.0	9.6	8.5	-	6.0	150.0	540	0.053	11.97	680	45.0
February	17.0	18.0	68.0	9.3	8.7	-	8.0	147.0	535	0.053	11.00	687	50.0
March	26.0	24.0	70.0	8.6	8.6	-	7.0	149.0	535	0.041	11.10	689	54.0
April	29.0	25.0	71.0	8.3	8.7	-	9.5	145.0	530	0.035	10.23	700	64.0
May	33.0	30.0	71.0	7.8	8.8	-	10.0	143.0	527	0.016	9.44	711	69.6
Average	27.625	25.104	58.45	7.862	8.308	4.388	8.033	140.833	556.708	0.078	11.685	703.333	59.458
Maximum	36.000	30.500	71.50	9.600	8.900	6.700	13.000	160.00	609.000	0.234	13.450	835.000	70.000
Minimum	17.000	16.000	40.00	6.800	7.600	2.500	2.000	127.00	527.000	0.016	9.000	640.000	50.000

Table-2: S.C.A. of Nandan Pahar Pond

	PD	Cl⁻	SiO_3-Si	Total Solid	PO_4-P	NO_3-N	HCO_3	CO_3	Free CO_2	pH	DO_2	Transparent
Water Temperature	-0.388	0.673	0.178	0.376	0.465	0.434	-0.379	-0.469	0.585	-0.515	-0.83	-0.505
	(1.976)	(3.885)	(0.895)	(1.908)	(2.465)	(2.259)	(1.922)	(2.497)	(3.389)	(2.808)	(7.676)	(2.751)
Transparency	0.928	-0.030	-0.890	-0.287	-0.776	-0.876	0.504	0.880	-0.848	0.935	0.607	
	(11.708)	(0.144)	(9.189)	(1.406)	(5.775)	(8.525)	(2.738)	(8.719)	(7.518)	(12.458)	(3.585)	
DO_2	0.456	-0.536	-0.275	-0.187	-0.500	-0.625	0.486	0.507	-0.694	0.638		
	(2.408)	(2.981)	(1.342)	(0.895)	(2.709)	(3.758)	(2.609)	(2.763)	(4.529)	(3.892)		
pH	0.892	0.073	-0.865	-0.257	-0.703	-0.891	0.504	0.904	-0.940			
	(9.279)	(0.342)	(8.118)	(1.249)	(4.641)	(9.249)	(2.738)	(9.971)	(12.992)			
Free CO_2	-0.778	0.016	0.717	0.132	0.629	0.904	-0.635	-0.765				
	(5.806)	(0.079)	(4.825)	(0.629)	(3.198)	(9.923)	(3.864)	(5.581)				
CO_3	0.028	0.146	-0.863	-0.415	-0.686	-0.710	+0.253					
	(6.902)	(0.697)	(8.018)	(2.142)	(4.433)	(4.729)	(1.228)					
HCO_3	0.437	-0.305	-0.366	-0.096	-0.418	-0.581						
	(2.282)	(1.506)	(1.840)	(0.455)	(2.159)	(3.354)						
NO_3-N	-0.806	0.020	0.763	0.005	0.691							
	(6.391)	(9.424)	(5.540)	(2.763)	(4.487)							
PO_4-P	-0.814	0.095	0.737	0.525								
	(6.675)	(0.448)	(5.125)	(2.894)								
Total Solids	-0.378	0.110	0.353									
	(1.919)	(0.521)	(1.772)									
SiO_3-Si	-0.923	-0.310										
	(11.308)	(1.533)										
Cl-	0.095											
	(0.51)											

N.B.: (i) Table value of t at 10%, 5% and 1% level of significance is 1.7171, 2.074 and 2.819 respectively.

(ii) Parenthesis value is observed t-value.

and varied from 2.5-6.7 ppm. It was absent during winter and summer season. Its presence during monsoon season may possibly be due to decreased rate of photosynthesis and increased rate of decomposition. (Sreenivasan *et al.,* 1977). Its absence during active period of phytoplankton growth may be attributed to rapid utilisation of free carbondioxide for their metabolic activities which led to its exhaustion.

Corbonate alkalinity was recorded during winter and summer and varied from 2-13 ppm. It was found maximum during January and minimum during May. Its biennial average was computed as 8.33 ppm. Simple correlation analysis revealed negative correlation with nutrients like silicate-silica, phosphate-phosphorus and nitrate-nitrogen.

Bicarbonate alkalinity fluctuated between 127-160 ppm maximum being in June and minimum in July 1996. Its average value was 140.833 ppm. Simple correlation analysis showed negative correlation with the nutrients (table-2).

Nitrate-nitrogen varied from 527μ g/litre to 609μ g/litre during the course of biennial investigation with an arithmatic mean of 556.708μ g/litre. During rainy season, nitrate-nitrogen content was detected to be higher in comparison to winter and summer season. This may be possibly due to slower rate of consumption and increased organic load from the surrounding along with rain water. Simple correlation analysis showed negative correlation with transparecy, dissolved oxygen, hydrogen ion concentration, carbonate and bicarbonate alkalinity, chloride content and phytoplankton density whereas positive correlation with water temperature, free carbondioxide, total solids, silicate-silica and phosphate-phosphorus (table-2).

Phosphorus is supposed to be one of the most important nutrients for algal growth. Its concentration ranged between 0.016-0.234 ppm with an arithmatic mean of 0.079 ppm. It showed strongly negative correlation with phytoplankton density, transparency and hydrogen ion concentration whereas positive correlation with nutrients.

Silicate-silica is an essential consituent of the wall structure of Diatoms. Its concentration varied from 9.00-13.45 ppm with an average of 11.685 ppm. Its concentration was estimated to be higher during monsoon when the density of Bacillariophytes was not much pronounced. This may be due to mixing of silica enriched, rain water and rain washings from the surrounding. In spite of high concentration of silica during monsoon months density of Diatoms was not so much pronounced due to high turbidity and cloudy sky condition. Simple correlation analysis revealed that it exhibited negative correlation with phytoplankton density, transparency, dissolved oxygen, hydrogen ion concentration, carbonate, bicarbonate, whereas positive correlation with total solid, water temperature, nitrate-nitrogen phosphate-phophorus.

Actually in an aquatic system the variation in phytoplankter density can not be fully correlated to any single factor, the influence of one factor can inhance or eclipse the effect of other. The phytoplankton abundance is the reflection of collective response of numerous factors. Multiple correlation and regression analysis (table-3), the testimony of the generalisation revealed the value of coefficient of determination as 0.948. It means 94.8 per cent variation in phytoplankton density is influenced by physico-chemical factors considered in this text and only 5.2 per cent by other factors not considered in the present investigation. Multiple regression equation may be utilised for predicting phytoplankton density at any time if we have the values of physico-chemical factors under study.

Table-3

Nandan Phar Pond		**Multiple regession of physico-chemical factors on phytoplankton standing crop.**
Maximum factors	R_2	multiple regression equation
$X_1, X_2, X_3, X_4, X_5, X_6, X_7$ $X_8, X_9, X_{10}, X_{11}, X_{12}$	0.948	Y = –21224 + –50.0761 × 1 + 58.28230 × 2 + –297.974 × 3 + 3151.057 × 4 + 169.2444 × 5 + –158.977 × 6 + –9.12774 × 7 + 6.62857910 × 8 + –4199.57 × 9 + 311.637 × 10 + –0.49948 × 11 + 5.951699 × 12

Standard error of Y-estimate = 442.5090

No. of observation = 24

Df = 11

Standard error of Co-efficients = 70.29277; 41.24866; 350.6187; 1880.414; 268.668; 72.40389; 18.38807; 13.70668; 2893.790; 401.7191; 4.2819; 34.57910.

N.B.: Y = Phytoplankton standing erop

X_1 = Water temperature

X_2 = Transparency

X_3 = Dissolved oxygen
X_4 = pH
X_5 = Free carbondioxide
X_6 = Carbonate
X_7 = Bicarbonate
X_8 = Nitrate-nitrogen
X_9 = Phosphate-phosphorus
X_{10} = Silicate-silica
X_{11} = Total solids
X_{12} = Chloride

References

Bharti, S.G. and Krishnamurthy, S.R. 1991. Benthic algae: Their ecological significance in river Kali, Dandeli, Karanakataka State, India. *Phykos.* **30(1** and **2): 153-163.**

Bose, S.K. and Goari, A.C. 1993. Diel variations of physico-chemical factors, plankton forms and their productivity in a fresh water tank of Dhanabad. *J. Water Biol.* **5(3): 203-208.**

Edmondson, W.T. 1970. Phosphorus, nitrogen and algae in lake Washington after division of Sewage. *Science*, **169: 690-691.**

Jana, B.B. 1979. Temporal plankton succession and ecology of tropical tank in West Bengal, India. *Int. Revue ges. Hydrobiol.* **64(5): 661-671.**

Mishra, G.P. and Yadav, A.K. 1978. A comparative study of physico-chemical characteristics of river and lake water in Central India. *Hydrobiol.* **59: 275-278.**

Padhi, S.B. 1995. Algal environment of plluted and unpolluted fresh water ponds. pp. **131-147.** In: Kargupta, A.N. and Siddiqui, E.N. (Eds.) *Alagal Ecologyl: An Overview.* International Book Distributors, Dehradun, India.

Prasad, B.N. and Suxena Manjula 1980. Ecological study of blue-green algae in the river Gomati, *Indian J. Env. Hlth.* **22(2): 151-168.**

Rana, B.C., Nirmal Kumar, J.I. and Sreenivas, S.S. 1995. Phytoplankton ecology of certain inland water bodies of Central Gujarat, India. In: Kragupta A.N. and Siddiqui, E.N. (Eds.) *Algal Ecolgy: An Overview*, International Distributors, Dehrandun, India. pp. **101-129.**

Reddy, P.M. and Vekateshwarlu 1986. Ecology of algae in the paper mill effluents and their impact on the river Tungabhadra. *J. Environ. Bio.* **76: 215-223.**

Saha, L.C. and Singh, N.K. 1980. Periodicity of algal flora in Bhagalpur ponds. *Acta. Eco.* **2, 3** and **4: 56-51.**

Saha, L.C., Choudary, S.K. and Singh, N.K. 1985. Factors affecting phytoplankton productivity and density in the river Ganges at Bhagalpur. *Geobios*. **12(2): 63-65.**

Seenayya, G. 1971. Ecological studies in the of certain ponds (fresh water) of Hydrabad, India part II. Phytoplankton-1 *Hydrobiologia*. **37: 58-68.**

Sharma, R.C. 1984. Potamological studies on lotic environment of the upland river Bhagirathi of Garwal, Himalaya. *Environment and Ecology* **2(4): pp. 239-242.**

Singh, C.S. 1962. Studies on the growth ad distribution of fresh water diatoms in relation to physico-chemical conditions of some ponds and rivers of Uttar Pradesh. Ph.D. Thesis, B.H.U., Varanasi.

Singh, N.K., Datta Munshi, J.S. and Bilgrami, K.S. 1983. A qualitative study of phytoplankton of the river Ganges at Bhagalpur, India. *Pol. Arco. Hydrobio.* **30(2): 81-87.**

Singh, N.K. 1993. Studies on density, productivity ad species composition of phytoplankton in relation to abiotic spectrum of the Ganges at Sahibganj. *J. Fresh Water Biol.* **5(1): 1-8.**

Sreenivasan, A., Sampath, V., Paramsivam, M. and Ananthanaryan, R. 1977. Pollution of river Cauvery from industrial and urban wastes. All India symposium on *Environ. Biol.* University of Kerala, Trivendrum.

Tanti, K.D. and Saha, K. 1993. Hydrobiological profiles along a thermal gradient of the hot springs of Rajgir (Bihar), India. *J. Fresh Water Biol.* **5(2): 107-117.**

Trivedi, R.K. and Goel, P.K. 1986. *Chemical and Biological Methods for Water Pollution Studies*. Environmental Publication, Karad, (India).

Welch, P.S. 1948. *Limnological Methods*. Blackiston Company, Philadelphia.

Zutschi, D.P. and Khan, M.A. 1977. Limnological investigations of two subtropical lakes. *Geobios*. **4(2): 45-48.**

❑❑❑

27

Symbiotic Association of Vesicular Arbuscular Mycorrhizal Fungi with Different Vegetable Crops

M.S. Darade and Dr. D.S. Mukadam

Department of Botany, Govt. Vidarbha Institute of Science and Humanities, Amravati–444 604 (M.S.), Department of Botany, Dr. Babasaheb Ambedkar Marathwada University, Aurangabad–431 004 (M.S.)

Introduction

VA-mycorrhizae are common and natural biotic components of terrestrial ecosystem. They are important group of a sexually reproducing soil borne fungi from the order Glomales of the class Zygomycetes. They form obligate symbiotic association with many angiospermic plants. The association of these fungi helps to maintain the general plant vigour under a variety of adverse and inhospitable ecological conditions (Trappe, 1981).

The infection of land plants with VAM fungi lead to a growth increase and increases the ability of plant to recover phosphorus from soil that results stimulated growth (Bagyaraj *et al.*, 1989).

Vegetables contribute the main source of fresh proteins and minerals in our diet. Daily consumption of the fresh vegetables is increasing day by day due to their awareness of for the importance of health. Considering the importance of VAM fungi and vegetables the present investigation was carried out.

Materials and Methods

The rhizosphere soil samples along with root systems were collected from different vegetable crops like Okra, cucumber, chilli, bottle gourd, tomato, bitter gourd and brinjal from Aurangabad (M.S.) region. The root system was duged and uprooted without damage. The soil particles adhered to the roots removed carefully and roots were washed with flooding tap water and preserved in FAA until further use.

The isolation of VAM fungi from soil were carried out by wet sieving and decantation technique suggested by Gerdemann and Nicolson (1963). 100 gm. shed dried soil sample were weighed and dissolved in 1000 ml tap water. The soil particles were allowed to settle at the bottom. The soil water mixture was passed through a stack of sieves. The residue left on sieves was washed with tap water and examined under stereoscopic microscope. The identification was done with the help synoptic keys of Trappe (1982) and manual of Schenck and Perez (1987).

The root infection and colonisation was assessed by technique of root clearing and staining suggested by Philips and Hayman (1976). The roots were cleared by 10 per cent KOH Solution, acidified with 5 N HCl and stained in 0.5 per cent Trypan blue.

The evaluation was carried out by using simplified slide technique suggested by Giovannetti and Mosse (1980). The root bits were taken on clean glass slide and observed for infection and VAM structures under compound microscope. The percentage root colonisation was calculated by the method of Read *et al.* (1976).

Result and Discussion

The vesicles, arbuscles, auxiliary cells, Intraradical and Extramatrical mycelium are the structures that are colonised as VAM fungal infecting structures.

The roots of different vegetables were found colonised by these structures and are considered to treat the root as infected.

The vesicles are small darkly staining globular or spherical structures formed. They arise as swellings terminaly or intercalary on fungal hyphae and gradually increase in size. They store oil and polyphosphate granules which are utilised by the plant under phosphorus deficiency conditions. The arbuscles are branched haustoria like structures formed early in the association resulting from repeated dichotomous branching of fungal hyphae inside the root cells that are thought to be sites of nutrient exchange. While external or extramatrical mycelium or hyphae explores the soil around roots travelling long distances in soil and extending beyond the depletion zone of nutrients. that absorbs the nutrients particularly phosphorus and make it available to the plant.

In many ecosystems these forms connecting channels between plants situated quite distantly and help in mineral flow. These hyphae due to their smaller diameter can reach areas where root hair connot. The structure of Intraradical and extramatrical mycelium were found in roots of all the vegetables (Table 1). The vesicles were found in the roots of Okra, bottle gourd, chilli and brinjal. The vesicles was globular as well as oval in shape formed in the parallely spread intraradical mycelium.

Table-1: Assessment of root infection in different vegetables

Name of Plant	Mycelium		Root colonising stricture		
	I.R.	E.M.	Vesicles	Arbuscles	Auxiliary cells
Abelmoschus esculentus L. (Okra)	+	+	+	+	–
Cucumis sativus L (Cucumber)	+	+	–	+	+
Momordica charantia (L bitter gourd)	+	+	–	+	+
Lagenaria siceraria (Molina) standl (bottle gourd)	+	+	+	+	–
Capsicum annum L. (Chilli)	+	+	+	–	–
Lycopersicum esculentum Mill (Tomato)	+	–	+	–	–
Solanum melongena L. (Brinjal)	+	+	+	–	–

I. R. = Intraradical mycelium

E.M. = Extramatrical mycelium

\+ = Present

– = Absent

The Arbuscles observed were dichotomously branched. The Auxilary cells were found to the extraradical soil exploring mycelium of the roots of cucumber and bitter ground. The cells were smooth and globular in shape and was formed in aggregate groups like grapes.

The vegetable plants were found susceptible to the infection caused by species of VAM fungi.

The maximum percentage of root colonisation was found in Okra (100%) and minimum in Brinjal (60%) (Table 2).

The root colonisation in vegetable plants were reported by Kavita *et al.* (2004) in chilli plant at nursery stage in the range of 52.83-96.48 per cent, Renu Sharma and P.C. Trivedi (1994) reported root colonisation in Brinjal in the range of 40-78.25 with different treatments of *Glomus fasciculatum* and *Glomus Mosseae*.

Table-2: Root colonisation in different vegetable plants

Name of the plant	Percent root colonisation
Abelmoschus esculentus L.	100
Cucumis Sativas L.	82
Momordica charantia L.	88
Lagenaria siceraria (Molina) standl	85
Capsicum annum L.	70
Lycopersicum esculentum L.	75
Solanum melongena L.	60

The root colonisation in plant helps to increase if plant growth and yield. Mohandas (1987) reported the growth response in tomato plant due to colonisation by *Glomus fasciculatum*. The increase in growth due to VAM clonisation was reported by Indi *et al.* (1990) in Brinjal and Konde and Patil (1990) in chilli.

The population of different VAM species were recorded in the rhizosphere soil of different vegetables (Table 3). The species of *Glomus* were recorded in the soil of Okra, bitter gourd, tomato brinjal and bitter gourd. The species of *Acaulospora*, found in Okra and chilli.

Table-3: Population of VAM fungi in the rhizosphere soil of vegetables

VAM fungi	I	II	III	IV	V	VI	VII
Glomus albidus walker and rhodes	+	–	+	–	–	+	+
Glomus fasciculatum Gerd. and Trappe	+	–	+	–	–	–	+
Glomus tortuosum Schenck and Smith	–	+	–	–	–	+	–
Glomus hoi Berch and Trappe	–	+	–	–	–	–	+

Glomus intraradices Schenk and Smith	+	–	–	–	+	–	–
Acaulospora mellea Schenck	+	–	+	–	–	–	–
Acaulospora rehmii Sieverd and Tora.	–	+	–	–	+	–	–
Acaulospora reticulata Rothell and Trappe	+	–	–	+	–	–	+
Gigaspora gigantea Nicol and Gerd.	–	–	+	–	+	–	–

I- *Abelmoschus esculentus,* II- *Cucumis sativus,* III- *Momordica charantia,* IV- *Lagenaria siceraria,* V- *Capsicum annum,* VI- *Lycopersicum esculentus,* VII- *Solanum melongena.*

Gigaspora gigantea was found in bottle ground and chilli. The maximum population of species of Glomus was found in the rhizosphere soil.

References

Bagyaraj, D.J., Leigh, H. and Wamer, M.J. 1989. (eds.) *Mycorrhiza in Tropical Rainforest Ecosystems.* Elsevier Sciences Publishers, Amsterdam. **537-546.**

Barea, J.M., Azon, R. and Azon-Aguilar, C. 1993. Mycorrhiza and crops, In: *Mycorrhizae Synthesis* (eds. Ingram, D.S. and Williams, P.H.) Academic Press, London, pp. **167-189.**

Gerdmann, J.W. and T.H. Nicolson 1963. Spores of mycorrhizal endogone species extracted from soil by wet sieving and decanting. *Trans. Br. Mycol. Soc.* **46: 235-244.**

Giovannetti, M. and Mosse, B. 1980. An evaluation of technique of measuring vesicular arbuscular mycorrhizal infection in roots. *New Phytol.* **84: 489-500.**

Indi, D.V., B.K. Konde and K.R. Sonar 1990. Response of brinjal genotypes in terms of dry weight and phosphorus uptake influenced by VAM inoculation. In: *Proc. First Nat. Conf. Mycorrhiza* held at Hissar. Feb. 14-16, pp. **116-117.**

Konde, B.K. and B.S. Patil 1990. N-P Uptake by chilli genotypes as influenced by VA-mycorrhizal inoculation, *Biol. Indica.* (1) **1: 43-49.**

Mohandas, S. 1987. Field response of tomato (*Lycopersicum esculentum* mill), pusa ruby to inoculation with a mycorrhizal fungus *Gloums fasciculatum* and with *Azotobactor vinelandi Pl. soil* **98(2): 295-297.**

Philips, J.M. and Hayman, D.S. 1970. Improved procedures for clearing root and staining parasitic roots and vesicular arbuscular mycorrhizal fungi for rapid assessment of infection. *Trans B. Mycol. Soc.* **55: 158-161.**

Read, D.S., Koucheki, H.R. and Hodgson, J. 1976. Vesicular arbuscular mycorrhiza in natural vegetation system. In the occurrence of infection. *New Pytol.* **77: 641-643.**

Trappe, J.M. 1981. Advances in food production system for arid and semi and lands (eds. Manassah, J.J. and Briskeg, E.J.), Academic Press, New York.

Trappe, J.M. 1982. Synoptic keys to the genera and species of zygomycetes mycorrhizal fungi. In: *Phytopathology*, 1982, **72(8): 1102-1108.**

Schenck, N.C. and Y. Perez 1987. *Manual for the identification of VA-mycorrhizal fungi*. pp. 191-**192.** Univ. of Florida, Gainevelle, Florida.

Kavitha, K., K.S. Meenakumari and P. Sivaprasad 2004. Standardisation of quantum and method of AMF inoculation for chilli nursery. In: *Indian J. of Microbiology,* Vol. 44, No. 2, June 2004, pp. **137-138.**

Sharma, R., P.C. Trivedi 1994. Interaction of root knot nematode *Meloidogynae incognita* and VA-mycorrhizae *Glomus fasciculatum* and *Gloums mosseae* on Brinjal (*Solanum melongena* L.) In: *J. Indian Bot. Soc., Vol.* 73 (1994) pp. **221-224.**

❑❑❑

28

Phytoextraction—A Green Technology for Metal Removal

A. Krishnaveni[1], A. Bharani[2] and R. Sudhagar[3]

1 and 2 Dept. of Environmental Science, Tamil Nadu Agricultural University, Coimbatore

3 Sugarcane Research Station, Melalathur, Vellore Dt.

Introduction

Heavy metals are ubiquitous environmental contaminants in an industrial society. Concerns over the possible health and ecosystem effects of heavy metals in soil have increased in recent years. Soil pollution by metals is essentially different persistence of metals in soil is reportedly much longer than in other compartments of biosphere. Once deposited on to the soil, certain metals such as Pb and Cr may be virtually permanent (Kabata Pendias and Adriano, 1995).

Phytoremediation of heavy metals from the environment serves as an excellent example of the process of plant-facilitated bioremediation and its role in removing environmental stress. Traditionally, when as area becomes contaminated with heavy metals the area must be excavated and the soil removed to a landfill site. (Lasat, 2000). This process is extremely expensive and therefore not entirely appealing in light of the recent discoveries regarding phytoremediation (Lasat, 2000). Analysts estimate that the cost of cleaning one hectare of highly contaminated land at a depth of one meter would range from $600,000 to $3,000,000 depending on the extent of the pollution and the toxicity of the pollutant. The cost of phytoremediation could be as much as 20 times less expensive, making this practice far less prohibitive than previous methods (Lasat, 2000). The ideal type of phytoremediator is a species that creates a large biomass, grows quickly, has an extensive root system, and must be easily cultivated and harvested (Clemens *et al.*, 2002). The only problem with this criterion is that natural phytoremediator often lack these qualities. Therefore, scientists have been forced to become very creative in the development of effective transgenic phytoremediator.

Toxic Heavy Metals

The focus of researchers on the phytoremediation of toxic metals in the environment has been considered an area of major scientific and technological progress and is a subject of contemporary relevance (Table 1). The widespread accumulation of heavy metals in soils is becoming a serious problem, as a consequence of industrial activity worldwide. Mining wastes, fertilisers, paper mills and toxic elements from atmospheric emissions, have all contributed to the continuous deposition and resulting accumulation of toxic metals in the environment. The concentrations of toxic metals in polluted soils are often hundreds of times greater than that required to exert a toxic effect on the majority of higher plants. Toxic metals can affect the biosphere for long periods of time and can be leached through the soil layers leading to the contamination of the water table. Consequently, the use of plants contaminated with high levels of heavy metals for food, might pose a serious risk to human and animals health (Wang *et al.*, 2003).

Table-1: Social and economic dimensions of toxic trace elements, metalloids and radionuclides and areas of scientific investigation (Prasad 2004a)

Area	Scope for detailed investigation
Ecotoxicology and ecophysiology	Nutrient availability efficiency and deficiency for optimisation of crop yield
Environmental chemistry waste management and mine reclamation	Bioremediation and restoration of metal contaminated and polluted ecosystems
Agricultural and nutritional sciences food industry	Plant and soil serve as vital links for supplementing nutrients on sustainable basis
Clinical biochemistry	Usage of radiolabelled antibodies and their disposal
Medicine and pharmacology	Use of plants as a source of certain disorders nutrients for
The atomic energy sector	The nuclear weapon testing (release of mainly ^{14}C, ^{137}Cs, ^{90}Sr and ^{95}Zr) production (release of mainly ^{137}Cs, ^{106}Ru, ^{95}Zr) and nuclear power production

Current Phytoremediation

Phytoremediation is the process that introduces plants into an environment and allows them to assimilate the contaminants into their roots and leaves. Such a process has been used to clean up heavy metal, pesticides and xenobiotics (Suresh and Ravishankar, 2004), organic compounds (Newman and Reynolds, 2004), toxic aromatic pollutants (Singh and Jain, 2003) and acid mine drainage (Archer and Caldwell, 2004). Interestingly, phytoremediation was recognised and documented by humans more than 300 years ago, however the scientific study and development of suitable plants was not conducted until the early 1980s (Lasat, 2000).

Phytoremediation is considered to be an environmentally friendly technology, that is a safe and also a cheap way to remove contaminants, in some cases doing the same job as a group of engineers for one-tenth of the cost. However, such technology cannot necessarily be effective all of the time or be used in all types of contaminated sites. If the contamination runs too deep, or the concentration of toxic compounds is too high, then plants alone cannot efficiently remediate the soil (Cunningham *et al.*, 1995).

Phytoremediation of Toxic Metals

Phytoremediation is characterised by the use of vegetative species for *in situ* treatment of land areas polluted by a variety of hazardous substances (Sykes *et al.*, 1999). Plants are especially useful in the process of bioremediation because they prevent erosion and leaching which can spread the toxic substances to surrounding areas (United States Environmental Protection Agency, 2001). There are several types of phytoremediation being used today. These include phytoextraction, which relies upon a plant's natural ability to take up certain substances (such as heavy metals) from the environment and sequester them in their cells until the plant can be harvested, phytodegradation, a means by which plants convert organic pollutants into a non-toxic form, phytostabilisation, where a plant releases certain chemicals that bind to the contaminant to make it less bioavailable and less mobile in the surrounding environment, and phytovolitisation, a process through which plants extract pollutants from the soil and then convert them into a gas that can be safely released into the atmosphere (Bentjen, 2002).

It is also known that some metal contaminated soils are difficult to remediate and such soils are usually excavated and land filled. Sites can also be treated by acid leaching, physical separation of the contaminant or electrochemical processes (Cunningham *et al.;* 1995). Moreover, the costs associated with soil remediation are variable and depend on the properties of the soil, the contaminants, site conditions and the volume of material to be remediated (Cunningham *et al.,* 1995). Considering these limitations, phytoremediation is one innovative approach that offers more environmental benefits and a cost effective alternative.

In phytoextraction and phytomining, accumulated toxic metals in plant tissues are harvested for metal recovery and reuse. Normally, the plants termed hyperaccumulators are preferably used, since they have the ability to withstand and build up high concentrations of metals, when compared to other plants. These plants can be processed to recover the metals accumulated during the phytoremediation process (figure 1). Although it is cheaper than the conventional methods, phytoremediation is not an easy technology that consists of simply planting and growing some hyperaccumulating plants in the metal polluted area (Alkorta *et al.,* 2004). It is in fact a highly technical strategy, requiring expert project designers with field experience that choose the proper species and cultivars for particular metals and regions. Research carried out with hyperaccumulator plant species has mainly focused attention on the physiological mechanisms by which the metal is taken up, transported and sequestrated, but little is known regarding the genetic basis of hyperaccumulation when compared with the genetic basis of metal tolerance. There is evidence for a quantitative genetic variation controlling the ability of these plants to hyperaccumulate between and within populations.

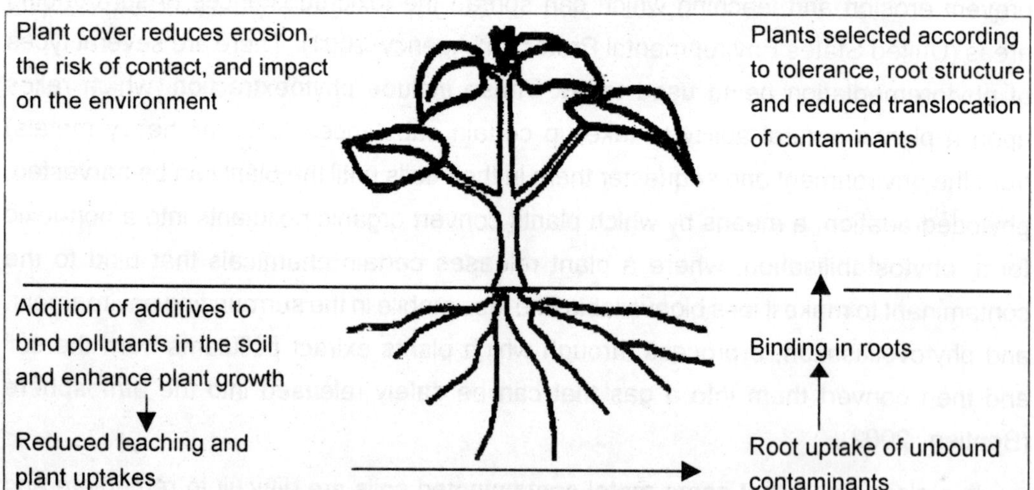

Fig. 1: Phytoremediation process

Transgenic Plants VS Natural Hyperaccumulators

Some plants and tree species are termed as "hyperaccumulators". Plants will extract and store extremely high concentrations (in excess of 100 times greater than non-accumulator species) of metallic elements (Lasat, 2000). Research has shown that these hyperaccumulators often do not exclude non-essential metals in the absorption process, thus resulting in plants that can extract high levels (1-2% of their biomass) of pollutants from contaminated soil (Lasat, 2000). It is believed that plants initially developed this ability to hyperaccumulate non-essential metallic compounds as a means of protecting themselves from herbivorous predators, who would experience serious toxic side effects from ingestion of the hyperaccumulator's foliage (Pollard *et al.,* 1997). Lead is one of the non-essential compounds hyperaccumulated by several species of plants including *Thlaspi rotundifolium* and *Brassica juncea*. The difficulty with these models, in terms of performing phytoremediation, is that they grow very slowly and have a very low biomass. Therefore, the bioremediation process would be extremely slow because the rate of bioremediation is directly proportional to growth rate and total amount of bioremediation is correlated with a plant's total biomass. Because no plant yet discovered meets the ideal criteria of an effective phytoremediator (fast growing, deep and extensive roots, high biomass, easy to harvest, hyperaccumulators of a wide range of toxic metals, it is necessary to introduce the hyperaccumulating genes into non-accumulator to make the plants better suited as agents in the phytoremediation process (Clemens *et al.,* 2002). The authors of a lead absorption study, Huang and Cunningham, cite corn as a perfect phytoremediator due to its large biomass, fast rate of growth, and the existence of extensive genomic knowledge of this crop. The presence of a body of knowledge regarding corn genetics would be helpful in the effective insertion of DNA that codes for hyperaccumulation of desired contaminant. Other potential model phytoremediators would include various varieties of transgenic trees. Trees are ideal in the remediation of heavy metals because the can withstand higher concentrations of pollutants due to their large biomass, they can accumulate large amounts of the contaminants in their systems because of their size, they can reach a huge area and great depths due to their extensive root systems, they can stablise an area and prevent erosion, and hence the spread of the contaminant, because of their perennial presence. They can also be easily harvested and removed from the area with minimal risk, effectively taking with them a large quantity of the pollutants that were once present in the soil. The introduction of hyperaccumulating genes – as well as genetic information that would better prepare these species to deal with diverse climatic conditions, up-regulate this hyperaccumulation, or increase the organism's growth rate and overall size – into these model agricultural species would undoubtedly accelerate the remediation process and result in faster removal of environmental stress from the ecosystem. Such metal accumulation and tolerance could be enhanced by over expressing natural or modified genes encoding antioxidant enzymes or those that are involved in the biosynthesis of glutathione and phytochelatins (figure 2).

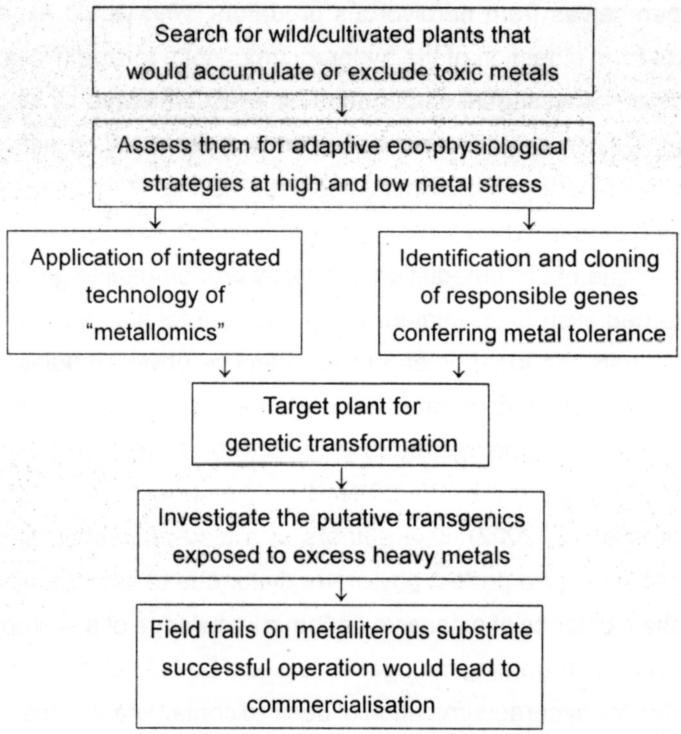

Fig. 2: The use of genetic engineering to breed plants with superior phytoremediation potential (Modified after Karenlampi *et al.*, 2000; Prasad, 2004b)

It is apparent that in order to breed plants with superior phytoremediation potential, it is necessary that these plants have a high rate of biomass production and are sufficiently strong and competitive in the climate they are to be used for phytoremediation (Pilon-Smits and Pilon, 2002). Moreover, plants suitable for phytoremediation should possess the ability to accumulate the target metal in the above-ground parts, and to tolerate the metal concentration that is to be accumulated (Karenlampi *et al.*, 2000). The over expression of a gene encoding a rate limiting gene product would be expected to lead to a faster overall rate of the pathway and to more efficient phytoremediation (Pilon-Smits and Pilon, 2002). However, it must be considered that metals rarely occur alone in the environment and an adaptive tolerance may be essential for several metals simultaneously (Karenlampi *et al.*, 2000).

Transgenic Plants for Toxic Heavy Metals

S-metabolism, glutathione and phytochelatins

Selenium (Se) is essential as a micronutrient for humans and animals but in toxic at medium to high concentration. Se occurs naturally in soils as selenate and selenite and often as a pollutant, following the industrial use of coal. Se and sulphur (S) have very similar chemical properties and their uptake and assimilation as selenate and sulphate, to incorporation in proteins as selenomethionine and methionine, proceed through common pathways and are catalysed by the same enzymes. A research group lead by Norman Terry and Elizabeth Pilon-Smits has considered the possibility of over expressing a range of genes encoding key enzymes in sulphur metabolism, with the aim of obtaining selenium tolerant plants.

Indian mustard (*Brassica juncea*) plants overxpressing ATP sulphurylase were shown to have higher shoot Se concentrations and enhanced Se tolerance compared to wild type when grown in the presence of selenate in either hydroponic systems or soil (Van Huysen *et al.*, 2004). Interestingly, young transgenic plants that were over expressing ATP sulphurylase, were more tolerant than the wild type to As (III), As (V), Cd, Cu, Hg, and Zn, but less tolerant to Mo and V (Wangeline *et al.*, 2004). Other transgenic Indian mustard plants over expressing cystathionine-□-synthase (CGS) showed a higher Se volatilisation rate, lower shoot Se levels, and higher Se tolerance than wild type grown in either soil or hydroponically (Van Huysen *et al.*, 2004).

Astragalus bisulcatus is a native plant that has the capacity to grow on Se containing soils and accumulate Se to high concentration but it has a slow growth rate. It has been proposed that in *A. bisulcatus* selenocysteine methyltransferase (SMT) specifically methylates selenocysteine (SeCys) to produce the nonprotein amino acid methyl-selenocysteine MetSeCys, which causes a reduction in the intracellular concentrations of SeCys and selenomethionine (SeMet), thus preventing their incorrect insertion into protein. Indian mustard plants over expressing the *A. bisulcatus* SMT gene, exhibited a greatly increased accumulation of MetSeCys and tolerance to Se compounds, in particular selenite (LeDue *et al.,* 2004).

Cd is a toxic element that normally occurs in low concentrations in soils, however the concentration can be significantly increased by activities such as zinc mining, iron foundries and the use of sewage sludge as fertiliser in agriculture (Zhao *et al.*, 2003). Cd may be detoxified in plants by a family of sulphur rich peptides termed phytochelatins (PCs) that are able to bind Cd and some other heavy metals (Cobbett and Goldsbrough, 2002). The peptides are structurally related to glutathione and contain a varying number (normally 2-5) of glutamate and cysteine, linked through the □-carboxyl group of glutamate. Phytochelatin synthase (PCS) has been characterised as a specific □-glutamyl cysteine dipeptidyl transpeptidase (EC 2.3.2.15) (Vatamaniuk *et al.,* 2004), which carries out the conversion of glutathione to PCs, and has been shown to be activated by Cd.

Various attempts have been made to increase the formation of PCs, by over expressing genes encoding enzymes that could stimulate the synthesis of cysteine and glutathione. Cysteine synthase catalyses the last step in the assimilation of sulphate into the amino acid. Transgenic tobacco plants over-expressing cysteine synthase in the cytosol, had elevated concentrations of PCs, were more tolerant to Cd, but did not accumulate the metal in the leaves (Harada *et al.,* 2001). In contrast, Dominguez-Solis *et al.* (2004) over expressed cysteine synthase in the cytosol of *A. thaliana.* One transgenic line was shown to be particularly resistant to Cd and to accumulate high concentrations in the leaves mainly in the trichomes. Transgenic tobacco plants over expressing cysteine synthase in either the cytosol or chloroplasts were more tolerant to metals such as Cd, Se and Ni. F_1 plants with expression on both the cytosol and chloroplast exhibited a higher tolerance than the other transgenic lines and accumulated Cd in the shoots (Kawashima *et al.,* 2004).

Sudhagar (2003) investigated the Cd tolerance limit for maize crop. The cadmium phytoremediation capacity of maize inbreds and the genetics of cadmium tolerance were investigated. Fifty-one maize genotypes with more divergence were utilised for the study. The cadmium critical concentration was fixed by studying the shoot and root development pattern of four randomly selected experimental genotypes. The 50 per cent reduction in shoot and root length were considered and critical concentration was fixed through SPSS analysis. All the genotypes were screened for cadmium tolerance at this critical concentration. Germination parameters, morphology and growth and varied biochemical parameters were utilised for isolating cadmium tolerant genotypes. The genotypes were scrutinised for the activity of H^+ATPase, nitrate reductase, superoxide dismutase, catalase, ascorbate peroxidase, monodehydroascorbate reductase, dehydroascorbate reductase, glutathione reductase and phytochelaythin sulfhydrill. Reduction in growth and biochemical parameters were observed invariably across the test genotypes. The inbreds with less and no reduction in morphology, growth traits and heavy metal induced enzymes were treated as tolerant genotypes. The genotypes with sharp decline in growth and related biochemical parameters under cadmium-imposed condition were considered as cadmium tolerant. The selected cadmium susceptible and tolerant inbreds were further test verified by isoenzyme and protein analysis. In the tolerant genotypes one additional catalase, peroxidase and poly phenol oxidase isoenzyme loci were observed. The tolerant genotypes manifested enhanced expression of 52 KDa protein that was absent or less expressed in susceptible genotypes. The tolerant and susceptible genotypes were crossed in line and tester mating design to obtain the genetic information of cadmium tolerance and other cadmium induced biometrical traits.

Indian mustard over expressing the *E. coli gshII* gene encoding □-glutamylcysteine synthetase (□-ECS) exhibited increased tolerance to Cd and had higher concentrations of PCs and □-GluCys, compared to the wild-type. When tested in a hydroponic system, □-ECS plants grew better and accumulated Cd in the shoots (Zhu *et al.,* 1999b). Transgenic Indian mustard containing the *E. coli gshII* gene encoding glutathione synthetase (GS) accumulated significantly more Cd than the wild type in the shoot and the plants showed enhanced tolerance to Cd at both the seeding and mature-plant stages (Zhu *et al.*, 1999a). Zhu *et al.* (1999b) concluded from the results obtained with the two type of transgenic plants, that under normal conditions that □-ECS limits the rate of glutathione and PC synthesis, but that in the presence of Cd □-ECS is activated and that GS becomes rate limiting. When grown in contaminated soil, the thiol-overproducing transgenic □-ECS and GS Indian mustard plants showed enhanced phytoextraction capacity, yet still produced the same shoot biomass as the wild type. As a result, the total shoot metal accumulation of the □-ECS and GS transgenics were

1.5-fold higher for Cd, and 1.5- to 2-fold higher for Zn. Furthermore, the □-ECS transgenics accumulated 2.4 to 3-fold more Cr, Cu, and Pb, relative to the wild type (Bennett *et al.,* 2003).

Following the positive results obtained with over expressing the enzymes involved in cysteine and glutathione synthesis, it was expected that over expressing PCS would be even more successful. However this did not prove to be the case for Cd toxicity. When PCS was over expressed in *A. thaliana*, the transgenic plants were shown to be hypersensitive to Cd and Zn but not to Cu (Lee *et al.,* 2003b). In a similar series of experiments, over expressing PCS in *A. thaliana*, Li *et al.* (2004) demonstrated that the transgenic plants were highly resistant to arsenic (As), accumulating 20-100 times more biomass on 250 and 300 μM arsenate, than the wild type. However, again the PCS plants were hypersensitive to Cd, despite the fact that they contained 2-6 fold higher concentrations of PCs. However, transgenic tobacco *Nicotiana glauca* expressing wheat PCS showed a slight increase in tolerance to Cd (Gisbert *et al.,* 2003). These somewhat surprising results were discussed in detail by Li *et al.* (2004) and they concluded that "much is still be learned about the processing of cadmium in Cd-PC petide complexes."

Metallothioneins

Metallothioneins (MTs) are low molecular mass cysteine rich proteins that were originally isolated as Cu, Cd and Zn biding proteins in mammals. There is now good evidence that four categories of these proteins occur in plants, which are encoded by at least seven genes in *A. thaliana* (Cobbett and Goldsbrough, 2002). A wide range of MT genes from various sources have been over expressed in plants, these include human, mouse, Chinse hamster and yeast. There was some variation in he range of Cd tolerance obtained. Metal uptake and accumulation in the shoot was not markedly altered, ranging from 70 per cent less to 30 per cent more than the wild type controls.

When a pea (*Pisum sativum*) MT gene *PsMTA* was expressed in *A. thaliana*, more Cu (several-fold in some plants) accumulated in the roots of transformed than of control plants (Evans *et al.,* 1992). Similarly when a type 2 MT gene, *tyMT*, cloned from Cattail (*Typha latifolia*), a wetland plant with constitutional tolerance, was introduced into *A. thaliana,* the transgenic plant showed an increased tolerance to both Cu^{2+} Cd^{2+}. The *A. thaliana* metallothionein proteins AtMT2a and AtMT3 were introduced as fluorescent protein-fused forms into the guard cells of *Vicia faba*. The MTs protected guard cell chloroplasts from degradation upon exposure to Cd, by reducing presence of reactive oxygen species. It was concluded that the Cd stays bound to the MT in the cytoplasm and is not sequestered into the vacuole, as occurs when Cd is detoxified by PCs (Lee *at al.*, 2004).

Transporters

There has been considerable interest in the possibility of manipulating transporters within plants in order to achieve different objectives, these include exclusion of a toxic metal ion, transporting the metal into the apoplastic space and transporting the metal into the vacuole where it would be less likely to exert a toxic effect (Tong *et al.*, 2004). In this article we will only cover a little of the extensive work that has been carried out.

Lee *et al.* (2003a) transformed *A. thaliana* plants with the *E. coli* gene, *ZntA*, which encodes a Pb(II)/Cd(II)/Zn(II) transporter. The *ZntA*-transgenic plants grew better than the wild-type plant in Pb, Cd and Zn-containing medium. The shoots of the transgenic plants exhibited decreased Pb and Cd contents, whereas transgenic protoplasts showed lower accumulation of Cd and faster release of pre-loaded Cd than wild-type protoplasts. Lee *et al.* (2003a) proposed that the *ZntA*-transgenic plants excluded the metal ions at the cellular level by pumping them from the plasma membrane to the extracellular space.

The over expression of the *AtHMA4* gene, which encodes an *A. thaliana* P-1B-ATPase Zn and Cd transporter, improved the growth of both primary and secondary roots in the presence of toxic concentrations of Zn, Cd and Co. A determination of metal content demonstrated that the over expressing lines, when exposed to toxic concentrations of Zn or Cd, translocated these metals at a greater extent to the shoot, compared to the control plants, an important criteria for phytoremediation. In contrast, the metal level was found to be rather similar in roots, indicating that the metal uptake by the roots compensated for the increased metal translocation to the shoot (Verret *et al.*, 2004).

The vacuole is generally considered to be the main storage site for metals in yeast and plant cells and there is evidence that phytochelatin-metal complexes are pumped into the vacuole. The best characterised of the known vacuolar transports and channels involved in metal tolerance is *YCF1* from *Saccharomyces cerevisiae*. *YCF1* is a MgATP-

Table-2: Advantages and limitations of some of the sub-processes of phytoremediation (Prasad, 2004b)

Advantage	Limitation
Phytoextraction	
The plant must be able to produce abundant biomass in short time. Fig. in green house experiments gold was harvested from plants (Anderson *et al.*, 1998)	Metal hyperaccumulators are generally show-growing and bio-productivity is rather small and shallow root systems. Phytomass after process must be disposed properly
Phytostabilisation	
It circumvents the removal of soil, low cost and is less disruptive and enhances ecosystem restoration/re-vegetation.	Often requires extensive fertilisation or soil modification using amendment, long-term maintenance is needed to prevent leaching.
Phytovolatilisation	
Contaminant/ Pollutant will be transformed in to less-toxic forms. e.g. elemental mercury and dimethyl selenite gas. Atmospheric process such as photochemical degradiation for rapid decontamination/transformation.	The contaminant of a hazardous metabolite might accumulate in vegetation and be passed on in later products such as friut or lumber. Low levels of metabolites have been found in plant tissue.
Phytofiltration/rhizofiltration	
It can be either in situ (floating rafts on ponds) or *ex situ* enginered tank systems); terrestrial or aquatic.	pH of the medium to be monitored contually for optimising uptake of metals; chemical speciation and interactions of all species in the influent need be understood; functions like a bioreactor and intensive maintenance is needed

energised glutathione S-conjugate transporter responsible for vacuolar sequestration of compounds after their S-conjugation with glutathione. Song *et al.* (2003) overexpressed the *YCF1* gene in *A. thaliana* and the *YCF1* proteins were found to be associated with the tonoplast and the plasma membrane. The vacuoles of the *YCF1*-transgenic plants exhibited a 4-fold higher rate of glutathione-Cd uptake than those of wild-type plants, indicating that expression of *YCF1* strongly increases Cd transport activity. The transgenic plants showed improved resistance to both Cd and Pb and elevated metal content, characteristics desirable for phytoremediation.

Other transporter proteins that could be of value include—the *A. thaliana* antiporter CAX2 (Hirshci *et al.*, 2000), LCTI, a nonspecific transporter for Ca^{2+}, Cd^{2+}, Na^{+} and K^{+} (Antosiewicz and Hennig, 2004), the *Thlaspi caerulesoens* heavy metal ATPase, TcHMA4 (Papoyan and Kochian, 2004), a novel family of cysteine rich membrane proteins that mediate Cd resistance in *A. thaliana* (Song *et al.*, 2004) and AtMRP3, an ABC transporter (Bovet *et al.*, 2005).

It can be seen from the above account of work that has been already carried out, that there are a large number of possibilities to improve the ability of plants to withstand the stress of high concentrations of toxic metals and also to accumulate them. Additional work not mentioned includes over expressing genes encoding—ACC deaminase (Grichko *et al.*, 2000), mercuric ion reductase, arsenate reductase aldose/aldehyde reductase (Hegedüs *et al.*, 2004) and enzymes of histidine biosynthesis (Wycisk *et al.*, 2004).

Clearly there may be some opportunity to combine some of these genes in fast growing species, but more importantly the plants need to be tested under carefully regulated environmental conditions.

Natural Hyperaccumulators

Natural metal hyperaccumulators plants can accumulate and tolerate greater metal concentrations in shoots than those usually found in non-accumulators, without visible symptoms. According to Baker and Brooks (1989), the minimum threshold tissue concentrations for Co, Cu, Cr, Pb or Ni hyperaccumulators should be 0.1 per cent dry weight, while for Zn or Mn the threshold is 1 per cent. Over 400 hyperaccumulator plants have been reported and include members of the Asteraceae, Brassicaceae,

Caryophyllaceae, Cyperaceae, Fabaceae, Flacourtiaceae, Lamiaceae, Poaceae, Violaceae, and Euphorbiaceae. The Brassicaceae is a very important group when heavy metal accumulation is concerned, with several being able to hyperaccumulate more than one metal (Prasad and Freitas, 2003).

Pistia stratiotes was examined in the presence of several distinct heavy metals separately, in order to determine the ability of the plant to tolerate and accumulate the metals tested and for eventual use for phytoremediation of waste water or natural water bodies polluted with these heavy metals (Odjegba and Fasidi, 2004). *P. stratiotes* exhibited different patterns of response to Ag, Cd, Cr. Cu, Hg, Ni, Pb and Zn, and although concentrations as high as 5 mM resulted in distinct levels of growth inhibition and biomass production, all the elements accumulated at high concentrations mainly in the root system. Furthermore, this plant species exhibited the highest tolerance index to Zn and the lowest to Hg (Odjegba and Fasidi, 2004).

Spartina plants have been shown to be 3-fold more tolerant to Hg than tobacco plants, due to an ability to absorb organic Hg and transform it into an inorganic form (Hg^+, Hg^{2+}). The inorganic Hg then accumulates in the underground part of the plants and is transferred back to the soil by diffusion and permeation, indicating that this species may be used in the phytoremediation of Hg polluted environment (Tian *et al.*, 2004). According to Bennicelli *et al.* (2004), the water hyperaccumulator fern *Azolla caroliniana* Willd. (Azollaceae) has the capacity to purify waters polluted by Hg and Cr by accumulation of these heavy metals in its tissues.

Helianthus annuus has been shown to concentrate Pb in the leaf and stem indicating that it has the prerequisites of a hyperaccumulator plant that could be used in the restoration of abandoned mines and factories sites contaminated with elevated Pb levels in the soil (Boonyapookana *et al.*, 2005). In an almost similar manner, *Hemidesmus indicus* has also been shown to be a Pb hyperaccumulating plat species, but the heavy metal was mainly accumulated in roots and shoots (Chandra *et al.*, 2005), exhibiting a high potential for use in the phytoremediation of industrial areas contaminated with this metal. The hyperaccumulator *Sesbania drummondii* is a plant species that has been shown to predominantly accumulate Pb as lead acetate in roots and leaves, although lead sulphate and sulphide have also been detected in leaves, whereas lead sulphide was detected in root samples (Sharma *et al.*, 2004). These results have indicated that *S. drummondii* is able to biotransform lead nitrate in the nutrient solution to lead acetate and sulphate in its tissues and the complexation with acetate and sulphate may be a lead detoxification strategy in this plant species (Sharma *et al.,* 2004).

Studies related to As accumulation in *Lemna gibba* have demonstrated that this plant species can be a preliminary bioindicator for As transfer from substrate to plants, indicating its use to monitor the transfer of arsenic from lower to higher trophic levels in abandoned mine sites. Moreover, *L. gibba* can be used for As phytoremediation of mine tailing waters because of its high accumulation capacity (Mkadawire and Dudel, 2005). *Pteris vittata* can hyperaccumulate As from naturally contaminated soils, but may be suitable for phytoremediation only in the moderately contaminated soil (Caille *et al.,* 2004). In addition to *P. vittata, P. cretrica, P. longifolia* and *P. umbrosa* are also able to hyperaccumulate As to a similar extent (Zhao *et al.,* 2002).

Broadhurst *et al.* (2004) have developed commercially viable phytoremediation/ phytomining technologies employing *Alyssum* Ni-hyperaccumulator species, where the majority of Ni is stored either in the leaf epidermal cell vacuoles, or in the basal portions of the numerous stellate trichomes. The metal concentration in the trichome basal compartment was the highest ever reported for healthy vascular plant tissue, approximately 15-20 per cent dry weight (Broadhurst *et al.,* 2004).

Solanum nigrum and *Conyza canadensis* have not only been shown to accumulate high concentration of Cd, but also to be tolerant to the combined action of Cd, Pb, Cu and Zn. The hyperaccumulator *Thlaspi caerulescens* J. and C. Presl. has a high Cd-accumulating capability, acquiring Cd from the same soil pools as non-accumulating species, by every efficient mechanisms (Schwartz *et al.,* 2003). Thus, *T. caerulescens* may be used as a tool to efficiently reduce the availability of Cd in soils, providing appropriate populations are used (Schwartz *et al.,* 2003). Moreover, *Arabis gemmifera* is a hyperaccumulator of Cd and Zn, with phytoextraction capacities almost equal to *Thlaspi caerulescens* (Kubota and Takenaka, 2003) and plants of the mined ecotype of *Sedum alfredii* have a greater ability to tolerate, transport, and accumulate Cd, when compared to the non-mined ecotype.

Understanding the mechanisms of rhizosphere interaction, uptake, transport and sequestration of metals in hyperaccumulator plants will lead to designing novel transgenic plants with improved remediation traits (Eapen and Souza, 2005). Moreover, the selection and testing of multiple hyperaccumulator plants could enhance the rate of phytoremediation, making this process a successful one for bioremediation of environmental contamination (Suresh and Ravisankar, 2004).

Advantages and Limitations of Phytoremediation

Several processes can limit the performance of plants in phytoremediation, such as the availability of the toxic metal ions in the soil for uptake by plant roots, the rate of uptake of the contaminants by plant roots, their translocation from roots to shoots and the extent of tolerance, or the rate of chemical transformation into less toxic, possibly volatile compounds (Prasad, 2003) (see Table 2).

Although the use of transgenic plants as phyroredïators is showing increasing potential for soils contaminated with toxic metals, possible risks should be considered, including the uncontrolled spread of the transgenic plants due to interbreeding with populations of wild relatives. Moreover, the transformation of metals into forms more bioavailable could increase the exposure of wildlife and humans to metals. However, the risk of metal accumulation in plant shoots that can be ingested by the wildlife, can be minimised by the reduction of the growth period and thus exposure of the transgenic plants. Furthermore, it is important that in the case of harvesting crop plants normally used to feed livestock and humans, the translocation of the toxic elements to the seeds is avoided at all costs.

Full utilisation of plant resources after they have been used for phytoremediation is an unsolved problem. Therefore the testing of the plant used in phytoremediation is necessary and may support their continuous use in contaminated soil. For instance, *Elshotzia argyi* flowers are used as fragrances and antiseptics due to the perfume ingredient and antibacterial components existing in their essential oils, and the analysis of these plants in Pb/Zn mined area, where they normally occur revealed that they can be safely exploited (Peng and Yang, 2005).

While phyotemediation processes hold a great promise as a way to remediate contaminated soils, there are disadvantages and limitations that must be carefully considered

Advantages of Phytoextraction

- Soil stabilisation.
- Aesthetically pleasing.

- Cost effective remedial alternative.
- Applicability to wide range of contaminants of varying concentration. These are subsequently rendered harmless or removed, with relatively low amounts of wastes being generated. Where metals are concerned, the contaminants can be recycled from the residues.
- Minimal environmental disturbance.
- Elimination of secondary air or water borne wastes.

Limitations to Phytoextraction

- Limited regulatory acceptance as a viable treatment alternative.
- Often long duration before there is clean up to acceptable threshold levels.
- Potential contamination of the food chain of organisms graze on the vegetation.
- Difficulty in establishing and maintaining vegetation in media containing toxic wastes.
- Potential for off-site migration and leaf transportation of metals to surface when using tree species.
- To minimise dispersal of seeds or pollen ensuring that metals are not mobilised down the soil profile.
- Preventing herbivores accumulating dangerous amounts of elements.

To overcome the Limitations

- The plants need to be harvested before they flower *i.e.* during a period of maximum biomass and element yield.
- Growth of non-crop or non-palatable species which might reduce the chances of biomagnification.

Future Needs

- Appropriate agronomic practices such as planting density, animal management, weed control and harvest methods to increase the biomass of hyperaccumulators to enhance phytoextraction of metals.
- Use of appropriate soil amendments to increase the bioavailability of metals to hyperaccumulators.

- Use of growth modifying hormones such as auxins and cytokinin as other compounds of accumulator plant with the aim of maximising removal.
- Increased production of hyperaccumulators by tissue culture techniques.
- Use of plant symbionts like bacteria or fungi (VAM) to enhance accumulation.
- Identification of gene or genes responsible for hyperaccumulation and insert them into a high biomass plant. None of the genes involved has been indentified and cloned.
- To identify the interactions of nutrients which affect the uptake of metals in the shoot.
- Elaboration of methods for the extraction of metals in hyperaccumulators.

Conclusions

It is clear that phytoremediation is not type, but hope as there are many examples that had the potential for improving environmental quality. Plants are very flexible and react quickly to environmental stress, but from a plant physiology point of view, what is the meaning of phytoremediation? Many plants are around us, but who will screen them? Exploiting the natural biodiversity is thus and important issue with choice of appropriate species for phytoremediation. Why should we go transgenic plants when natural variability is not yet explored? The economic utilisation of crops becomes more important than their heavy metal uptake ability or the time demand of phytoremediation.

It is evident that phytoremediation has benefits to restore balance to a stressed environment, but it is important to proceed with caution. The study and use of genetic modifications must be performed in order to determine the true costs and benefits of this technology to the ecosystem as a whole, before it is to be applied to a larger scale. Progress in the field of molecular genetics, will allow the analysis of metal hyperaccumulator plants and should provide new insights into metabolic detoxification processes and identify tolerance genes, thus providing considerable more information about the genomes of these model organisms.

Advances in other methods involving analysis of "omics" technologies could further reveal the non-targeted identification of all gene products in a specific biological sample, which could be followed by a refined analysis of quantitative dynamics in biological systems. The genomics can accelerate the discovery of genes that confer key traits, allowing their modification. In addition, metabolomics can provide biochemical and physiological knowledge about network organisation in plants subject to toxic metal stress, providing a much more detail understanding of the molecular basis of hyperaccumulation.

Approaches allowing recombination hotspots to be highlighted will further aid plant breeding efforts. Many signalling pathways and proteins can contribute to the cellular stress response, and the identification of the key ones within the stress response network is essential. The development of DNA and RNA microarray chip technologies in systematic genome mapping, sequencing, functioning and experimentation may allow the identification and genotyping of mutations and polymorphisms, allowing better insight into structure-function interaction of genome complexity under toxic metal stress.

Molecular genetics approaches such as insertion mutagenesis involving populations of T-DNA, can be used to identify genes involved in hyper-accumulation or transposon tagged plants screened to identify mutants impaired in the ability to accumulate metals (Pollard *et al.,* 2002). Recently, considerable progress has been made in identifying plant genes encoding metal ion transporters with important functions in cation transport and homeostasis (Papoyan and Kochian, 2004; Weber *et al.,* 2004). Modern molecular techniques, bioinformatics and computational techniques are effective tools for detailed structure-function genome analysis. It is clear that both fundamental and applied research must be carried out in association.

Phytoremediation technology is still in its early development stage and full scale applications are still limited. The results already obtained have indicated that the plants are effective and could be used in toxic metal remediation. Although it appears to be common sense among scientists, engineers, and regulators about the more widespread future use of this technique, it is important that public awareness about this technology is considered and clear and precise information is made available to the general public to enhance its acceptability as a global sustainable technology to be widely used.

References

Alkorta, I., Hernández-Allica, J., Becerril, J.M., Amezaga, I., Albizu, I., Garbisu, C. 2004. Recent findings on the phytoremediation of soils contaminated with environmentally toxic heavy metals and metalloids such as zinc, cadmium, lead, and arsenic. *Rev. Environs. Sci. Biotechnol.* **3: 71-90.**

Antosiewicz, D.M., Hennig, J. 2004. Overexpression of LCT_1 in tobacco enhances the protective action of calcium against cadmium toxicity. *Environ. Pollut.* **129: 237-245.**

Archer, M.J.G., Caldwell, R.A. 2004. Response of six Australian plant species to heavy metal contamination at an abandoned mine site. *Water Air Soil Poll.* **157: 257-267.**

Baker, A.J.M., Brooks, R.R. 1989. Terrestrial higher plants which hyperaccumulate metallic elements—a review of their distribution, ecology and phytochemistry. *Biorecovery* **1: 81-126.**

Bennett, L.E., Burkhead, J.L., Hale, K.L., Terry, N., Pilon, M., Pilon-Smits, E.A.H. 2003. Analysis of transgenic Indian mustard plants for phytoremediation of metal-contaminated mine tailings. *J. Env. Qual.* **32: 432-440.**

Bennicelli, R., Stepniewska, Z., Banach, A., Szajocha, K., Ostrowski, J. 2004. The ability of *Azolla caroliniana* to remove heavy metals (Hg(II), Cr(III), Cr(VI) from municipal waste water. *Chemosphere* **55: 141-146.**

Bentjen, Steve. Bioremediation and phytoremediation glossary. Published 1998.

Boonyapookana, B., Parakplan, P., Techapinyawat, S., DeLanue, R.D., Jugsujinda, A. 2005. Phytoaccumulation of lead by sunflower (*Helianthus annuus*), tobacco (*Nicotiana tabacum*), and vetiver (*Vetiveria zizanioides*). *J. Environ. Sci. Heal.* A **40: 117-137.**

Bovet, L., Feller, U., Martionia, E. 2005. Possible involvement of plant ABC transporters in cadmium detoxification: A cDNA sub-microarray approach. *Environ Int.* **31: 263-267.**

Broadhurst, C.L., Chaney, R.L., Angle, J.S., Maugel, T.K., Erbe, E.F., Murphy, C.A. 2004. Simultaneous hyperaccumulation of nickel, manganese, and calcium in *Alyssum* leaf trichomes. *Environ. Sci. Technol.* **38: 5797-5802.**

Caille, N., Swanwick, S., Zhao, F.J., McGrath, S.P. 2004. Arsenic hyperaccumulation by *Pteris vittata* from arsenic contaminated soils and the effect of liming and phosphate fertilisation. *Environ. Pollut.* **132: 113-120.**

Chandra Sekhar, K., Kamala, C.T., Chary, N.S., Balaram, V., Garcia, G. 2005. Potential of *Hemidesmus indicus* for phytoextraction of lead from industrially contaminated soils. *Chemosphere* **58: 507-514.**

Clemens, Stephan, Palmgren, Micheal, G., and Kramer, Ute. 2002. A long way ahead: understanding and engineering plant metal accumulation. *TRENDS in Plant Science* 7 (&), **309-314.**

Cobbett, C., Goldsbrough, P. 2002. Phytochelantins and metallothioneins: Roles in heavy metal detoxification and homeostasis. *Ann. Rev. Plant Biol.* **53: 159-182.**

Cunningham, S.D., Berti, W.R., Huang, J.W. 1995. Phytoremediation of contaminated soils. *TIBTECH* **13: 393-397.**

Dominguez-Solis, J.R., Lopez-Martin, M.C., Ager, F.J., Ynsa, M.D., Romero, L.C., Gotor, C. 2004. Increased cysteine availability is essential for cadmium tolerance and accumulation in *Arabidopsis thaliana*. *Plant Biotechnol. J.* **2: 469-476.**

Eapen, S., D'Souza, S.F. 2005. Prospects of genetic engineering of plants for phytoremediation of toxic metals. *Biotech. Adv.* **23: 97-114.**

Evans, K.M., Gatehouse, J.A., Lindsay, W.P., Shi, J., Tommey, A.M., Robinson, N.J. 1992. Expression of the pea metallothionein-like gene *PsMTA* in *Escherichia coli* and *Arabidopsis thaliana* and analysis of trace metal ion accumulation: implications for *PsMTA* function. *Plant Mol. Biol* **20: 1019-1028.**

Gisbert, C., Ros, R., DeHaro, A., Walker, D.J., Bernal, M.P., Serrano, R., Navarro-Aviñó, J. 2003. A plant genetically modified that accumulates Pb is especially promising for phytoremediation. *Biochem. Biophys. Res. Commun.* **303: 440-445.**

Grichko, V.P., Filby, B., Glick, B.R. 2000. Increased ability of transgenic plants expressing the bacterial enzyme ACC deaminase to accumulate Cd, Co, Cu, NI, Pb, and Zn. *J. Biotechnol.* **81: 45-53.**

Harada, E., Choi, Y.E., Tsuchisaka, A., Obata, H., Sano, H. 2001. Transgenic tobacco plants expressing a rice cysteine synthase gene are tolerant to toxic levels of cadmium. *J. Plant Physiol.* **158: 655-661.**

Hegedüs, A., Erdei, S., Janda, T., Tóth, E., Horváth, G., Dudits, D. 2004. Transgenic tobacco plants overproducing alfalfa aldose/aldehyde reductase show higher tolerance to low temperature and cadmium stress. *Plant Sci.* **166: 1329-1333.**

Hirschi, E.D., Korenkey, V.D., Wilganewski, N.I., Wagner, G.I. 2000. Expression of *Arabidopsis* CAX^2 in tobacco. Altered metal accumulation and increased manganese tolerance. *Plant Physiol.* **124: 128-133.**

Jansen, S., Broadly, M.R., Robbrecht, E., Smets, E. 2002. Aluminum hyperaccumulation in angiosperms: A review of its phylogenetic significance. *Bot. Rev.* **68: 235-269.**

Kabata Pendias, A. and Adriano, D.D. 1995. *Tace metals in soil amendments and environmental quality*. pp. **139-167** (eds. JE. Recheigh Lewis Publishers, New York).

Karenlampi, S., Schat, H., Vangronsveld, J., Verkleij, J.A.C., van der Lelie, D., Mergeay, M., Tervahatua, A.I. 2000. Genetic engineering in the improvement of plants for phytoremediation of metal polluted soils. *Environ. Pollut.* **107: 225-231.**

Kawashima, C.G., Noji, M., Nakamura, M., Ogra, Y., Suzuki, K.T., Saito, K. 2004. Heavy metal tolerance of transgenic tobacco plants over-expressing cysteine synthase. *Biotechnol. Lett.* **26: 153-157.**

Kubota, H., Takenaka, C. 2003. *Arabis gemmifera* is a hyperaccumulator of Cd and Zn. *Int. J. Phytoremediation* **5: 197-120.**

Lasat, M.M. 2000. Phytoextraction of metals from contaminated sites—A critical review of plant/ soil/metal interaction and assessment of pertinent agronomic issues. *J. Hazard Substance Res.* **2: 1-25.**

LeDuc, D.L., Tarun, A.S., Montes-Bayon, M., Meija, J., Malit, M.F., Wu, C.P., Abdel Samie, M., Chang, C.Y., Tagmount, A., DeSouza, M., Neuhierl, B., Bock, A., Caruso, J., Terry, N. 2004. Overexpression of selenocysteine methyltransferase in Arabidopsis and Indian mustard increases selenium tolerance and accumulation. *Plant Physiol.* **135: 377-383.**

Lee, J., Bae, H., Jeong, J., Lee, J.Y., Yang, Y.Y., Hwang, I., Martinoia, E., Lee, Y. 2003a. Functional expression of a bacterial heavy metal transporter in *Arabidopsis* enhances resistance to and decrease uptake to heavy metals. *Plant Physiol.* **133: 589-596.**

Lee, J., Shim, D., Song, W.Y., Hwang, I., Lee, Y. 2004. Arabidopsis metallothioneins 2a and 3 enhance resistance to cadmium when expressed in *Vicia faba* guard cells. *Plant Mol. Biol.* **54: 805-815.**

Lee, S., Moon, J.S., Ko, T.S., Petros, D., Goldsbrough, P.B., Korban, S.S. 2003b. Overexpression of *Arabidopsis* phytochelatin synthase paradoxically leads to hypersensitivity to cadmium stress. *Plant Physiol* **131: 656-663.**

Li, Y., Dhankher, O.P., Carreira, L., Lee, D., Chen, A., Schroeder, J.I., Balish, R.S., Meagher, R.B. 2004. Overexpression of phytochelatin synthase in *Arabidopsis* leads to enhanced arsenic tolerance and cadmium hypersensitivity. *Plant Cell Physiol.* **45: 1787-1797.**

Mkandawire, M., Dudel, E.G. 2005. Accumulation of arsenic in *Lemna gibba* L. (duckweed) in tailing waters of two abandoned uranium mining sites in Saxony, *Germany. Sci. Total Environ.* **336: 81-89.**

Newman, L.A., Reynolds, C.M. 2004. Phytodegradation of organic compounds. *Curr. Opin Biotech.* **15: 225-230.**

Odjegba, V.J., Fasidi, I.O. 2004. Accumulation of trace elements by Pistia straiotes: implications for phytoremediation. *Ecotoxicolgy* **13: 637-646.**

Papoyan, A., Kochian, L.V. 2004. Identification of *Thlaspi caerulescens* genes that may be involved in heavy metal hyperaccumulation and tolerance. Characterisation of a novel heavy metal transporting ATPase. *Plant Physiol.* **136: 3814-3823.**

Peng, H.Y., Yang, X.E. 2005. Volatile constituents in the flowers of *Elsholtzia argyi* and their variation: A possible utilisation of plant resources after phytoremediation. *J. Zhejiang Univ. Sci.* **6: 91-95.**

Pilon-Smits, E., Pilon, M. 2002. Phytoremediation of metals using transgenic plants. *Crit Rev. Plant Sci.* **21: 439-456.**

Pollard, A.J., Powell, K.D., Harper, F.A., Smith, J.A.C. 2002. The genetic basis of metal hyperacumulation in plants. *Crit. Rev. Plant Sci.* **21: 539-566.**

Prasad, M.N.V. 2004. Phytoremediation of metals in the environment for sustainable development. *Proceedings of the Indian National Science Academy* **70: 71-98.**

Prasad, M.N.V. 2003. Phytoremediation of metal-polluted ecosystems: hype for commercialisation. *Russ. J. Plant Physiol.* **50: 686-700.**

Prasad, M.N.V., Freitas, H. 2003. Metal hyperaccumulation in plants –Biodiversity prospecting for phytoremediation technology. *Electronic Journal of Biotechnology* **6: 275-321.** Online electronic journal.

Sharma, N.C., Gardea-Torresdey, J.L., Parsons, J., Sahi, S.V. 2004. Chemical speciation and cellular deposition of lead in *Sesbania drummondii. Environ. Toxicol. Chem.* **23: 2068-2073.**

Schwartz, C., Echevarria, G., Morel, J.L. 2003. Phytoextraction of cadmium with *Thlaspi caerulescens. Plant Soil* **24: 27-35.**

Singh, O.V., Jain, R.K. 2003. Phytoremediation of toxic aromatic pollutants from soil. *Appl. Microbiol. Biotech.* **63: 128-135.**

Song W.Y., Sohn, E.J., Martinoia, E., Lee, Y.J., Yang, Y.Y., Jasinski, M., Forestier, C., Hwang, I., Lee, Y. 2003. Engineering tolerance and accumulation of lead and cadmium to transgenic plants. *Nature Biotech.* **21: 914-919.**

Sudhagar, R. 2003. Cadmium tolerance in maize (*Zea mays* L.)-A genetic and biochemical perspective. Ph.D. Thesis submitted to the Tamil Nadu Agricultural University, Coimbatore.

Suresh, B., Ravishankar, G.A. 2004. Phytoremediation—A novel and promising approach for environmental clean-up *Crit. Rev. Biotechnol.* **24: 97-124.**

Sykes, Marguerite, Yang, Vina, Blankenburg, Julie, and Abu Bakr, Said 1999. Biotechnology: Working with nature to improve forest resources and products. *International Environmental Conference*, **631-637.**

Tian, J.L., Zhu, H.T., Yang, Y.A., He, Y.K. 2004. Organic mercury tolerance, absorption and transformation in Spartina plants. *Zhi Wu Sheng Li Yu Fen Zi Sheng Wu Xue Xue Bao* **30: 577-582.**

Tong, Y.P., Kneer, R., Zhu, Y.G. 2004. Vacuolar compartmentalisation: a second-generation approach to engineering plants for phytoremediation. *Trends Plant Sci.* **9: 7-9.**

United States Environmental Protection Agency. *A Citizen's Guide to Phytoremediation.*

Van Huysen, T., Terry, N., Pilon-Smits, E.A.H. 2004. Exploring the selenium phytoremediation potential of transgenic Indian mustard overexpressing ATP sulphurylase or cystathionine -□-syntheses. *Int. J. Phytoremed.* **6: 111-118.**

Vatamaniuk, O.K., Mari, S., Lang, A., Chalasani, S., Demkiv, Lo, Rea, P.A. 2004. Phytochelatin synthase, a dipetidyltransferase that undergoes multisite acylation with gamma-glutamylcysteine during catalysis—Stoichiometric and site-directed mutagenic analysis of *Arabidopsis thaliana* PCS_1-catalysed phytochelatin synthesis. *J. Biol. Chem.* **279: 22449-22460.**

Verret, F., Gravot, A., Auroy, P., Leonhardt, N., David, P., Nussaume, L., Vavasseur, A., Richaud, P. 2004. Overexpression of $AtHMA_4$ enhances root-to-shoot translocation of zinc and cadmium and plant metal tolerance. *FEBS Lett.* **576: 306-312.**

Wang, Q.R., Cui, Y.S., Liu, X.M., Dong, Y.T., Christie, P. 2003. Soil contamination and uptake of heavy metals at polluted sites in China. *J. Environ. Sci. Heal.* A **38: 823-838.**

Wycisk, K., Kim, E.J., Schroeder, J.I., Kramer, U. 2004. Enhancing the first enzymatic step in the histidine biosynthesis pathway increases the free histidine pool and nickel tolerance in *Arabidopsis thaliana. FEBS Lett.* **578: 128-134.**

Zhang, Y.W., Tam, N.F.Y., Wong, Y.S. 2004. Cloning and characterisation of type 2 metallothionein –like gene from a wetland plant, *Typha latifolia. Plant Sci.* **167: 869-877.**

Zhao, F.J., Dunham, S.J., McGrath, S.P. 2002. Arsenic hyperaccumulation of different fern species. *New Phytol.* **15: 27-31.**

Zhu, Y.L., Pilon-Smits, E.A.H., Jouanin, L., Terry, N. 1999a. Over expression of glutathione synthetase in Indian mustard enhance cadmium accumulation and tolerance. *Plant Physiol.* **119: 73-79.**

Zhu, Y.L., Pilon-Smits, E.A.H., Tarun, A.S., Weber, S.U., Jouanin, L., Terry, N. 1999b. Cadmium tolerance and accumulation in Indian mustard is enhanced by over expressing gamma-glutamylcysteine synthetase. *Plant Physiol.* **121: 1169-1178.**

□□□

29

Influence of Soil Salinity on Relative Biomass and Critical Limits of Growth in Selected Tree Species

B.R. Gurumurthy[1], M.B. Shivanna[2], S.P. Nataraja[1] and Narayan S. Mavarkar[1]

[1]College of Agriculture, Shimoga–577 204 and [2]Dept. Applied Botany, Kuvempu University, Jnana sahyadri–577 451, Karnataka

Introduction

Growth suppression in initiated at some threshold value of salinity, which varies with crop tolerance. (Anon, 1992). Very little work has been done on salt tolerance of perennial tree species. Biological amelioration of salt affected soils through afforestation has invited the attention of scientists and foresters from several years. Determination of salt tolerance of various tree species is of immense value before deciding their suitability for plantation on such sites, especially during establishment and early stage of growth when plants are too tender to bear soil salinity. The present study was conducted to find out the effect soil salinity on growth (biomass accumulation) and critical limits of salt tolerance of selected species.

Materials and Methods

Experiments were conducted in green house during 2000-01 and repeated in 2001-02 at Agriculture College, Shimoga. Six tree species—neem, jack, acacia, eucalyptus, pongamia and tamarind that are commonly grown in agro/social forestry were chosen for the study. Seeds were sown in small polybag (12 x 5 cm) containing potting mixture (red earth, farmyard manure and sand mixed in a ratio of 3:2:1) and placed in a green house with regular watering to raise healthy and uniform seedling. Saline soils were prepared by adding sodium chloride and calcium chloride to the potting mixture in a ratio of 2:1 (Fowler, 1991; Todd and Reed, 1998) to obtain EC of 4, 8, 12, 16 or 20 dsm^{-1}. Artificially prepared saline soils were filled into large polybag (40 x 15 cm) and four hundred and fifty-day-old-seedling raised in potting mixture were transplanted into these large polybag. Field capacity of the soil was calculated and each time only measured amount of water was applied to each bag, once in three days. Periodically samples were collected and biomass was recorded. The critical limit of salt tolerance of tree species was worked out on the basis of data of biomass accumulation. The salinity level, which caused 50 per cent reduction in biomass as compared to biomass content of control plant at a growth stage, has been considered as critical limit of growth. Also relative growth values were determined to compare the performance of tree species. Relative growth value is the percentage variation in the growth of plant over a period of time in a saline situation as compared to the growth at the time of treatment or that would be obtained under non-saline conditions. Determination of relative growth values aids in nullifying initial growth variations as the initial growth in different plant species is considered as 100. The data was subjected to statistical analysis as per the produce outlined by Gomez and Gomez (1984).

Results and Discussion

Relative Biomass Accumulation

There were no considerable differences in biomass production before treatment imposition but later on variations were evident in all the six tree species. The relative biomass contents on 180 DAT were 67, 42 and 37 in pongamia, neem and tamarind, respectively but they were 120, 60 and 20 in these crops on 360 DAT. This indicated that effect of soil salinity had a prolonged effect on these plant species. The increased relative biomass from 180 to 360 DAT in case of pongamia and neem indicated that the plants were developing tolerance to salinity. The recovery was slow in case of tamarind but mortality of seedling was not observed. Similarly, Evers *et al.* (1998) observed that salt stress induced biometric and physiological changes in *Solamum tuberosum*. In their experiment, potato grown *in vitro* showed decreased shoot and root length with increased salinity levels. In the present study, reduction in biomass could be attributed to reduction in leaf number of early senescence, dropping of leaves, reduced leaf production and retarded plant height. Even though biomass reduction was evident due to salinity stress when compared to control, a marginal increase was noticed in surviving plants after a short period of gap. As a result per gram increment in biomass was higher in treated plant, which further indicated that the plants were recovering from stress.

Table-1: Effect of soil salinity (EC 20) on relative biomass in tree species

DAT	Neem	Jack	Acacia	Ecualyptus	Pongamia	Tamarind
0	100	100	100	100	100	100
30	75	98	100	92	102	87
60	42	70	87	70	106	78
90	22	48	82	53	72	52
120	25	28	51	38	65	51
150	37	24	20	-	59	44
180	42	-	-	-	67	37
210	51	-	-	-	75	29
240	53	-	-	-	89	26
270	55	-	-	-	102	25
300	58	-	-	-	109	20
360	60	-	-	-	120	20

DAT- days after treatment

Critical Limits of Growth

Critical limit is the salinity level below which a reasonably satisfactory economic responses should be expected above which the probability of such response is low. Assessment of critical limits of salinity for various crops is essential for recommendation of plant species and for rational use of soil. According to the methodology proposed by FAO (Anon, 1992), 50 per cent reduction in biomass could be considered for defining a critical limit for different crop species. In the present study, among the different species tested pongamia appeared to be relatively tolerant to different levels of soil salinities as it is evident from the fact that all the salinity levels tired (EC 4 to 20) did not cause biomass reduction to a tune of 50 per cent as compared to other tree species at all four stages of observation. In case of neem the effect of salinity treatment was only for the period of 180 days. On the contrary in tamarind 50 per cent biomass reduction was noticed only after 120 days when the salinity was at and above EC 8. Mortality of neem, pongamia and tamarind was not noticed during the experimental period. The jack, acacia and eucalyptus seedlings recorded 50 per cent biomass reduction at the salinity of ≥ 8. And all the plants of these species completely died at the higher salinity ≥ 12. On the basis of present study pongamia could be considered as relatively saline tolerant, while neem and tamarind less tolerant types.

Table 2: Effect of soil salinities (EC 4 to 20) on critical limits in tree species

Crop	Days after treatment			
	120	180	270	360
Neem	≥8[1]	≥12	–[2]	—
Jack	≥12	≥8, M≥16	≥8, M≥16	≥8, M≥16
Acacia	–	≥8, M≥16	M≥12	≥8, M≥16
Eucalyptus	20	≥8, M≥16	M≥12	≥8, M≥16
Pongamia	–	–	–	–
Tamarind	–	≥8	≥8	≥8

[1]soil salinities in EC M-Mortality of plant

[2] Expected reduction was not evident.

References

Anonymous, 1992. CROPWAT a computer programme for irrigation planning and management. *FAO.*, Rome, p. **42.**

Evers, D.; Hemmer, K. and Hausman, J.F. 1998. Salt stress induced biometric and physiological changes in *Solauum tuberosum. Acta Physiologiae Plantarum.* **20: 3-7.**

Fowler, J.L. 1991. Interaction of salinity and temperature on the germination on crambe. *J. Agron.* **83: 169-172.**

Gomez, A.K. and Gomez, A.A. 1984. *Statistical Produce for Agricultural Research,* 11 John Wiley and Sons, Inc., New York- 10158. pp. **464-470.**

Todd, N.M. and Reed, D.W. 1998. Characterising salinity limits of New Guinea impatiens in recirculating sub-irrigation. *Journal of American Society of Horticultural Science.* **123: 156-160.**

❑❑❑

30

Effect of Growth Regulators on Chlorophyll Content, Growth and Yield Parameters of Berseem (*Trifolium alexandrinum* L.) under Punjab Condition

J.S. Kang, Avtar Singh and Jyoti Gulati
Department of Agronomy and Agrometeorology,
Punjab Agriculture University, Ludhiana, Punjab

Introduction

Application of wide variety of both naturally occurring and synthetic chemical growth regulators has been extensively used in order to ascertain their beneficial effect on plants. A number of synthetic growth retardant have been discovered and proved to be of considerable importance in agriculture (Yadava *et al.*, 1984).

Gibberelic acid, Succinic acid, Mepiquat chloride have been reported to be playing a significant role in seed production of Persian clover (*Trifolium resupinatum* L.) (Kang *et al.*, 2002). But the information on growth, flowering behaviour chlorophyll content in lacking which is also a tool for judging the crop productivity. Accordingly, the present investigation was undertaken to explore the possibility of improving seed yield in tetraploid berseem (*Trifolium alexandrinum* L.) crop in relation to leaf growth, dry matter accumulation, flowering behaviour and photosynthetic pigment.

Materials and Methods

A field trial was carried out at Punjab Agricultural University, Ludhiana during

2001-2002 and 2002-2003. The experimental soil was loamy sand, low in organic carbon and available nitrogen medium in available phosphorus. Berseem variety BL–42 was sown on October 5. The first cutting was taken around 55 days and subsequent cutting were taken at 25/30 days interval. Flowering behaviour was noted from four spots of 2 x 2 feet in each plot. Various growth regulators like Gibberelic acid (GA), Mepiquat chloride (MC), Succinic acid SA), Garden glow (GG) (mixture of essential plant nutrients) and water spray as control were tested. First spray of growth chemicals was given at 30 days after sowing and second spray was given at the time of initiation of flowering.

Leaf growth in term of length, breadth, dry matter accumulation in leaves and photosynthetic pigments (Chlorophyll) were recorded at 3-sucessive cut. Pigments were extracted by aqueous acetone (Mackinney, 1941) and concentration was measured in duplicate samples at 660 and 640 mμ wavelength. The fractions of pigment were calculated from the absorption data using the equation of Comar (1942). Days to complete flowering were recorded keeping the initiation of flowering as starting first day (day one). Number of heads /sq.feet were recorded one month before harvest from specific spots. Total seed yield was recorded after threshing of the harvested produce and the percentage increase or decrease over control was also worked out.

Results

Average length and breadth of fully expanded terminal leaf lamina in untreated and growth regulator treated plants are presented in Table 1. It was observed that leaves treated with Mepiquat chloride were shorter and narrower than those in control whereas Gibberelic acid treated plants showed increasing trend in length and breadth of leaves with the increase in concentration up to 40 ppm. Observation from Succinic acid treated plots showed that with increase in concentration the leaf length and breadth rapidly decreased. Garden glow positively also affected the leaf growth.

Percentage of dry matter accumulation in leaves treated with Gibberelic acid (GA), Succinic acid (SA), Garden glow (GG) was recorded higher as compared to control (Table 1). However, Mepiquat Chloride (MC) up to 400 ppm improved the production but thereafter it showed declining trend. GA, SA and GG treatments were superior than control. The effect of treatment was not proportional to the concentration used.

Table-1: Effect of growth regulators on leaf growth and dry matter (%) of berseem leaves (means of 4 cuts)*

Treatments	Leaf length (cm)	Leaf breadth (cm)	Dry matter (%)
Gibberelic Acid (ppm)			
20	5.30	2.12	15.68
30	5.38	2.19	16.24
40	5.59	2.32	16.98
Succinic Acid (ppm)			
400	5.30	2.12	16.19
500	5.16	2.00	15.81
600	5.10	1.96	15.67
Mepiquat Chloride (ppm)			
400	5.02	1.91	15.98
500	4.87	1.89	15.07
600	4.21	1.72	14.10
Garden Glow (5g/l)	5.45	2.19	16.32
Water Spray (control)	5.06	2.00	15.09
CD at 5%	0.23	0.11	0.39

* based on the 2 years means data

Data on the photosynthesis pigments (Chlorophyll and its fractions) of berseem leaves under the various treatments are presented in Table 2. Total chlorophyll content increased with application of GA from 20 to 40 ppm. However, it decreased with increasing concentration of Succinic acid (SA) and Gibberelic acid (GA). Except 600 ppm of MC, all the treatment of chemicals induced more total chlorophyll content then control. Highest chlorophyll–a content (0.823) was observed from plots treated with Garden g low and among the chemicals, GA (40 ppm) treated plots gave the maximum contents (0.780). Overall, chlorophyll-a was more than then chlorophyll–b, marked differences were also noticed in the ratio of chlorophyll–a and chlorophyll–b.

Data on the time (days) taken for compete flowering after last cut is presented in table 3. It was observed that the number of days taken for complete flowering varied from 17 to 26 under different treatment. The flowering was advanced under the application of GA and SA and it was delayed under the Mepiquat chloride spray and garden glow. However, under Garden glow the differences were minor when compared with GA and SA.

Table-2: Effect of growth regulators on photosynthetic pigments of berseem leaves (means of 4 cuts)*

Treatments	Chlorophyll-a (mg/l) fresh wt.	Chlorophyll-b (mg/l) fresh wt.	Total Chlorophyll (mg/l) fresh wt.	Ratio (a:b)
Gibberelic Acid (ppm)				
20	0.768	0.469	1.237	1.63
30	0.776	0.495	1.271	1.56
40	0.780	0.521	1.301	1.49
Succinic Acid (ppm)				
400	0.812	0.483	1.295	1.68
500	0.791	0.480	1.271	1.64
600	0.725	0.456	1.181	1.58
Mepiquat Chloride (ppm)				
400	0.708	0.455	1.163	1.55
500	0.695	0.463	1.158	1.50
600	0.597	0.403	1.000	1.48
Garden Glow (5 g/l)	0.823	0.480	1.303	1.71
Water Spray (control)	0.693	0.349	1.042	1.98

Seed yield increased (Table 3) significantly with the application of GA, SA and garden glow over the control. However seed yield was not influenced with the application of Mepiquat chloride and it was lower than the control under all the treatments of MC. The increase in seed yield was 23.57, 25.71, 31.96, 24.46, 35.89, 39.28 and 18.21 per cent higher under the application of GA, SA and garden glow at 20, 30, 40 ppm, 400, 500, 600 ppm and 5g/l, respectively over the control. This might be due to the more flower heads under all the treatment over the control. Data on the seed yield indicated that application of GA, SA and Garden Glow were found effective to enhance the seed yield of berseem.

Discussion

Mepiquat chloride (MC) induced shorter and narrow leaves than those in control but Gibberelic acid treated plots showed increasing tend in length ad breadth of leaf and no phytotoxic symptoms were observed on the foliage. The results in conformity with earlier reports on other plants (Kang *et al.*, 2002; Brar *et al.*, 1998; Jain and Yadava, 1980; Pandey *et al.*, 2001) observed reduction in plant height, leaf growth with IAA and CCC in rice crop. MC reduced the length and breadth of the leaves but increased the dry matter production in comparsion to control where length and breadth from GA and SA treated plots increased with increasing dry matter, therefore it is conceivable that Mepiquat chloride, cause cells to be smaller with a more dense cytoplasm. The viewpoint has been somewhat substantiated by an increase in chlorophyll content with more dry matter accumulation from Gibberelic acid and Succinic acid treated plants.

Minimum time to reach the complete flowering stage was observed under GA and SA treated plots and the corresponding data on seed yield showed higher yield from GA and SA treated plots over the control. Higher seed yield from GA and SA treated plots over the control might be due to the mobilisation of metabolites from the vegetative to the reproductive part which resulted in higher seed yield. The treatments which took more days to reach complete flowering might have been having a slow pace of metabolic reaction due to stress caused by chemical spray as seen from the data on leaf length, breadth and chlorophyll content and therefore less of the metabolites reached the sink site leading to many more days from attaining the complete flowering stage. The delay in complete flowering and with more rise in temperature during April and May months there is drastic reduction in the number of flower heads/sq.ft and ultimately lesser seed yield as seen from the data from Table 3.

Mepiquat chloride (400 and 500 ppm) had increased the chlorophyll—'a' content when compared with control but at the highest concentration (600 ppm) the chlorophyll-'a' were lower as compared to control. Chlorophyll-'b' contents were higher at all levels of MC treated plots over control. However, in spite of this, plots treated with MC took more days to reach complete flowering and which might have been having a slow pace of metabolic reaction due to stress caused by chemical spray as seen from data on leaf length, breadth and chlorophyll content and therefore less of the metabolites reached the sink site.

Table-3: Effect of growth regulators on complete flowering, number of heads/sq/ft and seed yield. (Mean of 2 years)

Treatments	Days to complete flowering after spray at initiation stage	Number of flower heads (heads/sq.ft.)	Seed yield (Q/ha)	Percentage Increase/ Decrease over control
Gibberelic Acid (ppm)				
20	18	290	6.92	+ 23.57
30	17	295	7.04	+ 25.71
40	18	308	7.39	+ 31.96
Succinic Acid (ppm)				
400	20	290	6.97	+ 24.46
500	17	312	7.61	+ 35.89
600	18	314	7.80	+ 39.28
Mepiquat Chloride (ppm)				
400	26	231	5.36	– 4.28
500	23	220	5.10	– 8.92
600	24	209	5.16	– 7.85
Garden Glow (5 g/l)	21	286	6.62	+ 18.21
Water Spray (control)	22	236	5.60	–
CD At 5%	–	35	0.84	–

Wasnik and Bagga (1996) were of the view increased chlorophyll content condition in chick pea could not be harnessed because of lack of any increase in sink size. The delay in flowering and rise in temperature during April and May months caused drastic reduction in the number of flower heads/sq. ft. which ultimately caused lesser seed yield. Similar findings have been reported by Kang *et al.* (2002), Kang and Tiwana (2001), Pandey *et al.* (2001), Bidlack and Roxton (1995), Mohamed *et al.* (1992), and Dhasmana and Aggarwal (1991).

References

Bidlack, J.E. and Roxton, D.R. 1995, Chemical regulators of yield, growth and digestibility of alfalfa. *J. Plant Growth Regulators.* **14: 1-7.**

Brar, Z.S., Singh, M. and Singh, G. 1998. Effect of planting dates and growth regulators on production of mung beam. *J. Res. Punjab Agric. Univ. Ludhiana.* **25(4): 515-520.**

Comar, C.L. 1942. Chloroplast substance of spinach leaves. *Bot. Gaz.*, **104: 122-127.**

Dhasmana, R. and Aggarwal, A.K. 1991. Effect of foliar spray of GA, IAA and 2, 4-D on production of clover (*Trifolium alexandrinum* L.). *Res. and Dev. Rep.* **8: 39-46.**

Jain, N. and Yadava, R.B.R. 1980. Effect of Phospon-D on seed yield and quality of cowpea (*Vigna unguiculata* L.) plants. *Seed Res.* **8(1): 42-45.**

Kang, J.S., Tiwana, M.S. and Thind, I.S. 2002. Seed yield response to growth regulators in Persian clover (*Trifolium resupinatum* L.) *J. Res. Punjab Agric. Univ. Ludhiana.* **39(3): 334-336.**

Mohamed, N.A., Ahmed, M.K., Abed, E.L., Halim, A.Z. and Moausa, M.E. 1992. Effect of Alar and micronutirents on yield and morphological characters of berseem. *Egyptian J. Agric. Res.* **70(4): 177-190.**

Pandey, A.L., Tripathi, R.S. and Yadava, R.S. 2001. Effect of certain growth substances on growth, yield and quality of rice (*Oryza sativa* L.). *Indian J. Agric. Res.* **35(2): 118-120.**

Xu, X. and Taylor, H.M. 1982. Increase in drought resistance of cotton seedlings treated with Mepiquat chloride. *Agron. J.* **84: 569-74.**

Wasnik, K.G. and Bagga, A.K. 1996. Effect of Mepiquat chloride on growth and yield of chickpea (*Cicer arietinum* L.). *Indian J. Plant Physiol.* **1(4): 239-242.**

❑❑❑

31

Plant Biodiversity with Cultural Perspectives of Two Sacred Groves at Vanur Taluk of Villupuram District of Tamil Nadu

D. Kadamban and R. Maheshwari

Department of Botany, K.M. Centre for P.G. Studies, Pondicherry–605 008

Introduction

Sacred groves are patches of natural vegetation preserved by ancient societies on religious and cultural traditions. They exemplify the perceived interlink between man and his natural environment as well as his ecological prudence (Gadgil and Vartak, 1975). They have been described variously as natural museums of giant trees, treasure houses of threatened species, dispensaries of medicinal plants, regulators of water sheds, recreation centres for urban life, variable gardens for botanists, gene banks, economic species, paradise for nature-lovers and labortatory for environmentalists (Gadgil and Vartak, 1975; Vartak, 1983; Vartak and Kumbhonjaker, 1984; Vartak *et al*., 1986; Somashekar, 1998; Ramanujam, 2000).

A comprehensive treaty on the distribution, historical background, cultural and religious connection and conservation strategies of the grove was brought out by Chandran, Gadgil and Hughes (1998).

Sacred groves have been reported from all parts of world wherever indigenous societies existed. Gadgil and Vartak (1976; 1981) reported hundreds of sacred groves on the Western Ghats in Maharashtra and Karnataka. These scared groves are very characteristic of the agricultural landscape. Reports poured in from different parts of North India-the Aravalli Hills of Rajasthan, Gond region of Madhya Pradesh, plains of

Bihar and Garhwal range of the Himalayas. Simultaneously, groves from Uttra and Dahshina Kannada of Karnataka, north and south Kerala and also in Tamil Nadu were reported by different groups. The traditional and cultural aspects like taboos, and beliefs, folklores and rites and rituals were behind the conservation and protection of each and every grove. These patches of vegetation were preserved as "Amara vana, Venu vana, Salai vana, Ashoka vana, Kadmaba vana, Vilva vana" etc. and named after the dominant species.

In India, sacred groves occur under a variety of ecological situations. They have evolved under resource-rich situations. Floristically, sacred groves are considered as patches of climax vegetation of the respected areas except in the case of some memorial groves where the species are selectively introduced (Gadgil and Vartak, 1981). They represent a variety of formations ranging from typically evergreen to dry-deciduous forest types of Himalayan ranges, Western and Eastern Ghats, swamps of the coastal plains along the western coast, bases of thorny scrubs in Aravally ranges and scrub woodlands of Coramandel coastal plains, corresponding to the different climatic zones.

Kanjilal and Das (1934) were pioneers in the study of nature conservation based on religious beliefs. Floristic richness and the biocultural perspectives sustaining them have been documented in 11 important groves (Kadamban, 1998). Encouraged by the Plant diversity and the conservation significance, the groves along the Coramandel coast covering Cuddalore, Villupuram districts of Tamil Nadu and Union Territory of Pondicherry were studied (Kadamban and Ramanujam, 1999a, b; Ramanujam and Kadamban 1999a, b). Saxena *et al.* (1998) reviewed the cultural and religious perspectives associated with sacred groves. The religious perspectives on the conservation of plant diversity was studied by Praveen Kumar (1999) and Ramanujam and Kadamban (2001) in Cuddalore and Pondicherry region respectively. Lalitha (2003) and Maheswari (2004) have studied the link between cultural perspectives and plant diversity in the groves of Villupuram District. The plant biodiversity of sacred groves alone was studied by Meher-Homji (1986), Sethi (1993), Parthasarathy and Karthikeyan (1997), and King (1997).

The sacred groves harbours majestic shrubs, trees and lianas besides rare, endangered and rediscovered taxa. Balachandran and Gastmans (1997) identified *Derris ovalifolia* from Puthupet (near Pondicherry). *Dipterocarpus indicus*, a majestic evergreen tree of Western ghat (Chandran and Gadgil, 1993). Chandrasekara and

Sankar (1998) reported 20 per cent of endangered and 40 per cent of medicinal plants in Kerala groves. Raj and Sukumaran (1991) have studied 43 "miniature temples" from Southern district of Tamil Nadu and found them harbouring rare, endemic and endangered plants.

At low elevation of Maharashtra sacred groves are dominated by Hirada tree (*Terminalia chebula*) which is a source of tannin. A small grove at village Tunbad in Srivardhan taluk in the same state shelters a gregarious specimen of *Entada phaseoloides* var. *pusaetha* of Mimosoideae (Fabaceae). The bark of this leguminous liane is the only source of medicine in 40 km radius for treating snakebite in cattle (Vartak *et al.*, 1986). A gigantic specimen of *Michelia champaca,* a sacred tree worshipped the Maharaja of Mysore, is part of a grove in the Biligiriranagam hills of Mysore district (Gadgil and Vartak, 1976). A WWF (1996) survey revealed the existence of over 750 sacred groves in 23 districts of Andhra Pradesh. About 134 species of medicinal, rare and endemic plants. A mammoth liane of *Derris scandes* (> 45 cm gbh) and *Combretum albidum* (35 cm gbh) were found in Omiper and Oorani respectively (Ramanujam and Kadamban, 1999b).

Materials and Methods

The study area is located in the Vanur taluk of Villupuram Revenue district of Tamil Nadu which includes Thriuchitrambalam and Vanur. The former is situated at north of the village, 9 km from Pondicherry on the high way of Pondy – Tindivanam. The later is also situated at north of the Village on Pondy to Mailam hill road, 15 km away from Pondicherry.

A typical maritime tropical climate with a dissymmetric rainfall regime prevails in the study area. The weather is generally humid and hot for most part of the year. The south-west monsoon contributes an average of 356.6 mm of annual rain whereas the north-east monsoon contributes 317.8 mm rainfall. The minimum temperature was 18°C in January and the maximum was 41°C in May. The relative humidity average to 69 per cent. In the Vanur taluk of Villupuram district, sandy, clay (black), red loam-laterite and alluvial soil types were recorded.

Information about the societal composition of the study area was collected from the revenue authorities and traditional heads of the respective region. The traditions, beliefs, taboos, restrictions and folklores pertaining to each temple were collected from the villagers and cross checked with the devotees. The behaviour of the devotees inside the temple during worship and festivals were observed closely for corroboration, the field visits were timed accordingly.

Floristic composition of each grove was intensively studied during field visits in different seasons. The flowering twigs were collected and identified using the regional and state floras (Gamble, 1957; Matthew, 1983).

Observations

Thiruchitrambalam Grove

Thiruchitrambalam is a small village of 105 households and has a population of 420 people. The village is dominated by Naidu and Yadhava community. The sacred grove which is about 2.0 ha., situated at north of the village in the midst of agricultural fields.

The Presiding deity of the grove is Ponniamman and is sheltered within a temple of brick and mortar. In front of the temple, there are cult spot and metallic trident with lemon. A temple for Lord Shiva with modern architecture is erected by filling the lake with soil, found near the grove is an evident of modernism disturbing the water resource cycle. A lake found on the eastern side of the temple serves as water source.

Rites and Ritual

People regularly worship the deity by performing aradhana with flowers, and fruits like banana, lime with sweetened rice (pongal) on every Friday. During festivals people leave terracotta images of horses and toy cradles as votive offerings. Animals like goats, fowls are sacrificed to Amman.

Taboos and Beliefs

The people believe that the Amman has more power and protect them. There is no severe restriction recorded here other than that women are not permitted during menstruation time into the sanctum. At the time of epidemics like dysentry and cholera, people from the suffering villages organise a special pooja seeking salvation. People believe that lighting lamps made of rice flour on every Friday make early marriage for unmarried and child birth for married woman. People believe that tieing the placental chord of the cattle on the branches of the banyan tree would bestow the cattles with good amount of milk.

Folklore

The goddess, Ponniamman patrols the village usually at midnight, and she gets angry if she saw anybody out of their house at that time. Beneath Amman's idol there is a Panchalokam *i.e.* an idol made of 5 metals. One day a priest stole the Panchalokam idol. As soon as he reached home with the idol he got paralysed and died. Amman punished the priest with a curse of not having any future generation in the family. So people are prevented from desecrating the beliefs and the vegetation.

Floristic Composition

The sacred grove of Tiruchitrambalam, measuring about 2.0 ha., has a total of 123 species, representing 110 genera and 47 families. Among them, herbs accounted for 39 (31.7 %) species from 35 genera of 18 families; Shrubs were 24 (19.5 %) from 24 genera of 16 families; Trees were 27 (21.95 %) species from 25 genera of 19 families; climbers were 32 (26.02 %) species from 30 genera of 19 families (Figure 2b); there is only one parasite *Dendrophyhoë falcata* of Loranthaceae, an evergreen parasitic shrub on *Ficus religiosa*. (Table 1).

Vegetation Analysis

The vegetation analysis shows that the family Euphorbiaceae represented the maximum number (9) of taxa. There were 6 species each from Acanthaceae and Asteraceae. The five families, *viz.* Caesalpiniaceae, Convolvulaceae, Fabaceae, Mimosaceae and Moraceae had donated 5 species each. There are 9 families each with four species and 3 families with 3 species each. The dominant genus is Ficus with 4 species; *Cassia, Ipomoea* and *Solanum* were represented by 3 species, in each.

Barringtonia acutangula of Lecythidacea an ecologically valuable tree, has a maximum growth with a gbh of 385 cm; *Butea monosperma* a member of Fabaceae with a hob of 261 cm, was heavily cut down for the leaves which are used as plates to serve food. *Derris scandens* of Fabaceae a liana recorded 88 cm (ghb) from the total of 2 branches. They are the robust specimens of the grove. *Maeruva oblongifolia* of Capparaceae a rare climber reported in Villupuram District. *Jatropha curcus* is a hilly species recorded from this grove. *Argyreia cymosa*, *Cryptolepis grandiflora* and *Hewitia sublobata* reported from this grove are uncommon in distribution. Medicinally important plants like *Crateva adansonii, Rauvolfa tetraphylla, Desmodium gangeticum, Plumeria rubra* have attained their full growth in this grove..Northern part of the grove is cleared and *Eucalyptus* plantation is raised by TANCOPF.

Vanur Grove

Vanur is a village of 150 households with a population of 500. The grove measuring 1.5 ha. Stands amidst the paddy fields and Casuarina and mango plantations. The lake situated on the Western side serves as the water reservoir. The grove is situated on the northern side of Pondicherry- Mailam state highway. It belongs to Vanur taluk of Villupuram district. The village is dominated by Vanniar community followed by Yadhava, Udayar, Scheduled caste, and Mudaliar.

The presiding deity is Patchaivazhi Amman. The deity is sheltered within a temple made of brick and cement. There is a pali peedam in front of the deity. The temple is surrounded mostly by *Azadirachta indica* trees. There are tridents standing in front of trees. The yellow, red and blue cloths offered by devotees were tied around each tree to protect the trees from the cutters.

Apart from main deity, a termite mound found in front of the temple under the neem and palm tree is also worshipped. A trident with a garland of lemon found near the puthu is worshipped. The turmeric tied with yellow rope believed as "tali" are seen hanging on the neem tree for seeking early marriage. And also, a cloth and a stone within, was found hanging from the branches of the tree for seeking early child birth.

Rites and Rituals

People visit regularly on Friday, Sunday and Tuesday. The villagers, who had migrated to different area, try to come and participate in the annual temple festivals regularly during which urani pongal is made in the month of Aadi to please the goddess. The first born child in the family is brought to the temple for tonsuring of head and ear boring. During important occasions they offer cloth, bangles, terracotta images of horses and toy cradles as votive offering. Animals like goats, fowls are offered sacrificially to the Goddess.

Taboos and Beliefs

The people have deep faith on the power of Patchaivazhi Amman. The scheduled caste people and women during menstruation period are restricted from entering the Grove. Lopping are axing of the groves was totally banned. Footwear is not permitted inside the grove. People also followed Viradh/fasting prior to ceremonial visits to the grove.

Folklore

There is a folklore behind the Goddess "Patchaivazhi Amman" at Vanur. Her name is "Patchai Valaiyalkutchi Amman."

About 14 generations back, an unmarried girl wanted to water bangles on her hands and she asked her parents. They refused regularly, so she went out of the house. The parents and the villagers could not find her. Menawhile, one day a bangle vendor heard the sound of a lady from the forest, "asking bangles for her hand"; without answering he went back to his home. On another day, while he was crossing the same forest he heard the same voice from the termite mound. Two bare hands suddenly appeared from the termite mound before him and asked to put the bangles on her hand. The vendor also wore it and he asked for the cost of bangles. She told him to get it from her house. He went and explained the events to her parents and collected the money. Later on, they came over the spot and requested her to come out. She told them that she got a suiable place and likes to live there and also asked them to build a temple over it. Thereafter, a temple was constructed and named as "Patchai Valayalkutchi Amman" temple.

Floristic Composition

The sacred grove of Vanur has a total of 87 species representing 80 genera and 39 families. Of these, herbs accounted 33 (37. 93%) species from 31 genera of 19 families. Shrubs were represented by 21 (24.14%) species from 20 genera and 15 families. Trees had 14 (16.1%) species from 14 genera and 11 families. Climbers were 19 (21.8%) belonging to 18 genera and 16 families. There is no parasite reported from this grove. (Table 1).

Vegetation Analysis

The maximum number of species (7) was represented from Acanthaceae. There are 5 species each from Convolvulaceae and Rubiaceae; Amaranthaceae, Asclepidaceae, Febaceae, Malvaceae and Mimosaceae had four each.

Table-1: Plant diversity of Two Sacred Groves at Vanur Taluk, Villumpuram district, Tamil Nadu

Sl. No.	Family Name	Species Name	Habit	Present study	
				TC	VN
1	Acanthaceae	*Asystasia gangetica* (L.) T. Anderson	Herb	+	
2	Acanthaceae	*Blepharis maderaspatensis* (L.) Roth.	Herb		+
3	Acanthaceae	*Dipteracanthus prostratus* (Poiret) Nees	Herb	+	+
4	Acanthaceae	*Ecbolium ligustrinum* (Vahl) Vollesen	Herb	+	
5	Acanthaceae	*Elytraria aculais* (L. F.) Lindan	Herb	+	+
6	Acanthaceae	*Hygrophila schulii* (Buch. Ham.) M. R. et S.M. Almida	Herb	+	+
7	Acanthaceae	*Justicia adhatoda* L.	Shrub		+
8	Acanthaceae	*Justicia prostrata* (C. B. Clarke) Gamble	Herb	+	+
9	Acanthaceae	*Justicia transquebariensis* L.	Herb		+
10	Agavaceae	*Sansevieria roxburghiana* Schultes et Schultes f.	Herb		+
11	Alangiaceae	*Alangium salviiforlium* (L. f.) Wangerin	Tree	+	+
12	Amaranthaceae	*Achyranthes aspera* L.	Herb	+	+
13	Amaranthaceae	*Aerva lanata* (L.) Juss.	Herb	+	+
14	Amaranthaceae	*Almania nodiflora* (L.) R. Br.	Herb		+
15	Amaranthaceae	*Alternanthera sessilis* (L.) R. Br. ex DC.	Herb	+	+
16	Amaranthaceae	*Amaranthus viridis* L.	Shrub		+
17	Apocynaceae	*Carissa spinarum* L.	Shrub		+
18	Apocynaceae	*Ichnocarpus frutescens* (L.) R. Br.	Climber	+	
19	Apocynaceae	*Plumeria rubra* L.	Tree	+	
20	Apocynaceae	*Rauvolfia tetraphylla* L.	Shrub	+	
21	Arecaceae	*Borassus flabellifer* L.	Tree	+	+
22	Arecaceae	*Phoenix pusilla* Gaertn.	Shrub		+
23	Arecaceae	*Phonenix sylvestris* (L.) Roxb.	Tree	+	

24	Aristolochiaceae	*Aristolochia indica* L.	Climber		+
25	Asclepiadaceae	*Calotropis gigantea* R. Brown	Shrub	+	+
26	Asclepiadaceae	*Calotropis procers* (Ait.) R. Br.	Shrub		+
27	Asclepiadaceae	*Pergularia daemia* (Forsskal) Chiov.	Climber	+	
28	Asclepiadaceae	*Tylophora indica* (Burm. f) Merr.	Climber	+	+
29	Asclepiadaceae	*Wattakata vilubuilis* (L. f.) Staf	Climber	+	+
30	Asparagaceae	*Asparagus racemosus* Willd.	Climber		+
31	Asteraceae	*Ageratum conyzoides* L.	Herb	+	
32	Asteraceae	*Grangea madaraspatana* (L.) Poiret	Herb	+	
33	Asteraceae	*Parthenium hysterophorus* L.	Herb	+	
34	Asteraceae	*Synedrella nodiflora* (L.) Gaertner	Herb	+	
35	Asteraceae	*Tridax procumbens* L.	Herb	+	+
36	Asteraceae	*Vernonia cinerea* (L.) Less.	Herb	+	+
37	Bombacaceae	*Ceiba pentandra* (L.) Less.	Tree	+	
38	Boraginaceae	*Cordia dichotoma* L.	Tree	+	
39	Boraginaceae	*Carmona retusa* (Vahl) Masam.	Shrub	+	+
40	Boraginaceae	*Coldenia procumbens* L.	Herb	+	
41	Boraginaceae	*Heliotropium indicum* L.	Herb	+	
42	Cactaceae	*Cereus pterogonus* La Marie	Shrub		+
43	Cactaceae	*Opuntia dillenii* (Ker Gawler) Haw.	Shrub	+	+
44	Caesalpiniaceae	*Bauhinia tomentosa* L.	Shrub	+	
45	Caesalpiniaceae	*Cassia fistula* L.	Tree	+	
46	Caesalpiniaceae	*Cassia occidentalis* L.	Herb	+	+
47	Caesalpiniaceae	*Cassia tora* L.	Herb	+	
48	Caesalpiniaceae	*Senna auricultа* Roxb.	Shrub	+	+
49	Capparaceae	*Cadaba fruticosa* (L.) Druce	Shrub	+	+
50	Capparaceae	*Capparis sepiaria* L.	Shrub	+	
51	Capparaceae	*Capparis zeylanica* L.	Climber		+
52	Capparaceae	*Crateva adansonii* DC. Var. Odora	Tree	+	+
53	Capparaceae	*Maeruva oblongigolia* (Forsskal) A. Rich	Climber	+	
54	Caricaceae	*Carica papaya* L.	Tree	+	

55	Cleomaceae	*Cleome viscosa* L.	Herb	+	
56	Combretaceae	*Combretum ovalifolium* Roxb.	Climber	+	
57	Convolvulaceae	*Argyreia cymosa* (Roxb.) Sweet	Climber	+	
58	Convolvulaceae	*Evolvulus alsinodies* (L.) L.	Herb		+
59	Convolvulaceae	*Hewittia sublobata* (L. f.) Kuntz	Climber	+	
60	Convolvulaceae	*Ipomoea fistulosa* (Choisy) D. Austin	Shrub	+	+
61	Convolvulaceae	*Ipomoea obscura* (L.) Ker Gawler	Climber	+	
62	Convolvulaceae	*Ipomoea speiaria* J. Koening ex Roxb.	Climber		+
63	Convolvulaceae	*Ipomoea staphylina* Roemer et Schultes	Climber	+	
64	Convolvulaceae	*Merremia emarginata* (Burm. f.) Hallier. f	Herb		+
65	Convolvulaceae	*Merremia tridentata* (L.) Hallier. f.	Climber		+
66	Cucurbitaceae	*Coccinia grandis* (L.) Voigt	Climber	+	+
67	Cucurbitaceae	*Luffa cylindrica* (L.) Roemer	Climber	+	
68	Euphorbiaceae	*Acalypha indica* L.	Herb	+	
69	Euphorbiaceae	*Bridelia retusa* Willd.	Tree	+	
70	Euphorbiaceae	*Croton bonplandianus* Baill.	Herb	+	+
71	Euphorbiaceae	*Jatropha curcus* L.	Shrub	+	
72	Euphorbiaceae	*Phyllanthus amarus* Schum, et Thonn.	Herb	+	
73	Euphorbiaceae	*Phyllanthus reticulatus* Poiret	Shrub	+	+
74	Euphorbiaceae	*Securinega leucopyrus* (Willd.) Muell. Arg.	Shrub	+	+
75	Euphorbiaceae	*Tragia involucrata* L.	Climber	+	
76	Euphorbiaceae	*Tragia plukenetii* R. Sm.	Climber	+	
77	Fabaceae	*Abrus precatorius* L.	Climber		+
78	Fabaceae	*Alysicarpus monilifer* (L.) DC.	Herb		+
79	Fabaceae	*Butea monosperma* (Lam.) Taubert	Tree	+	+
80	Fabaceae	*Derris scandens* (Roxb.) Benth.	Climber	+	
81	Fabaceae	*Desmodium gangeticum* (L.) DC.	Herb	+	
82	Fabaceae	*Rhynchosia minima* (L.) DC.	Climber	+	
83	Fabaceae	*Tephrosia purpurea* (L.) Pers	Herb		+
84	Fabaceae	*Teramnus labialis* (L. f.) Spengel	Climber	+	
85	Lamiaceae	*Anisomeles indica* (L.) Kuntz	Herb	+	

86	Lamiaceae	*Anisomeles malabarica* (L.) R. Br. Ex Sims	Shrub		+
87	Lamiaceae	*Leucas aspera* (Willd.) Link	Herb	+	
88	Lamiaceae	*Leucas diffusa* Benth.	Herb		+
89	Lamiaceae	*Leucas lavundulifolia* Sm.	Herb	+	+
90	Lecythidaceae	*Barringtonia acutangula* (L.) Gaertner	Tree	+	+
91	Loranthaceae	*Dendrophthoe falcata* (L. f.) Ettingsh	Parasite	+	
92	Lythraceae	*Ammania baccifera* L.	Herb	+	
93	Lythraceae	*Lawsonia inermis* L.	Shrub	+	
94	Malvaceae	*Abutilon indicum* (L.) Sweet	Shrub		+
95	Malvaceae	*Hibiscus ovalifolius* (Forrskl) Vahl	Herb	+	
96	Malvaceae	*Pavonia zeylanica* (L.) Cav	Herb	+	
97	Malvaceae	*Sida acuta* Burm.f.	Herb	+	+
98	Malvaceae	*Thespesia populnea* (L.) Solex. Corr.	Tree		+
99	Malvaceae	*Urena lobata* L. ssp. *Sinuata* (L.) Borssum Waalkes	Shrub	+	+
100	Maliaceae	*Azadirachta indica* A. Juss.	Tree	+	+
101	Menispermaceae	*Cissampelos pareira* L.	Climber	+	
102	Menispermaceae	*Cocculus hirsutus* (L.) Diels	Climber		+
103	Menispermaceae	*Pachygone ovata* (Poiret) Hook. f. et Thomson	Climber	+	
104	Menispermaceae	*Tiliaceora acuminata* (Lam.) Miers	Climber	+	
105	Menispermaceae	*Tinospora cordifolia* (willd.) Hook. f. and Thomson	Climber	+	
106	Mimosaceae	*Acacia auriculiformis* A. Cunn. Ex. Benth.	Tree	+	
107	Mimosaceae	*Acacia caesia* (L.) Willd.	Climber	+	+
108	Mimosaceae	*Albizia amara* (Roxb.) Boivin	Tree		+
109	Mimosaceae	*Albizia lebbeck* (L.) Benth.	Tree	+	
110	Mimosaceae	*Prosopis juliflora* (Sw.) DC.	Tree	+	+
111	Mimosaceae	*Samanea saman* (Jacq) Merr.	Tree	+	+
112	Molluginaceae	*Mollugo pentaphylla* L.	Herb		+
113	Moraceae	*Ficus amplissima* Smith	Tree	+	

114	Moraceae	*Ficus benghalensis* L.	Tree	+	+
115	Moraceae	*Ficus hispida* L. f.	Shrub	+	
116	Moraceae	*Ficus religiosa* L.	Tree	+	
117	Moraceae	*Plecospermum spinosum* Trecul	Climber		+
118	Moraceae	*Streblus asper* Lour.	Tree	+	+
119	Myrtaceae	*Eucalyptus macarthurii* H. Deane et Maiden	Tree	+	
120	Nyctaginaceae	*Boerhavia diffusa* L.	Herb	+	+
121	Oleaceae	*Jasminum angustifolium* (L.) Willd. var. *angustifolium*	Climber	+	+
122	Oleaceae	*Jasminum angustifolium* (L.) Willd. var *sessiliflorum* (Vahl) P.S. Green	Climber		+
123	Onagraceae	*Ludwigia perennis* L.	Herb	+	
124	Passifloraceae	*Passiflora foetida* L.	Climber	+	
125	Pedaliaceae	*Martynia annua* L.	Herb		+
126	Pedaliaceae	*Pedalium murex* L.	Herb	+	+
127	Periplocaceae	*Cryptolepis grandiflora* Wight	Climber	+	
128	Periplocaceae	*Hemidesums indicus* (L.) R. Br.	Climber		+
129	Rhamnaceae	*Zizyphus oenoplia* (L.) Miller	Climber	+	+
130	Rubiaceae	*Canthium coromadenlicum* (Burm. f.) Alston	Shrub	+	+
131	Rubiaceae	*Morinda pubescens* Sm.	Tree	+	+
132	Rubiaceae	*Pavetta indica* L.	Shrub	+	+
133	Rubiaceae	*Psilanthus wightianus* (Wight et Arn.) J. Leroy	Shrub	+	+
134	Rubiaceae	*Spermacoce hispida* L.	Herb		+
135	Rutaceae	*Atalantia monophylla* (L.) Corr. Serr.	Tree	+	+
136	Rutaceae	*Glycosmis mauritiana* (Lam.) Yuich Tanaka	Shrub	+	+
137	Rutaceae	*Limonia acidissima* L.	Tree	+	
138	Rutaceae	*Toddalia asiatica* (L.) Lam.	Climber	+	+
139	Sapindaceae	*Allophyllus serratus* (Roxb.) Kurz	Shrub	+	
140	Sapindaceae	*Cardiospermum helicacabum* L. var. *microcarpum* (Kunth) Blume	Climber	+	

141	Sapindaceae	*Dodonea viscosa* L. f.	Shrub	+	
142	Sapindaceae	*Lepisanthes tetraphylla* (Vahl) Radlk	Tree	+	
143	Scrophulariaceae	*Scoparia dulcis* L.	Herb	+	+
144	Solanaceae	*Solanum melongena* L. var. *insanum* (L.) Prain	Herb	+	
145	Solanaceae	*Solanum torvum* Sw.	Shrub	+	
146	Solanaceae	*Solanum trilobatum* L.	Climber	+	
147	Sterculiaceae	*Waltheria indica* L.	Herb		+
148	Tiliaceae	*Grewia rhamnifolia* Heyne ex Roth	Climber	+	+
149	Tiliaceae	*Triumfetta rhomboidea* Jacq.	Herb		+
150	Verbenaceae	*Lantana camera* L.	Shrub	+	
151	Verbenaceae	*Phyla nodiflora* (L.) E. Greene	Herb		+
152	Verbenaceae	*Stachytapheta jamaicensis* (L.) Vahl	Herb		+
153	Verbenaceae	*Vitex negundo* L.	Shrub	+	+
154	Violaceae	*Hybanthus enneaspermus* (L. f.) Muell.	Herb	+	+
155	Vitaceae	*Cayretia pedata* (Lour) A. L. Juss ex Gagnepain	Climber	+	
156	Vitaceae	*Cissus quadrangularis* L.	Climber	+	+
157	Zygophyllaceae	*Tribulus terrestris* L.	Herb	+	

The maximum number of herbs reported from Acanthaceae and Amaranthaceae was 6 and 4 species, respectively. *Tylophora indica* and *Wattakaka volubilis* are the two medicinally important climbers reported from Asclepiadaceae and *Jasminum angustifolium* with two varieties *viz. angustifolium* and *sessilflorum* are the two climbers represented by Oleaceae. Asclepiadaceae, Caesalpinanceae, Euphorbiaceae, Moraceae and Rubiaceae are the families contributing two shrubs each. *Albizia amara, Prosopis juliflora* and *Samanea saman* are the three tree species reported from Mimisaceae. *Martynia annua, Senna auriculata, Pavetta indica* and *Vitex negundo* are the medicinally important shrubs of this grove.

Discussion

Balick and Cox (1997) identified the rapid erosion of biodiversity and suggested that focusing on culture and tradition of societies would be more rewarding by revealing people's perceptions of nature and natural resources. Chandran and Hughes (1997) traced the transformation of temples in the groves of the Uttra Kannada hills which can be recaptured in the study site also. The groves analysed in the present study share one of those attributes: Termite mound worship is found in both Thiruchitrambalam and Vanur grove, with offerings of terracota horses is the first object of veneration. With the advent of modernism people gave more importance to deity forgetting the vegetation and constructed ornate temples. This is evident from the temples of Lord Shiva of Thiruchitrambalam.

The interface between the people and the grove has blossomed into the belief system/cultural traditions. The diverse form of rites, rituals and festivals prove the unflinching faith of the people in the gods. The variety of the cultural traditions and rituals are a measure of the intimacy of diverse sections of the society on which the groves survive. People gather at special/annual occasions to praise/please the goddess by making Urani pongal, appeal for salvation of epidemic diseases it suggests that the groves serve as centres of socialisation among the assorted groups and nurtures harmony between the factions. Krishnan (2004) reported the impressive gathering of hundreds of Nadhaswaram musicians during the annual festival at Thennampakkam grove which is an evidence for such socialisation. Malhotra and Das (1997) observed the gathering of different ethnic groups around the groves in Kolhapur district of Maharashtra and Minapore district of West Bengal to renew their contact and reassert the group identities similarly. Whereas the anthropological dimensions are impressive and intense, it must be admitted that undue emphasis on the construction of modern temple by clearing vegetation will only amplify the sociological dimensions diverting the focus from the ecological, botanical and conservational values.

A total of 157 species representing 133 genera from 51 families are recorded from the study area of 3.5 ha. Among them herbs accounted for 55 (33.03%) species from 51 genera from 25 families; shrubs were 31 (19.75%) from 30 genera of 19 families, trees were 28 (17.83%) from 25 genera of 19 families; Climbers were 42 (26.78) from 38 genera of 22 families; theres is only one Parasite *Dendrophthoe falcata* of Loranthaceae an evergreen parasitic shrub of *Ficus religiosa*. Out of which 127 species representing 110 genera of 47 families are recorded from Thiruchitrambalam grove and 87 species representing 80 genera of 39 families are from Vanur grove. The structure of vegetation of both groves resembles a forest patch, thus befitting the definition of formal grove.

The dissimilarities in the distribution of taxa in two sites, despite very little difference in climate conditions might be due to the site heterogeneity as suggested by Murphy and Lugo (1986). Kadamban (1998) reported 279 species of angiosperms representing 228 genera and 78 families from seven groves of same forest type found in Cuddalore, Villupuram and Pondicherry. A TDEF type fond at Puthupet by Sethi (1993) had recorded 51 species of 46 genera of 30 families of woody taxa. Parthasarathy and Karthikeyan (1997) reported 42 species, 37 genera of 26 families from Thirumanikkuzhi, Cuddalore district of Tamil Nadu. Praveen Kumar (1999) reported 111 species from 103 genera of 53 families from South Arcot district of Tamil Nadu. Lalitha (2003) reported 188 species from 158 genera of 60 families. Krishnan (2004) reported 200 species from 176 genera and 62 families from the groves of Pondicherry and its environs.

Generally, monsoon forests are rich in trees and climbers as exemplified by 26 tree species and 6 climbers recorded in a 5 ha. grove at Maharashtra (Vartak and Gadgil, 1972). In an isolated Kan forest of Uttara Kannada, Chandran and Gadgil (1993) estimated about 60 tree species. But species richness is generally low in a tropical dry evergreen forest (Murphy and Lugo, 1986). The mixed deciduous forest in Peninsular region contain about 250 tree species of which 150 are recorded in Mudumalai Wild Life Sanctuary. Sukumar *et al.*, (1992) enumerated about 30 tree species in a typical 1 ha plot in such a forest site. The tree species inventory is bound to decline in areas where chance factors of human impacts have played a diachronically dominant role (Joshi *et al.*, 1997). In the present study, the maximum number of tree species (27) was from Thiruchitrambalam whereas only 14 were recorded from Vanur. This pattern can be correlated to the advantage of disadvantage of the location facilitating with human population and their impacts.

Thiruchitrambalam grove is rich in species diversity with 123 species in 110 genera and 47 families with close canopy of trees with dense underground vegetation whereas Vanur grove has only 87 species in 80 genera and 39 families with less tree coverage.

Joshi and Gadgil (1991) described the sacred groves as refugia as they contained plants selectively sheltered within them. Magnificent liana, *Derris scandens* (88 cm gbh) of Fabaceae, *Barringtonia acutangula* (385 cm gbh) of Lecythidaceae and *Butea monosperma* (261 cm gbh) of Fabaceae are of cultural and ecological important plants reported from Thiruchitrambalam grove.

Most of the rare, endangered and endemic plant species of Meghalaya are now confined to the sacred groves (Haridasan and Rao, 1985). The rare and rediscovered entity was reported earlier by various authors in these regions (Krishnamurty 1960; Gadgil and Vartak 1976; Balasubramainan 1977; Balachandran and Gastmans, 1997; Chandran, 1997; Kadamban, 1998, Praveen Kumar, 1999; Lalitha, 2003; Krishnan, 2004). *Jatropha curcas* of Euphorbiaceae, a hilly species was recorded in the present study. The rare species like *Crateva adansonii* and *Maeruva oblongifolia* of Capparaceae are reported from Thiruchitrambalam grove.

The description of Champion and Seth (1968) about the tropical dry evergreen forest types holds good with the present study. The maximum number of tress and shrubs were recorded as evergreen. But he *Eucalyptus* plantation raised by TANCOPF, of course reduces the groundwater level as well as the ground vegetation. Since the sacred groves of the study sites are amidst the paddy field, sugarcane, casuarina plantation and mango orchard, the encroachment was very much observed.

Brandis and Grant (1868) described the sacred groves as 'the traditional form of forest protection'. The factors contributing the decline the sacred groves were gathered by Chandran and Hughes (1997). In Vanur, the tank foreshore area vegetation was reduced by the construction of the temples. Rapid industrialisation, westernisation, urbanisation, modernisation of temple are the key factors for the depletion of the plant genetic resources in the groves of the study area.

A redeeming feature emerging from this study is that despite acculturation, demographic pressures and economic allurements, the rural India is still vibrant with ancient values and not inclined to break it links to nature.

References

Balachandran, N and Gastmans, W. F. 1997. A complete description of Rare and Endemic Derris ovalifolia (Wight et Arn.) Benth. (Fabaceae), Rediscovered from Pondicherry, *South India. J. Econ. Tax. Bot.* Vol. 21.

Balasubramanian, K. 1977. Biotaxonomical studies on the Kurumburam Reserve Forests, Coromandel Coast, Ph.D. Thesis, Annamalai University, Annamalai Nagar, India.

Balick, M.J. and Cox, P.A. 1997. Ethnobotanical Research and Traditional health Care in Developing Countries. In: *Medicinal Plants for forest conservation and health care.* (Schimincke, K.H., Bodeker, F. and Vantamme, P. (eds.), GIFTS of Health, F.A.O. of UN, Rome).

Brandis, D. and Grant, I.P. 1868. Joint Report No. 33, (11th May, 1868), on *The Kans in the Sorab taluka.* Forest Department, Shimoga, Karnataka, India.

Champion, H.G. and Seth, S.K. 1968. *Forest Types of India.* Manager of Publications, Delhi.

Chandran, M.D.S. 1997. On the Ecological History of the Western Ghats. *Curr. Sci.*, **73: 146-155.**

Chandran, M.D.S. and Gadgil, M. 1993. Kans-safety forests of Uttara Kannada. In: *Proc. IUFRO Forest History Group Meeting on Peasant Forestry.* Brandil, M. (ed) Freiburg, Germany, Sept, 2-5, 1991, pp. **49-57.**

Chandran, M.D.S. and Gadgil, M. and Hughes, J.D. 1998. Sacred groves of the Western Ghats of India. In: *Conserving the Sacred for Biodiversity Management* (eds. Ramakrishnan, P.S., Saxena, K.G. and Chandra Shekera, U.M.). Oxford IBH Publishing Co. Pvt. Ltd., New Delhi. pp. **3-16.**

Chandran, M.D.S. and Hughes, J.D. 1997. The sacred groves of Soutl India: Ecology, Traditional Communities and Religious change. *Social Compass*, **44: 413-427.**

Chandra Shekara, U.M. and Sanker, S. 1998. Structure and Functions of Sacred Groves: Case Studies in Kerala. In: *Conserving the Sacred for Biodiversity Management* (eds. Ramakrishana, P.S., Saxena, K.G. and Chandra Shekara, U.M.) Oxford IBH Publishing Co. Pvt. Ltd., New Delhi. pp. **323-336.**

Gadgil, M. and Vartak, V.D. 1975. Sacred Groves of India: A Plea for continued conservation. *J. Bombay Nat. Hist. Soc.*, **72: 314-320.**

Gadgil, M. and Vartak, V.D. 1976. The Sacred Groves of Western Ghats in India. *Econ. Bot.*, **30: 152-160.**

Gadgil, M. and Vartak, V.D. 1981. Sacred groves of Maharashtra: An Inventory. In: *Glimpses of Ethno Botany*, Jain, S.K. (ed.), Oxford University Press, Mumbai, pp. **279-294.**

Gamble, J.S. 1925-57. *Flora of the Presidency of Madras.* Vol. I-III.

Haridasan, K. and Rao, R.R. 1985. *Forest Flora of Meghalaya* Vol. 1 Bishen Singh, Mahendra Paul Singh, Dehradun.

Joshi, N.V. and Gadgil, M. 1991. On the role of refugia in promoting prudent use of biological resource, *Theor. Popul. Biol.* **40: 211-229.**

Joshi, N.V., Suresh, H.S., Daharaja, H.S. and Sukumar, R. 1997. The Spatial organisation of Plant Communities in a Deciduous Forest: A Computational-geometrybared analysis. *J. Indian Inst. Sci.* **77: 365-375.**

Kadamban, D. 1998. Biocultural Perspectives and Plant diversity of Scared Groves and Traditional Medical Knowledge in Pondicherry Environs. Ph.D. Thesis, Pondicherry.

Kadamban, D. and Ramanujam, M.P. 1999a. Medicinal herbs used by Traditional Medical Practioners of Pondicherry Region in curing/treating injuries and poisonous bites: In: *Seminar on Modern Trends in Plant Sciences.* A.V.C. College, Mayiladuthurai, Tamil Nadu. March 4-6 Abstract, pp. **77-78.**

Kadamban, D. and Ramanujam, M.P. 1999b. Sacred Groves of Pondicherry bioregion. In: *National Symposium on Emerging Trends in Plant Sciences.* Department of Botany, Sri Venkateswara University, Tirupathy. March 15-17, Abstract, pp. **41-42.**

Kanjilal and Das, 1934. Cited from Suting, W. 1984. Natural conservation based on Religious beliefs I.F.S. Dissertation- Indian Forest College, Dehra Dun.

King, O.I. 1997. *Botanical and Socio-Cultural Studies on Sacred Groves in Cuddalore District, Tamil Nadu*, M. Phil. Dissertation, University of Madras. (Dept of Botany, Madras Christian College, Chennai), India.

Krishnamoorthy, K. 1960. Myristica swamps in the Evergreen Forests of Travancore. *Indian For.* **86:** pp. **314-315.**

Krishnan, V. 2004. Plant Biodiversity and Biocultural Perspectives of sacred groves in Pondicherry and its Environs. Ph.D. Thesis, Pondicherry University, Pondicherry.

Lalitha, R. 2003. Biocultural Perspectives and Plant Biodiversity of Four sacred Groves of Villupuram District of Tamil Nadu. M.Phil. Thesis, Manonmaniun Sundaranar University. Tirunelveli.

Malhotra, K.C. and Das, K. 1997. Interface between faunal biodiversity and cultural heritage in south west Bengal, in India. In: Fujiki, N. and Macer, R.J. (Eds.). *Bioethics in Asia, Eubois Ethics Institute*, Japan. pp. **338-345.**

Maheswari, R. 2004. Sacred Groves of Vanur taluk in Villupuram District of Tamil Nadu: Plant Biodiversity and People- Grove interface. M.Phil. Dissertation, Pondicherry University.

Mathew, K.M. 1983. *Flora of the Tamil Nadu Carnatic.* Vol. 3.

Meher-Homji, V.M. 1986. *Puthupet: A Sacred termite-mound protects a forest Blackbuck*, II: **1-4.**

Murphy, P.G. and Lugo, A.E. 1986. Ecology of tropical dry forest. *Annual Review of Ecology and Systematics*, **17: 67-88.**

Parthasarathy, N. and Karthikeyeyan, R. 1997. Plant Biodiversity inventory and conservation of two tropical dry evergreen forests on the coromandel coast, South India. *Biodiversity and Conservation*, **6: 1063-1083.**

Praveenkumar, C. 1999. Plant Biodiversity and Biocultural traditions of four sacred groves of composite South Arcot district of Tamil Nadu. M.Phil. Dissertation, Pondicherry University, Pondicherry.

Raj, A.D.S. and Sukumaran, S. 1997. Observations on the sacred groves of South Tamil Nadu. In: Abstracts of *National Symposium on Natural Resources Management Systems*. 26 Feb. St. Joseph College, Thiruchirapalli, India.

Ramanujam, M.P. 2000. Conservation of environment and Human rights: Sacred groves in cultural connections to Biodiversity, *PRP Journal of Human Rights*. **4: 34-38.**

Ramanujam, M.P. and Kadamban, D. 1998. Role of Scared Groves in Cultural and Biological diversity Conservation. In: *Sixth International Congress of Ethno botanical Dialogue between Cultures*. Forging meaningful partnership. New Zealand. November 23-28. Abstract, p. **31.**

Ramanujam, M.P. and Kadamban, D. 1999a. Sacred Groves: A fading legacy of biodiversity Conservation. In: *Seminar on Modern Trends in Plant Sciences*. A.V.C. College, Mayiladuthurai, Tamil Nadu. March 4-6. Abstract, pp. **76-77.**

Ramanujam, M.P. and Kadamban, D. 1999b. Rare, Endangered and Threatened plants occurs in the sacred groves of Pondicherry bioregion. In: *National Symposium on Emerging Trends in Plant Sciences*. Department of Botany, Sri Venkateswara University, Tirupathy. March 15-17. Abstract, p. **43.**

Ramanujam, M.P. and Kadamban, D. 2001. Plant Biodiversity of two tropical dry evergreen forests in the Pondicherry region of South India on the role of belief system in their conservation. *Biodiversity and Conservation* **10: 1023-1217.**

Saxena, K.G. Rao, K.S. and Maikhuri, R.K. 1998. Religious and Cultural Perspective of Biodiversity Conservation in India. A Review. In: *Conserving the Sacred for biodiversity Management*. (eds. Ramakrishnan, P.S., Saxena, K.G. and Chandra Shekara, U.M.). Oxford IBH Publishing Co. Pvt. Ltd. New Delhi. pp. **153-162.**

Sethi, P. 1993. Phytosociology of Tropical Dry Evergreen Forest Patch in the Puthupet Sacred Grove, Coromandel Coast, Tamil Nadu. MS. Thesis, Pondicherry University, Pondicherry.

Somashekar, B.S. 1998. Treasure House in Trouble. *Amruth*, **2(5): 3-7.**

Sukumar, S. *et al.* 1992. Long-term monitoring of Vegetation in Tropical Deciduous forest in Mudumalai, Southern India. *Current Science*. **62: 602-616.**

Vartak, V.D. 1983. Observation on rare, imperfectly known and endemic plants in the sacred groves of Western Maharashtra. In: *An Assessment of Threatened Plants of India*, Jain, S.K. and Rao, R.R. (eds.). Botanical Survey of India, Howrah. pp. **169-178.**

Vartak, V.D. and Gadgil, M. 1972.Dev-Rahati: an ethno-obtanical study of tracts of forest preserved on grounds of religious beliefs: *Proc. Indian Science Congress*, Sixtieth session (Abstracts).

Vartak, V.D. and Kumbhojakar, M.S. 1984. Notes on the lianas of some sacred groves in Western Maharashtra. *Biovigyanam*, **11:** pp. **214-215.**

Vartak, V.D., Kumbhojakar, M.S. and Vandana, D. 1986. Sacred Groves- a sanctuary for lofty trees and lianas. In: Proc. *Seminar on Ecodevelopment of Western Ghats*, Kerala Forest Research Institute, Peechi. pp. **55-59.**

World Wild Fund for Nature – Andhra Pradesh, 1996. *Sacred and Protected Groves of Andhra Pradesh*. WWF–AP State Office, Andhra Pradesh.

❑❑❑

32

Influence of Planting Techniques and Nitrogen Management on Root Density and Yield Attributes of *durum* Wheat (*Triticum durum* desf.)

Navneet Kaur Brar and D.S. Kler

Department of Agronomy and Agrometeorology, Punjab Agricultural University, Ludhiana–141 004, Punjab

Introduction

Durum wheat occupies 8 to per cent area in Punjab covering about 250 to 300 thousand hectares with a production of about 800 to 1000 thousand mt annually (Anon, 1995). *Durum* wheat plays an ecologically important role in providing biodiversity to check the appearance of insect pests and diseases. The make *durum* wheat competitive in international market it becomes imperative to improve its productivity and quality by manipulating agronomy of the crop through variation in planting methods and nitrogen management with the emphasis on use of organic manures. Keeping these points in view the present study was conducted to evaluate the influence of various nitrogen management levels on bed planted *durum* wheat.

Materials and Methods

The experiment was conducted at the Punjab Agricultural University, Ludhiana during 2002-03 on a loamy sand irrigated soil. The soil of the experimental field tested low in available nitrogen and medium in available phosphorous and potash. The 16 treatment combinations (15 bed planted + 1 flat sown) consisting of 3 nitrogen management levels [120 kg N (Rec.); 120 kg N + 15 t FYM and 150 kg N ha^{-1}] with 5 times of N application [1/2 at sowing (S) + 1/2 at crown root initiation (CRI) (Rec.); 1/3 at CRI + 1/3 at tillering (T); 1/4 at S + 1/2 at CRI + 1/4 at T; 1/3 at S + 1/3 at CRI + 1/3 at T + 3% urea spray (US) at flowering (F) and 1/4 at S + 1/2 at CRI + 1/4 at T + 3% US at F] to bed sown wheat including 1 recommended treatment of flat sown *durum* wheat (Rec.) and an extra treatment where whole N (120 kg ha^{-1}) was applied at sowing to bed planted wheat for comparison were replicated thrice in a randomised block design having gross plot size of 6.00 m x 4.05 m while net plot measured 5.00 m x 2.70 m. A seed rate of 100 kg ha^{-1} of *durum* wheat (cv. PDW 274) was used. Two rows per bed (67.5 cm) were sown and row to row spacing of 22.5 cm was kept for flat sowing. All other economic management practices of the crop were uniformly adopted. The crop was sown on November 9, 2002 and harvested on April 24, 2003. The data on yield attributes and yield were recorded at harvest.

Data on root density were taken at boot stage at different rooting depths using the following formula:

$$\text{Root density } (\text{g m}^{-3} \text{ of soil}) = \frac{\text{Total root weight in particular depth increment (g)}}{\text{Total soil volume from which roots were collected } (\text{m}^3)}$$

All interaction effects were found to be non-significant therefore the mean effects are discussed in results.

Results and Discussion

Root density: Root development pattern plays as important role in uptake of water and nutrients and ultimately the growth and yield of crop. Root density of *durum* wheat was studied at boot stage from soil samples taken by keeping the plant stump in the centre.

The data on root density in Table-1 revealed that it was higher in upper layers of 0-15 and 15-30 cma and thereafter decreased with depth. On an average the root density was higher in bed planted crop than flat sown crop throughout the depth. The more favourable existing soil conditions for root growth under bed planting promoted higher root growth. Better soil structure and more favourable conditions for root growth occurred under raised bed farming. Less soil strength (30%) in uncompacted ridges

Table-1: Effet of nitrogen management and time of nitrogen application on root density at boot stage in bed planted *durum* wheat

Treatment	Root density (gm^{-3})						
	Depth						
	0-15	15-30	30-60	60-90	90-120	120-150	150-180
Nitrogen management (ha-1)							
120 kg N	1336.2	234.1	114.6	92.6	74.5	51.2	14.2
120 kg N + 15 t FYM	1445.2	248.1	120.8	100.2	79.0	54.7	22.7
150 kg N	1434.7	212.5	118.6	97.9	76.8	50.2	18.5
Time of N application							
1/2S + 1/2CRI	1299.3	232.2	114.6	94.9	72.5	50.9	17.7
1/3S + 3CRI + 1/3T	1431.0	230.9	117.9	98.4	77.9	52.8	18.1
1/4S + 1/2CRI + 1/4T	1431.2	232.5	118.0	97.4	77.9	52.4	18.8
1/3S + 1/3CRI + 1/3 T + 3% US	1433.1	230.7	119.9	97.5	77.7	52.0	18.0
1/4S + 1/2CRI + 1/4 T + 3% US	1434.3	231.8	119.6	96.8	78.0	52.0	18.2
Flat	1298.6	226.9	110.9	88.6	60.1	30.0	10.2
Bed (S)	1290.5	230.2	113.5	91.8	71.2	49.2	13.5

S= Sowing, CRI = Crown root initiation, T = tillering, US = Urea spray at flowering stage.

than on flat lands has been reported by Tisdall and Hodgson, 1990. Similar results have been reported by Deol in 2001.

With increase in the nitrogen levels the root density showed an increasing trend throughout the rooting depth upto 180 cm soil depth. The highest values of root density were obtained at 120 kg N + 15 t FYM followed by 150 kg N ha^{-1} and lowest for 120 kg N ha^{-1}. The increase in root density with increase in nitrogen levels have been reported (Narang *et al.*, 1992; Deol 2001; Kumar 2002). This was due to beneficial effect of FYM on soil structure, which in turn increased the infiltration and retention of water, improved soil aeration, root penetration and moderated soil temperature (Negi and Mahajan, 2000).

Among the split application of nitrogen the root density increased with increase in number of splits. It was lower for two splits than three splits. However as root density was taken at boot stage before urea spray at flowering so effect of urea spray was not observed. Similar results have been reported by Kumar in 2002.

Spikelength: Spikelength may serve as reliable criterion to assess crop being an indicator of spikelets and number of grains. Bed planting showed higher spikelength than flat sown crop though the differences were nonsignificant.

Nitrogen management did not influence significantly the spikelength though with increase in nitrogen level, spikelength tended to increase (Table-2). The trend of increasing year length with the application of FYM was also recorded by Sushila and Giri in 2002. The difference in the spikelength was not significant with time of nitrogen application. Application of whole N at sowing to bed planted wheat also resulted in comparable spikelength. The results corroborate the earlier findings of Singh (1996); and Uppal (1999).

Number of spikelets per spike: Spikelets per spike in wheat is an important growth index of grain yield. The data on spikelets per spike are presented in Table-2. The differences were observed nonsignificant between the bed planted and flat sown crop.

Among the nitrogen levels, 120 kg N + 15 t FYM ha^{-1} produced the highest spikelets per spike, though it was statistically on par with 120 kg and 150 kg N ha^{-1}. The nitrogen dose of 120 kg N + 15 t FYM ha^{-1} resulted in 6.0 per cent and 3.1 per cent higher spikelets per spike over 120 kg N ha^{-1} and 150 kg N ha^{-1} respectively. The results were in accordance with that of Nehra *et al.* in 2001.

Table-2: Effect of nitrogen management and time of nitrogen application on yield attributing characters in bed planted *durum* wheat

Treatments	Spike length (cm)	Number of spikelets per spike	Number of grains per spike	1000 grain weight (g)
Nitrogen management (ha^{-1})				
120 kg N	7.4	14.17	43.81	40.29
120 kg N + 15 t FYM	7.8	15.08	46.81	42.65
150 kg N	7.7	14.61	45.90	39.79
CD (P = 0.05)	**NS**	**NS**	**NS**	**NS**
Time of N application				
1/2S + 1/2CRI	7.6	14.24	42.06	43.14
1/3S + 1/3CRI + 1/3T	7.6	14.60	43.91	41.09
1/4S + 1/2CRI + 1/4T	7.6	14.57	45.33	40.41
1/3S + 1/3CRI + 1/3 T +3% US	7.7	14.80	48.00	40.41
1/4S + 1/2CRI + 1/4 T + 3% US	7.7	14.80	48.00	39.83
CD (P = 0.05)	**NS**	**NS**	**NS**	**NS**
Interaction	**NS**	**NS**	**NS**	**NS**
Flat (Rec.)	6.8	12.10	34.33	38.51
Bed (S)	7.3	14.00	43.90	43.50

The effect of the time of nitrogen application was nonsignificant. Almost comparable values were obtained with application of whole N at sowing on beds. Similar results were reported by Singh (1996).

Grains per year: The number of grains per spike is an important character influencing the final yield of crop. The number of grains per year did not differ significantly between the bed planted and flat sown crop.

The data on grains per year on bed planted crop (Table 2) revealed that with increase in nitrogen level there was not significant increase in grains per year, maximum being in case of 120 kg N + 15 t FYM ha^{-1} which was 6.4 per cent and 1.9 per cent higher than 120 kg and 150 kg N ha^{-1} respectively. The findings were in accordance with those of Thakur *et al.*, 1999; Nehra *et al.*, 2000; and Singh and Agarwal, 2001. Various times of nitrogen application did not differ significantly in the number of grains per year. Almost similar data on grains per year were obtained by applying whole N at sowing on beds. Marginal increase in number of grains with three equal splits over two split application of nitrogen were earlier recorded by Singh (1996); and Uppal (1999).

Thousand grain weight: The test weight is an important and reliable index of grain yield. The data on thousand grain weight are given in Table-2. There is no significant difference between the tests weight of bed planted and flat sown crop.

Thousand grain weight did not differ significantly due to nitrogen management levels. However, as the nitrogen levels increased from 120 kg to 150 kg N ha^{-1} there was a tendency of decrease in thousand grain weight. The nitrogen level of 120 kg N + 15 t FYM ha^{-1} increased the thousand grain weight and the increase was 5.5 per cent and 6.7 per cent over 120 kg and 150 kg N ha^{-1} (Thakur *et al.*, 1999) reported that rest weight was significantly higher with application of 10 tonnes FYM ha^{-1} over no FYM.

The time of application of nitrogen did not influence the 1000 grain weight although with increase in number of splits thee was tendency of decrease in thousand grain weight. However thousand grain weight on beds where whole N was applied at sowing was slightly higher than two splits although there was no difference statistically. The results are in accordance with Singh (1996); and Uppal (1999).

Grain yield: The ultimate effect of experimental variables is reflected in the final yield of the crop and thus it is a major criterion to identify the efficiency of various treatments in a given situation. The efficiency of different factors can be judged mainly by the addition they make to yield which is the ultimate aim of the field research investigations. The data on the grain yield as affected by various treatments are given in Table-3. The grain yield of bed planted *durum* wheat was statistically on par with the flat sown crop.

Nitrogen management levels did not influence the grain yield significantly. However, the application of combined use of 15 t FYM with 120 kg N ha^{-1} as fertiliser in bed planted wheat tended to increase the grain yield over both 120 kg N ha^{-1} (4.21 q ha^{-1} or 9.9%) and 150 kg N ha^{-1} (3.35 q ha^{-1} or 7.9%) as fertiliser. This increase in the yield was due to cumulative increased effect of various yield attributing characters in 120 kg N + 15 t FYM ha^{-1} treatment (Negi and Gulshan, 2000) also reported 10.5 per cent increase in wheat yield with FYM application over no FYM application. Similar results were reported by Nehra *et al.* (2001); Thakur *et al.* (1999) and Singh and Agarwal (2001). Time of nitrogen application did not influence the grain yield significantly and all treatments were at par statistically. Similar results were reported by Robinson *et al.* (1979); Dhuka *et al.* (1991); Mairona *et al.* (1992); Singh (1996); and Uppal (1999).

Harvest index: The data regarding harvest index were presented in Table 3. The harvest index values for bed planted and flat sown crop were almost same. Among nitrogen management treatments the highest harvest index value was obtained with 120 kg N + 15 t FYM ha^{-1} (0.380) followed by 120 kg N ha^{-1} (0.379) and lowest was recorded in 150 kg N ha^{-1} (0.372). Thus showing that increasing the level of nitrogen through FYM manure increased or atleast maintained harvested index when compared with 120 kg N ha^{-1} whereas when nitrogen level was increased only through fertiliser at 150 kg N ha^{-1} the harvest index decreased because there was more increase in straw yield than grain yield. Slow availability of nitrogen from FYM improved the yield attributes and thus grain yield.

Table-3: Effect of nitrogen management and time of nitrogen application on grain yield and harvest index in bed planted *durum* wheat

Treatment	Grain yield (q ha^{-1})	Harvest index
Nitrogen management (ha^{-1})		
120 kg N	38.33	0.379
120 kg N + 15 t FYM	42.54	0.380
150 kg N	39.19	0.372
CD (P = 0.05)	**NS**	-
Tine of N application		
1/2S + 1/2CRI	38.33	0.373
1/3S + 1/3CRI + 1/3T	39.00	0.372
1/4S + 1/2CRI + 1/4T	39.78	0.379
1/3S + 1/3CRI + 1/3 T + 3% US	41.12	0.378
1/4S + 1/2CRI + 1/4 T + 3% US	41.88	0.383
CD (P = 0.05)	**NS**	-
Interaction	**NS**	-
Flat (Rec.)	**39.53**	**0.377**
Bed (S)	**38.31**	**0.373**

Time of N application did not influence the harvest index value and 3% urea spray tended to increase the harvest index marginally. When whole dose of N was applied at sowing on beds almost similar harvest index value was obtained as compared with two split application on beds and flat.

The performance of *durum* wheat with respect of level of nitrogen management and time of nitrogen application under bed and flat sowing was statistically on par between each other.

Under bed sown, management of nitrogen through integration of chemical fertiliser with FYM (120 kg N + 15t FYM ha^{-1}) tended to improve yield attributed and yield over alone chemical fertiliser in *durum* wheat.

Reference

Deol, K.S. 2001. Studies in nitrogen requirement in relation to method of planting of late sown wheat *(Triticum aestivum* L.). M.Sc. Thesis, Punjab Agricultural University, Ludhiana, India.

Dhuka, A.K.; Sadaria, G.S.; Patel, J.C.; and Vyas, M.N. 1991. Response of irrigated late sown wheat to nitrogen application. *Indian J. Agron.* **36: 276-77.**

Kumar, S. 2002. Studies on nitrogen management in bed planted timely sown wheat *(Triticum aestivum* L.) M.Sc. Thesis, Punjab Agricultural University, Ludhiana, India.

Mairona, M.; Rizzo, V.; and Bari, V.D.I. 1992. Effects of straw incorporation depth and time of nitrogen application on yield of *durum* wheat in continuous cropping. *In: Proceedings of second congress of the European Society for Agronomy.* Warmick University, 23-28 August, 1992. Field Crop. Abstracts, 46: Entry No. 07, 1993.

Narang, R.S.; Brar, S.S.; and Grewal, D.S. 1992. Tillage management of crops in different cropping system/situations. *Final Technical Report.* ICAR ad hoc scheme. Department of Agronomy, Punjab Agricultural University, Ludhiana, pp. **14-15.**

Nehra, A.S.; Hooda, I.S. and Singh, K.P. 2001. Effect of integrated nutrient management on growth and yield of wheat (*Triticum aestivum). Indian J. Agron.* **46(1): 112-17.**

Negi, S.C.; and Mahajan, G. 2000. Effect of FYM, planting methods and fertiliser levels on rainfed wheat. *Crop. Res. Hisar* **20: 534-36.**

Robinson, F.E.; David, C.; and William, F.L. 1979. Nitrate fertiliser timing, irrigation on protein and yellow berry in *durum wheat. Agron. J.* **71: 304-08.**

Singh, R.; and Agarwal, S.K. 2001. Growth and yield of wheat (*Triticum aestivum*) as influenced by levels of farmyard manure and nitrogen. *Indian J. Agron.* **46(3): 462-67.**

Singh, K. 1996. Effect of irrigation regime and nitrogen levels on growth, yield and quality or *durum* wheat *(Triticum durum* desf.) M.Sc. Thesis, Punjab Agricultural University, Ludhiana, India.

Sushila, R.; and Giri, G. 2000. Influence of farmyard manure, nitrogen and biofertilisers or growth, yield attributes ad yield of wheat (*Triticum aestivum*) under limited water supply *Indian J. Agron.* **45(3): 590-95.**

Thakur, S.S.; Pandey, I.B.; and Mishra, S.S.; 1999. Effect of organic manures, fertiliser level and seed rate on yield and quality of late sown wheat (*Triticum aestivum*). *Indian J. Agron.* **44: 754-59.**

Tisdall, J.M.; and Hodgson, A.S. 1990. Ridge tillage in Australia: A review. *Soil Tillage Res.* **18: 127-44.**

Uppal, R.S. 1999. Effect of levels of NPK and time of nitrogen application on yield and quality of *durum* wheat *(Triticum durum* Desf.). Ph.D. Dissertation, Punjab Agricultural University, Ludhiana, India.

❑❑❑

33

People-grove Interface in the Plant Conservation of Agasampattu Sacred Grove, Union Territory of Pondicherry

D. Kadamban and K. Meerabay

Department of Botany, K.M. Centre for P.G. Studies, Pondicherry–605 008

Introduction

Sacred groves are patches of natural vegetation preserved by ancient societies on religious and cultural traditions. They exemplify the perceived interlink between man and his natural environment as well as his ecological prudence (Gadgil and Vartak, 1975; Ramanujam and Kadamban, 1998; Ramanujam, 2000). Sacred goves have widely been recorded from many packets in Asia and Africa (Frazer, 1922).

Gadgil and Vartak (1976) have pioneered such studies in India in the Western Ghats of Maharashtra State. Tamil Nadu, Meher- Homji (1986) first reported a patch of natural vegetation receiving protection due to presence of deities in Puthupattu, about 13 km north of Pondicherry on the East Coast road to Marakkanam and Chennai. Even though the main deity Manjineewara Aiyanar with images of horse and other minions is present, the *puthu* (a termite mound with a snake hole) is the primary place of worship, hence the name Puthupattu. The 20 ha. of green vegetation is still under protection on the basis of religious faith. It is the relict of forest with 104 woody species in 44 families (Sethi, 1993). Parthasarathy and Karthikeyan (1997) described the Aiyanar temples, plant biodiversity and vegetations pattern of two sacred groves, Kulaindaikuppam and Thirumanikkuzhi in Cuddalore district of Tamil Nadu.

Sacred groves have been widely regarded as relicts of past vegetation but have several other attributes too: natural museum of giant trees, treasure house of threatened species, dispensaries of medicinal plants, regulators of watersheds, recreation centre for urbanites, experimental garden for botanist, gene bank of economic species, shelters of rare taxa, a paradise for nature lovers and laboratory for conservation biologist (Vartak, 1983; Vartak and Kumbhojaker, 1984; Vartak *et al.*, 1986; Ramakrishnan *et al.*, 1998; Somasekar, 1998; Kadamban and Ramanujam 1999; Ramanujam and Kadamban, 1999).

A miniature sacred grove on the Passumari hill top, near Vedanthangal Bird Sanctuary in Chenglepattu district of Tamil Nadu with 110 flowering plants in 40 families shows the floristic richness (Maheswaran *et al.*, 1995). Balasubramainam and Rajasekaran (1996) of the Salim Ali Institute of Ornithology and Natural History have surveyed 23 sacred groves of Coimbatore forest division. They enumerated 82 woody angiosperms belonging to 39 families, besides 21 species of birds and three species of mammals.

Ramanujam *et al.* (2004), have enumerated more than 500 groves from a segment of Coromandel coast covering Cuddalore Villupuram, district of Tamil Nadu and Union Territory of Pondicherry. Floristic richness and the biocultural perspectives sustaining them have been analysed in 37 important groves. The study was encouraged by the plant diversity and the conservational significance of groves as refugium, relict and museum of giant specimens (Kadamban, 1998; Ramanujam and Kadamban, 1999). Another intensive survey enlisted Aiyanar temple groves within Pondicherry region. (Krishnan, 2004).

Sacred groves have existed in India from time immemorial as patches of densely wooded area, venerated on religious grounds. Sacred groves have preserved many rare, endangered and endemic wild plant species, many f which hold potential benefit to man in medicine, agriculture and industry. In fact, sacred groves represent the ancient Indian way of *insitu* conservation of genetic diversity. These groves are, therefore, valuable gene pools and the first major effort is to recognise and conserve them.

Methodology

Study Area: Sacred groves are the remnant patches of Tropical Dry Evergreen Forest vegetation described by Champion and Seth (1968) found along the Coromandel Coast of South India. The sacred grove at Agasampattu village situated one km. north of Pondicherry Mailam road was selected for study on the basis of belief system linked to the biodiversity.

Climate and Soil: A typical tropical climate with a dissymmetric rainfall regime prevails in the study area. The weather is generally humid and hot for most part of the year. The South-West monsoon contributes to the average of 356.6 mm of annual rain whereas the North-East monsoon contributes 317.8 mm. The minimum temperature of 18°C in January and the maximum of 41°C in the month of May were recorded. The soil is of clay (Black) type.

Biocultural perspectives: Details of the topography, status of vegetation and the evolution of the temple complex were recorded by the direct observation during field visits. Information on the religious/cultural traditions followed in the sacred grove village was recorded from the priest(s) as well as from the responsible villages. The traditions, beliefs, taboos, restrictions and folklores pertaining to the temples were collected from the villagers and cross checked from the devotees.

Though discussion, observations ad formal of informal interviews are adopted to elicit information, the data were immediately transferred on the questionnaire (Kadamban, 1998), which closely follows the format of World Wide Fund for Nature and Nature Conservation (WWF, 1996).

Floristic Survey: Floristic composition of the grove was intensively studied during field visits. The flowering/fruiting twigs were collected and identified using the regional and state floras (Gamble, 1957; Matthew, 1983).

Results and Discussion

Agasampattu is a small village of 167 households with a population of 500. The grove is dedicated to Aiyanar ad measured about 0.8 ha. The village is dominated by Naidu community. The main deity is Aiyanar, whose image is carved along with his two wives—Poorani and Porkalai—on a monolithic plate; it lies open under *Ficus benghalensis* tree. Images of terracotta horses and elephants of different times with soldiers are also found arranged in rows in front of the main deity. One of the characteristic feature of the sacred grove is the presence of a spring/a stream/a pond/ or a pool adjacent to a stand of trees. It might have been a compelling reason for humans to establish settlements close to these sacred groves (Chandran and Hughes,

1997). This uniqueness was shared by the presence of the pond in the centre of the grove as conserved by Godbole *et al*., (1998).

Taboos and Beliefs

The villagers consider Aiyanar as the only guardian deity for their village and the surrounding villages too. The whole grove is sanctified and people did not use the footwear inside the grove. Women entry is restricted during menstruation period. Tree felling and fire wood collection are completely banned.

Rites and Rituals

People visit the temple regularly on every Monday and the annual festival falls on the first Monday of the Tamil month, Aadi. People offer flowers, fruits, cooked sweetened rice, cloth, terracotta horses, granite elephants, granite horses as votive offerings during important occasions as a mark of fulfilling their vows. One full day event was celebrated within the grove. During that occasion, people will carry head-loads of ingredients for making pongal and arrive in a procession accompanied by playing of musical instruments. Though the women of every family father together and make pongal, the virgin or the unmarried girls make lamps out of rice flour. They offer it to Aiyanar and pray for early marriage.

Floristic composition

The sacred grove of Agasampattu, measuring about 0.8 ha., harbours a total of 105 species representing 92 genera and 40 families (Table 1). Of these, herbs accounted for 33 species (31%) from 30 genera of 14 families and shrubs were 27 species (26%) from 23 genera and 19 families. The trees are represented by 15 species (14%) belonging to 15 genera of 12 families. The climbers accounted for 29 species (28%) from 24 genera and 14 families (Table 2). The gigantic specimens found in the grove include *Mitragyna parivfolia* (gbh 195 cm) of Rubiaceae and *Ficus benghalensis* (gbh 300 cm) of Moraceae. *Dendrophthoe falcata* of Loranthaceae was the only parasitic species found growing on *Albizia lebbeak* trees.

Generally, monsoon forests are rich in trees and climbers as exemplified by 26 tree species and 6 climbers recorded in a 5 ha. grove at Maharashtra (Vartak and Gadgil, 1972). The description of Champion and Seth (1968) about the tropical dry evergreen forest types holds good with the present study. But species richness is generally low in a Tropical Dry Evergreen Forest (Murphy and Lugo, 1986). The tree species inventory is bound to decline in areas where chance factors or human impacts have played a diachronically dominant role (Joshi *et al.,* 1997). Trees had 15 species of 15 genera belonging to 12 families proving this as the rich diversity conserved with the culture, faith and belief. More number of herbaceous (33) species has indicated the heavy disturbance. The controversial results, 15 tree species and 33 herbaceous species is emphasising the recent disturbances.

Table-1: Plant diversity of Agasampattu grove

S. No	Species Name	Family Name	Habit
1	*Andrographisalata* (Vahl) Nees	Acanthaceae	Herb
2	*Elytraria aculais* (L.f.) Lindan	Acanthaceae	Herb
3	*Ecbolium ligustrinum* (Vahl) Vollesen	Acanthaceae	Herb
4	*Rugia pectinata* C. B. Clarke	Acanthaceae	Herb
5	*Dipteracanthus patulus* (Jacq.) Nees	Acanthaceae	Herb
6	*Justicia prostrata* (C.B. Clarke) Gamble	Acanthaceae	Herb
7	*Alangium salviiforlium* (L. f.) Wangerin	Alangiaceae	Tree
8	*Achyranthes aspera* L.	Amaranthraceae	Herb
9	*Aerva lanata* (L.) Juss.	Amaranthraceae	Herb
10	*Alternanthera sessilis* (L.) R. br. Ex DC.	Amaranthraceae	Herb
11	*Gompherena celosioides* C. Martius	Amaranthraceae	Herb
12	*Ichnocarpus frutescens* (L.) R. Br.	Apocynaceae	Climber
13	*Carissa spinarum* L.	Apocynaceae	Shrub
14	*Rauvolfia teraphylla* L.	Apocynaceae	Shrub
15	*Theriophonum minutem* (Willd.) Baillion	Araceae	Herb
16	*Amorphophallous sylvaticus* (Roxb.) Kuth	Arecaceae	Herb
17	*Phoenix pusilla* Gaertn.	Arecaceae	Shrub
18	*Phoenix sylvestris* (L.) Roxb.	Arecaceae	Tree
19	*Calotropis gigantea* R.Brown	Asclepiadaceae	Shrub
20	*Pentatropis capensis* (L. f.) Bullock	Asclepiadaceae	Climber
21	*Tylophora asthmatica* (L. f.) wight et A rn.	Asclepiadaceae	Climber
22	*Wattakaka volubilis* (L.f.) Stapf	Asclepiadaceae	Climber
23	*Blumea mollis* (D. Don) Merr.	Asteraceae	Herb

24	*Parthenium hysterophorus* L.	Asteraceae	Herb
25	*Tridax procumbens* L.	Asteraceae	Herb
26	*Vernonia cinerea* (L.) Less	Asteraceae	Herb
27	*Carmona retusa* (Vahl) Masam.	Boraginaceae	Shrub
28	*Heliotropium indicum* L.	Boraginaceae	Herb
29	*Opuntia dillenii* (Ker Gawler) Haw.	Cactaceae	Shrub
30	*Cassia fistula* L.	Caesalpiniaceae	Tree
31	*Senna auriculata* (wallich) Gamble	Caesalpiniaceae	Shrub
32	*Ehretia pubescens* Benth.	Cordiaceae	Shrub
33	*Capparis brevispina* L.	Capparaceae	Climber
34	*Capparis divaricata* Lam.	Capparaceae	Shrub
35	*Capparis zeylanica* L.	Capparaceae	Climber
36	*Maeruva oblongigolia* (Forsskal) A. Rich.	Capparaceae	Climber
37	*Maytenus emarginata* (willd) Ding Hou	Celastraceae	Shrub
38	*Coccinia grandis* (L.) Voigt	Cucurbitaceae	Climber
39	*Luffa cylindrica* (L.) Roemer	Cucurbitaceae	Climber
40	*Kedrostis foetidissima* (Jacq.) Cogn.	Cucurbitaceae	Climber
41	*Croton bonplandianus* Baill.	Euphorbiaceae	Herb
42	*Euphorbia hirta* L.	Euphorbiaceae	Herb
43	*Euphorbia serpans* Kunth.	Euphorbiaceae	Herb
44	*Euphorbia rosea* Retz.	Euphorbiaceae	Herb
45	*Jatropha glanduiifera* L.	Euphorbiaceae	Shrub
46	*Jatropha gossypifolia* L.	Euphorbiaceae	Shrub
47	*Jatropha tanjorensis* Ellis and Saroja.	Euphorbiaceae	Shrub
48	*Tragia involucrata* L.	Euphorbiaceae	Climber
49	*Tragia plukenetii* R.Sm.	Euphorbiaceae	Climber
50	*Abrus precatorius* L.	Fabaceae	Climber
51	*Butea monosperma* (Lam.) Taubert	Fabaceae	Tree
52	*Canavalia cathartica* L.	Fabaceae	Climber
53	*Crotaloria angulata* Miller	Fabaceae	Herb
54	*Desmodium gangeticum* (L.) DC.	Fabaceae	Herb
55	*Pongamia pinnata* (L.)	Fabaceae	Tree
56	*Pseudarthiria viscida* (L.) Wight et Arn.	Fabaceae	Herb
57	*Rhynchosia minima* (L.) DC.	Fabaceae	Climber
58	*Tephrosia purpurea* (L.) Pers	Fabaceae	Herb

59	*Teramnuslabialsi* (L.f.) Sprengel	Fabaceae	Climber
60	*Flacourtia indica* (Busm. F.)	Flacourtieae	Shrub
61	*Diospyros ferrea* (willd.) Bakh.	Ebenceae	Shrub
62	*Disopyros montana* Raxb.	Ebenceae	Tree
63	*Cureuligo orchiodes* Gaernter	Hypoxidaceae	Herb
64	*Anisomeles indica* (L.) Kuntz	Lamiaceae	Herb
65	*Leucas lavundulifolia* Sm.	Lamiaceae	Herb
66	*Dendrosphthoe falcate* (L.f.) Ettingsh.	Loranthaceae	Parasite
67	*Sida cordata* (Burm.f.) Borssum Waalkes	Malvaceae	Herb
68	*Azadirachta indica* A. Juss.	Meliaceae	Tree
69	*Cocculus hirsutus* (L.) Diels	Menispermaceae	Climber
70	*Tillacora acuminata* (Lam.) Miers	Menispermaceae	Climber
71	*Tinospora cordifolia* (willd.) Hook.f. and Thomson	Menispermaceae	Climber
72	*Acacia caesia* (L.) Willd.	Mimosaceae	Climber
73	*Albizia labbeck* (L.) Benth.	Mimosaceae	Tree
74	*Prospois juliflora* (Sw.) DC.	Mimosaceae	Tree
75	*Ficus benghalensis* L.	Moraceae	Tree
76	*Ficus hispida* L. f.	Moraceae	Shrub
77	*Streblus asper* Lour.	Moraceae	Tree
78	*Jasminum angustifolium* (L.) Willd. var. angustifolium	Oleaceae	Climber
79	*Biophytum sensitivum* D.C.	Oxalidaceae	Herb
80	*Hemidesmus indicus* (L.) R. Br.	Periplocaceae	Climber
81	*Plumbago zeylanica* L.	Plumbagonaceae	Shrub
82	*Polygala arvensis* Willd.	Polygalaceae	Herb
83	*Vetiveria zizanoides* (L.) Nash	Poaceae	Herb
84	*Zizyphus oenoplia* (L.) Miller	Rhamaceae	Climber
85	*Benkara malabarica* (Lam.) Tiru	Rubiaceae	Shrub
86	*Conthium coromandelicum* (Burm. f.) Alston	Rubiaceae	Shrub
87	*Mitragyna purvifolia* (roxb.) Korth var Parvifolia	Rubiaceae	Tree
88	*Morinda pubescens* Sm.	Rubiaceae	Tree
89	*Oldenlandia biflora* L.	Rubiaceae	Herb
90	*Pavetta indica* L.	Rubiaceae	Shrub
91	*Psilanthus wightianus* (Wight et Arn.) J. Leroy	Rubiaceae	Shrub
92	*Atalantia monophylla* (L.) Corr. Serr.	Rutaceae	Tree

93	*Glycosmis mauritiana* (Lam.) Yuich. Tanaka	Rutaceae	Shrub
94	*Limonia acidissima* L.	Rutaceae	Tree
95	*Cardiospermum helicacabum* L. var. microcarpum (Kunth) Blume	Sapindaceae	Climber
96	*Solanum torvum* Sw.	Solanaceae	Shrub
97	*Solanum trilobatum* L.	Solanaceae	Climber
98	*Grewia bracteata Heyne* ex Roth	Tiliaceae	Climber
99	*Grewia hirsute* vahl.	Tiliaceae	Shrub
100	*Gmelina asiatica* L.	Verbenaceae	Shrub
101	*Lantana camera* L.	Verbenaceae	Shrub
102	*Vitex negundo* L.	Verbenaceae	Shrub
103	*Cayretia pedata* (Lour) A. L. Juss ex Gagnepain	Vitaceae	Climber
104	*Cissus setosa* wallich L.	Vitaceae	Climber
105	*Cissus quandrangularis* L.	Vitaceae	Climber

Table-2: The hierarchical analysis of taxa

Parametes	Name of the grove
	Agasmpattu
No. of Species	105
No. of Genus	92
No. of Family	40
Life Form	
Herb	33
Shrub	27
Tree	15
Climbers	29
Parasite	1

Joshi and Gadgil (1991) described the sacred groves as refugia as they contained plants selectively sheltered within them. Magnificent tree species like *Ficus benghalensis* of Moraceae and *Mitragyna parvifolia* of Rubiaceae also strengthening the disturbance as well as showing the age/climax condition of the forest.

Jatropha curcas of Euphorbiaceae, a hilly species was recorded from this grove next to Maheswari (2004). The presence of such species agrees with the view of

Pushpangadan *et al.* (1998) that the sacred groves are considered to be the patches of virgin forest prescribed and protected by the local people on religious ground.

Vegetation analysis: Taxonomically, Fabaceae is the largest family with 10 species followed by Euphorbiaceae with nine species; seven species in Rubiaceae; six species in Acanthaceae; four species each in Amaranthaceae, Asclepidaceae, Asteraceae, and Capparaceae; three species each from Apocynaceae. Menispermaceae, Rutaceae and Verbenaceae. Maximum number of species is contributed by the genera *Capparis, Jatropa* and *Euphorbia* with three each. *Tragia, Diospyros, Ficus*, *Grewia* and *Cissus* had two each.

Acanthaceae was the largest herbaceous family with six species. Amranthaceae, Asteraceae, Euphorbiaceae, Fabaceae had four species of herbs each. The shrubs were represented Apocynaceae, Euphorbiaceae and Rubiaceae with two species each. Rubiaceae had maximum of three tree species followed by Fabaceae, Rutaceae and Mimosaeae had two tree species each. The climbers were mainly represented from Fabaceae (4), Asclepidaceae, Capparaceae and Vitaceae with three species each. In addition to this, there were two climbers from Cucurbitaceae.

Vetiveria zizioides belonging to Poaceae was found growing abundantly around the pond. The margin of the grove had dense vegetation compared to the centre of the grove, it is occupied by a pond which is serving the water source of the grove. *Atalantia monophylla*, *Capparis is brevispina, Capparis sepiaria, Maytenus emarginata* are bushy species found growing along with *Jatropha glandulifera, Jatropha gossypifolia* and *Pentatropis capensis.*

Medicinally valuable plants like *Biophytum sensitivum, Canthium coromandelicum, Desmodium gangeticum, Rauvolfia tetraphylla, Senna auriculata, Solanum trilobatum, Streblus asper, Vitex negundo* are present in the grove. The taxa *Andrographis alata, Jatropha curucs, Rugia pectinata* are uncommon in distribution, and are well represented in this grove. The gigantic specimens of the grove are *Mitragyna parvifolia* (195 ghb) of Rubiaceae, *Ficus benghalensis* (300 cm ghb) of Moraceae.

Chandran and Hughes (1997) enumerated the characteristic of typical sacred groves. The grove analysed in the present study share several of those attributes—an amorphous stone encarved with the images of Aiyanar, Porkalai and Poorani found openly in the grove. It is similar to Vellaripattu (Lalitha, 2003) and Suriyanpet (King,

1997) mimicking the primitive pattern of worship. People perform various kinds of rites and rituals during organised festivals with votive offerings of terracotta images in gigantic forms related to Hindu philosophy of birth, death and rebirth (eternity of life). Special poojas are conducted to celebrate important functions in their families like tonsuring of head of the first born baby of the family, marriages etc.,

The role of belief systems in upholding conservation ethics has been the subject of several texts (Nair, 1987; Jain, 1981; 1983). The cult practices and folklore were prevalent in the sacred groves of Western Ghats (Vartak and Gadgil, 1981). Significantly, majority of the population from the sacred grove villages were belonged to socially and /or educationally depressed strata. Slackening of faith and belief among the people is seen in the form of clear felling and cutting of trees, firewood collection, encroachment of sacred grove area by agricultural land, housing sites etc. These events negate the statement of Gadgil (1995), *i.e.* biodiversity is the corner stone of lie security.

Brandis (1897), described the sacred grove as 'the traditional form of forest protection'. The factors contributing to decline the sacred groves were gathered by Chandran and Hughes (1997).

Usually, the erosion of sacred grove is due to anthropogenic pressures like collection of fire wood, medicinal plant, grazing of plants by cattle, encroachment of sacred grove land by agriculturist of nearby areas. Owing to anthropogenic pressure, the grove is shrinking day-by-day. Such beliefs, taboos, restrictions, folklores and the rituals were practised and passed on. It is therefore tempting to agree with Alcorn (1990) that the prescriptions and the local rules limiting the extractions of natural resources, prevent their indiscriminate destruction. Whatever be the affinities, the sacred groves are conserved ecologically and environmentally by the ethics of the ancestors. The need of the time is to take necessary action to protect these sacred groves, otherwise we lose priceless heritage. The God father created the nature for the beloved son/daughter (people) to enjoy them instead of destruction, we have to mind it.

Though the sacred groves are conserved ecologically and environmentally by the ethics of the ancestors, we should seek the cooperation of the village community by providing better opportunities for their livelihood to protect the sacred land from deterioration.

References

Alcorn, 1990. Cited from Berkes *et al.*, 1995.

Balasubramaniyam, P. and Rajasekaran, A. 1996. An Inventory of sacred groves of Coimbatore forest division, Tamil Nadu. In: *Abstracts of Seminar on Sacred Groves of India-Their Biodiversity Conservation*, Apr. 21 Hyderabad, India.

Brandis, D. 1897, *Indian Forest*, Oriental Institute, Pune, India.

Champion, H.G. and Seth, S.K. 1968. *Forest Types of India*. Manager of Publications, Delhi.

Chandran, M.D.S. and Hughes, J.D. 1997. The sacred groves of South India: Ecology, Traditional Communities and Religious change, *Social Compass*, **44: 413-427.**

Frazer, J.G. 1922. *The Golden Bough*. A Study in Magic and Religion. Macmillan, London.

Gadgil, M. 1995. Traditional Conservation Practices. *Encyclopedia of Env. Biol.,* **3: 423-425.**

Gadgil, M. and Vartak, V.D. 1975. Sacred Groves of India: A Plea for Continued Conservation. *J. Bombay Nat. Hist. Soc.*, **72: 314-320.**

Gadgil, M. and Vartak, V.D. 1976. The Sacred Groves of Western Ghats in India. *Econ. Bot.*, **30: 152-160.**

Godbole, A., Satve, A., Prabhu, A. and Sarnaik, J. 1998. Role of Sacred Groves in biodiversity conservation with local people's participation: A case study from Ratnagiri District, Maharashtra. In: *Conserving the Sacred for Biodiversity Management* (eds. Ramakrishnan, P.S., Saxena, K.G. and Chandra Shekara, U.M.). Oxford IBH Publishing Co. Pvt. Ltd., New Delhi. pp. **233-246.**

Gamble, J.S. 1925-57. *Flora of the Presidency of Madras*. Vol. I-III.

Jain, S.K. 1981. *Glimpses of Ethnobotany*. Oxford University Press, Mumbai.

Jain, S.K. 1983. Rare and endangered species: Observation of rare, imperfectly known endemic plants in the sacred groves of Western Maharashtra. In: *An Assessment of Threatened Plants of India*. Jain, S.K. and Rao, R.R. (eds.) Botanical Survey of India, pp. **169-178.**

Joshi, N.V. and Gadgil, M. 1991. On the role of refugia in promoting prudent use of biological resource, *Theor. Popul. Biol.*, **40: 211-229.**

Joshi, N.V., Suresh, H.S., Dattaraja, H.S. and Sukumar, R. 1997. The spatial organisation of plant communities in a Deciduous Forest: A computational–geometry-based Analysis. *J. Indian Inst. Sci.* **77: 365-375.**

Kadamban, D. 1998. Biocultural perspectives and Plant diversity of Sacred Groves and traditional medical Knowledge in Pondicherry Environs. Ph.D. Thesis, Pondicherry University, Pondicherry.

Kadamban, S. and Ramanujam, M.P. 1999. Sacred Groves of Pondicherry bioregion. In: *National Symposium on Emerging Trends in Plant Sciences*. Department of Botany, Sri Venkateswara University, Tirupathy. March 15-17. Abstract pp. **41-42.**

King, O.I. 1997 Botanical and Socio-Cultural Studies on Sacred Groves in Cuddalore District, Tamil Nadu, M. Phil. Dissertation, University of Madras. (Dept of Botany, Madras Christian College, Chennai), India.

Krishnan, V. 2004. Plant Biodiversity and Biocultural Perspectives of Sacred Groves in Pondicherry and its Environs. Ph.D. Thesis, Pondicherry University, Pondicherry.

Lalitha, R. 2003. Biocultural Perspectives and Plant biodiversity of Four Sacred Groves of Villupuram District of Tamil Nadu. M.Phil. Thesis, Manonmaniun Sundaranar University, Tirunelveli.

Maheswaran, B., Dayanadan, P., and Narasimhan, D. 1995. Miniature Sacred grove near Vedanthagnal Bird Sanctuary. In: *Abstracts of 2nd congress on Traditional Science and Technology of India*, Bio-3, Dec. 26-31, Chennai, India.

Maheswari, R. 2004. Sacred Groves of Vanur taluk in Villupuram District o Tamil Nadu: Plant Biodiversity and people- Grove interface. M.Phil. Dissertation, Pondicherry University.

Matthew, K.M. 1983. *Flora of the Tamil Nadu Carnatic*. Vol. 3.

Meher-Homji, V.M. 1986. Puthupet: A Sacred termite-mound protects a forest. *Blackbuck* II: **1-4.**

Murphy, P.G. and Lugo, A.E. 1986. Ecology of tropical dry forest. *Annual Review of Ecology and Systematics*, **17: 67-88.**

Nair, P.K.B. 1987. *Religion, Mythology and Ecosystem*. Report submitted to the Ministry of Environment and Forests, New Delhi.

Parthasarathy, N. and Karthikeyan, R. 1997. Plant Biodiversity inventory and conservations of two tropical dry evergreen forests on the coromandel coast, South India. *Biodiversity and Conservation,* **6: 1063-1083.**

Pushpangadan, P., Rajendraprasad, M. and Krishnan, P.N. 1998. Sacred Groves of Kerala – A synthesis on the state of art of knowledge. In: *Conserving the Sacred for Biodiversity Management*. (eds. Ramakrishana, P.S., Saxena, K.G. and Chandra Shekara, U.M.). Oxford IBH Publishing Co. Pvt. Ltd., New Delhi. pp. **193-210.**

Ramakrishana, P.S. 1998. Conserving the Sacred: Epilogue. In: *Conserving the Sacred for Biodiversity Management* (eds. Ramakrishana, P.S., Saxena, K.G. and Chandra Shekara, U.M.). Oxford IBH Publishing Co. Pvt. Ltd., New Delhi. pp. **457-458.**

Ramanujam, M.P. 2000. Conservation of environment and Human rights: Sacred groves in cultural connections to biodiversity, PRP *Journal of Human Rights*. **4: 34-38.**

Ramanujam, M.P. and Kadamban, D. 1998. *Role of Sacred Groves in Cultural and Biological Diversity Conservation.* In: Sixth International Congress of Ethno botanical Dialogue between Cultures. Forging meaningful partnership. New Zealand. November 23-28. Abstract, p. **31.**

Ramanujam, M.P. and Kadamban, D. 1999. *Rare, Endangered and Threatened plants occurs in the sacred groves of Pondicherry bioregion.* In: National Symposium on Emerging Tends in Plant Sciences. Department of Botany, Sri Venkateswara University, Tirupathy. March 15-17. Abstract, p. **43.**

Ramanujam, M.P. Kadamban, D., Kumaravelu, G., Krishan, V., Ganesan, T., Devaraj, P., 2004. *Flora of Sacred Groves of Pondicherry Region (A Pictorial Catalogue)*, Pub. – Department of Forest and Wild Life, Govt. of Pondicherry.

Sethi, P. 1993. *Phytosociology of a Tropical Dry Evergreen Forest patch in the Puthupet Sacred Grove, Coromandel Coast, Tamil Nadu.* MS. Thesis, Pondicherry University, Pondicherry.

Somashekar, B.S. 1998. Treasure House in Trouble. *Amruth*, **2(5): 3-7.**

Vartak, V.D. 1983. Observation on rare, imperfectly known and endemic plants in the sacred groves of Western Maharashtra. In: *An Assessment of Threatened Plants of India*, Jain, S.K. and Rao, R.R. (eds.). botanical Survey of India, Howrah, pp. **169-178.**

Vartak, V.D. and Gadgil, M. 1972. Dev-Rahati: An ethno-botanical study of tracts of forest preserved on grounds of religious beliefs; *Proc. Indian Science Congress,* Sixtieth session (Abstracts).

Vartak, V.D. and Gadgil, M. 1981. Studies on Sacred Groves along the Western Ghats from Maharashtra and Goa: Role of belief and Folklores' In: *Glimpses of Indian Ethno Botany*, Jain, S.K (ed.), 1981. Oxford university Press, Mumbai, pp. **272-278.**

Vartak, V.D. and Kumbhojakar, M.S. 1984. Notes on the lianas of some sacred groves in Western Maharashtra. *Biovigyanam*, **11:** pp. **214-215.**

Vartak, V.D., Kumbhojakar, M.S. and Vandana, D. 1986. Sacred Groves- a sanctuary for lofty trees and lianas. In: Proc. *Seminar on Ecodevelopment of Western Ghats*, Kerala Forest Research Institute, Peechi, pp. **55-59.**

World Wild Fund for Nature- Andhra Pradesh, 1996. *Sacred and Protected Groves of Andhra Pradesh.* WWF – AP State Office, Andhra Pradesh.

❑❑❑

34

Electrophoretic Protein Patterns in Different Body Parts of Edible Oyster from Ratnagiri Coast of Maharashtra

G.D. Suryawanshi, K.B. Shejule and Z.A. Langekar*
Dept. of Zoology, Dr. B.A.M. University, Aurangabad
* Naik Sea Food Ltd., Ratnagiri

Introduction

Apart from biochemical analysis, the protein patterns can be detected by electrophoretic techniques in which molecular sizes and molecular weights of a single protein can also be determined. Perusal of literature reveals the reports showing detection of the protein patterns from molluscs. Reports indicate detection of protein patterns in relation to the closely related species, genetic differentiation on the species inhabiting in the different environments, and also isolation and characterisation of proteins. Sreenivasan (1984) studied the protein pattern of the ovary, developing embryo and free veliger in the snail *Cerithium coralium*. Rodriguez *et al.* (1989) studied electrophoretically the variations between two oyster populations of *Crassostrea carteziensis* from Mexican coast. Brock *et al.* (1990) determined the intergenetic distance between *Ostrea crassostrea* and *Saccostrea* by means of immuno electrophoresis. Sarver and Boundnslager (1991) also studied electrophoretically the genetics of California population of the blue mussel. Brandan *et al.* (1992) separated a high molecular weight proteoglycon from mollusc, *C. concholepas*. Nagawanshi (1997) studied the effect of cerebralectomy on protein profile of gonad and hepatopancreas of a freshwater *L. marginalis*. Therefore, the aim of this study is to develop and utilise electrophoretic method on west coast bivalve molluscs for identification of individual species when morphological traits are insufficient for proper classification.

Materials and Methods

Crassostrea cattuekensis were collected with help of commercial fishermen from Bhatye estuary at Ratnagiri coast in monsoon (July/August). Soon after the fishing, they were brought to the laboratory and the shells were brushed to clean the fouling biomass and mud. They where then stocked in continuous aerated filtered seawater pumped in the laboratory from the estuary for 24 to 48 hours for depuration.

The oysters (80-84 mm shell length) were shucked for their different body parts. The 300 mg of each tissue *viz.* mantle, gonad, gill, hepatopancreas, adductor muscle and siphon were homogenised in 0.1M phosphate buffer of pH 7.5. The homogenised tissues were centrifuged at 15000 rpm in a refrigerated centrifuge at 4°C for 10 minutes. The sample was dialysed in cellophane membrane sac within 10-12h in 0.1M, phosphate buffers. After the dialysis the protein content was estimated according to Lowry *et al.* (1951). The electrophoretic system of Genei (Bangalore) dual model was used to determine the protein profile. A discontinuous polyarylamide slab gel electrophoretic systems was prepared according to Davis (1964). The 20 –μ1 samples were loaded in each well on the 10 per cent gel. Simultaneously, standard proteins of different molecular weight markers were loaded on the 10 per cent separate gel. The current of 150V and 30-40 mA was supplied. The electrophoresis was carried out at room temperature. Proteins were stained with amido black stain in a place of commassie brilliant blue because the latter one failed to stain a few protein fractions. The numbers, thickness and length of the bands were counted. The relative mobility values (REM) with the molecular weights (MW) of specific proteins used were determined using regression equation.

Results and Discussion

Nowadays, the highly versatile polyacrylamide gel electrophoresis technique has become the choice of priority for the biochemical researchers in attempts to characterise the protein components present in the bivalve for further biotechnological studies. During electrophoresis of the basic method with denatured proteins using SDS-PAGE has been described in many books and research publications. In the present study when denatured by heating in the presence or excess SDS and thiol reagent (usually 2-mercaptoethanol). Most polypeptides bind SDS in a constant weight ratio such that they have essentially identical charge densities and migrate in polyacrylamide gels of the correct porosity according to polypeptide size. Under these conditions a plot of $\log_{10}$ Polypeptide molecular weight versus relative mobility reveals a straight-line relationship. In the present study showed four different protein molecular weight markers were used as A, B, C and D (Figure 3) which ranged between 14,300 to 97,400, 29,000 to 205,000, 3,000 to 43,000 and 3,000 to 205,000 Daltons respectively. Protein molecular weight marker A showed six bands, B showed five bands, C with five bands and D with six bands. Bands with molecular weight 97,400 and 66,000 Da. were visible in marker A, B and D. Bands with molecular weights 43,000 and 29,000 Da. were visible in all the four types of markers. Bands with molecular weight 201,100 and 14,300 Da. were visible in the markets A, C, and D. Protein molecular weight marker A and D showed six bands. The marker A showed clear visible bands ranging from 14,300 to 97,400 Da. Therefore, for determination of protein profiles in the oysters protein molecular weight marker A was selected as a standard marker, however in this attempt a few proteins of high and low molecular weight from the body parts crossed the range and no further attempt was made for detailed study within the range of standard molecular weight B or D. The protein content in siphon was 0.342, mantle 0.175, gill 0.40, gonad 0.346, adductor muscle 0.415 and hepatopancreas 0.03 mg/20 μl of the tissue homogenate. The maximum value of protein content was recorded in adductor muscle and the minimum in hepatopancreas. The electrophoretic pattern of proteins, *i.e.* protein bands, their relative mobility values (REM) and molecular weight (MW) are presented in Table 1. With regarded to their REM values the bands were named as serially 1 to 13. The siphon exhibited eight different molecular weight protein fractions, of which three bands were dark and others were light. In mantle the total number of bands observed were ten, two bands were light and others were dark. In gill out of total eight bands two were dark and rest were light. In gonad all the ten bands were light. In adductor muscle maximum number of thirteen bands were observed, two bands were dark and others were light. Hepatopancreas showed four protein fractions of which one band was light and others were dark. From the Table, it was revealed that amongst the different body parts the REM values increased with decrease in MW of the proteins. It was also apparent that proteins with specific MW were present probably with tissue specificity and were not common in all the tissues, for example, with MW 37,020 and

Table-1: Relative mobility values and molecular weight of potein in different body parts of *C. cattucknesis*

Tissues		1	2	3	4	5	6	7	8	9	10	11	12	13
Siphon	REM values	0.031	0.091	0.113	0.151	0.316	0.354	0.444	0.512					
	Molecular weight	108,555	94,922	90,365	83,002	57,388	52,712	43,101	37,020					
Mantle	REM values	0.023	0.113	0.151	0.369	0.452	0.474	0.542	0.564	0.602	0.625			
	Molecular weight	110,515	90,365	83,002	50,973	42,337	40,304	34,618	32,956	30,270	28,753			
Gill	REM values	0.031	0.151	0.316	0.369	0.452	0.527	0.542	0.587					
	Molecular weight	108,555	83,002	57,388	50,973	42,337	35,799	34,618	31,303					
Gonad	REM values	0.121	0.151	0.316	0.369	0.407	0.444	0.512	0.527	0.572	0.602			
	Molecular weight	88,763	83,002	57,388	50,973	46,820	43,101	37,020	35,799	32,371	30,270			
Adductor	REM values	0.023	0.106	0.151	0.452	0.489	0.504	0.542	0.572	0.602	0.677	0.925	0.933	0.970
	Molecule weight	110,512	91,791	83,002	42,337	38,974	37,688	34,618	32,371	30,270	25,596	14,699	14,438	13,291
Hepatop-anareas	REM values	0.031	0.151	0.369	0.452									
	Molecular weight	108,555	83,002	50,973	42,337									

52,712 in siphons, MW 46,820 in gonad, MW 28,753 and 40, 304 in mantle, MW 31,303 in gill, and MW 25,596 in adductor muscle.

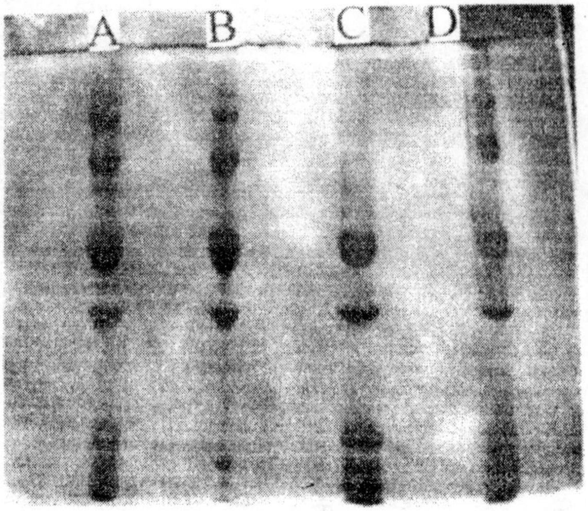

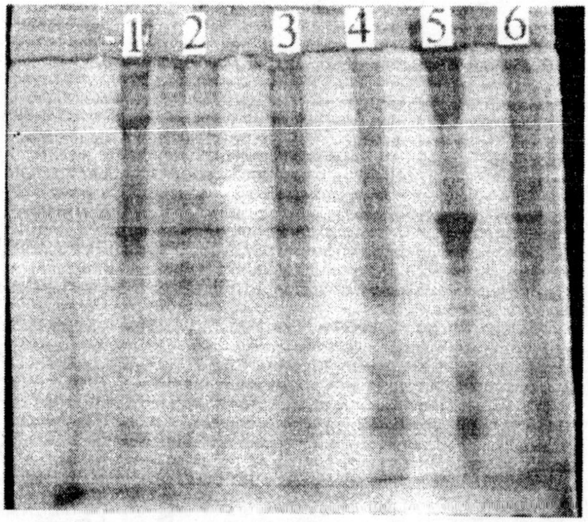

The photographs showing protein profile using standard molecular weight from body parts of oyster in Monsoon

A. Protein Molecular weight marker 14 to 100 kd

B. Protein Molecular weight marker 29 to 200 kd

C. Protein Molecular weight marker 3500 to 43000 kd

D. Protein Molecular weight marker 3500 to 205000 Da

1. Siphon 2. Mantle 3. Gill 4. Gonad 5. Adductor muscle 6. Hepatopancreas.

The approach is, therefore, to electrophorese a set of market polypetides of known molecular with and use the distance migrated by each to construct a standard curve from which the molecular weight and the sample polypetides can be calculated based on their mobility under the same electrophoretic conditions. As mentioned earlier studies on elctrophoretic protein patterns in different animals of most economic importance are receiving much attention. According to Leaback (1976) the fast moving fractions have lower molecular size than those of slow moving fractions and probably simpler molecular constitution. Decrease in the number of protein fractions and simplification of molecular structure might be due to functional requirements for that stage of development in the life cycle. The bivalve molluscs remained an under exploited source of edible protein and hence greater attention needs to be given to their genetic diversity in order to improve amino acid profile and eliminate the antinutritional factor present in them. Further, we studied that distinct protein pattern in terms of the number of bands and also in terms of presence of absence of certain REM values and MWs of poteins. The adductor muscle showed presence of maximum thirteen protein fractions and hepatopancreas showed minimum four protein fractions. The rhythmic activity of adductor muscles in bivalves is well documented and these bivalves in coordination with the external environment open the bivalves for respiration and feeding (Morton, 1983). The oysters are sedentary bivalves throughout their life and must open and close the valves in relation to existing external environment for respiration and feeding. In Bhatye estuary *C. cattuckensis* inhabits at the depth of 5 to 7 meters and among from the river mouth where these submerged populations experience very low salinities during low tides in monsoon, unfavourable for even shell valve opening. Thus these oysters became opportunistic in sense that with the rise in salinity the stimulus to open the valves is reached. In this context for rapid and slow rhythmic activity that it is most likely to have different proteins fraction to a maxima compared to other body parts. the most sensing organ to the external environment for the oysters to open shell-valves is considered mantle, which bear tentacles for rapid extension and contraction during sensing. This tissue of physiological importance also showed maximum protein fractions. The probable reason accounting to minima of different proteins in hepatopancreas amongst different body parts may be that these oysters receive only a limited time for feeding in monsoon. The ingested food from hepatopancreas upon digestion is probably accounting for the energy demand of the oysters to maintenance metabolism and spawning. Thereby, the depot tissue like hepatopancreas, which carries function of digestion and assimilation under the conditions of sample preparation during electrophoretic study, could have lost some proteins, which had likely accumulated for minimum protein fractions compared to other tissues. It was also apparent that proteins with specific MW are present in specific physiological important tissues and the variations in the protein profile amongst some tissues like hepatopancreas and gonad may be brought about due to environmental variables suitable/unsuitable to the oyster's healthy life.

The reports on the electrophoretic protein fractions from bivalve molluscs are scantly. Moreover, similar results along the Mexican coastline sixteen species of oysters belonging to *Ostrea* and *Crassostrea* have been reported by Rodriguez *et al.* (1989). Further, they studied differences between the electrophoretic patterns of two-oyster population of *Crassostrea corteziensis* from those brackish and hyper saline environments in order to follow some of the possible variations in the adaptive genetic expression. Suzuki (1990) studied haemolymph proteins of the pearl oyster, which were separated by SDS and gel filtration into approximately thirty bands ranging from 10-200 kDa in non-reducing condition. Thus the pearl oysters haemolymph found abundant in relatively in low molecular weight proteins of 10-30 kDa. An electrophoretic study of regional variations of protein patterns of different body parts of the edible oyster *C. madrasensis* was made by Samuel and Thangavelu (1991). They have reported the different tissues of the same species from areas suggests that the samples from Ennore, Muttukadu and oyster farm possess distinct protein patterns in terms of the number of bands and also in presence or absence of certain protein fractions. These variations may be brought about by the difference in the environment especially salinity and exposure of the natural beds during low tide and the resultant physiological stress. Nagawanshi (1997) found decreased protein fractions of different soft tissues of the field population of the freshwater bivalve *L. corrianus* upon removal of cerebral ganglia.

References

Brandan, E. 1992. A high molecular weight proteoglycon in differentially expressed during development of molluscs *Concholepas concholepas* (Mollusca: Gastropoda: Muricidae) *J. Expt. Zool.*, **264: 363-371.**

Brock, V. 1990. Intergeneric distances between Ostrea, *Crassostrea* and *Saccostrea*, studied by means of crossed immuno-electrophoresis. *Mar. Ecol. Progress Series*, **68(1-2): 59-63.**

Davis, B.J. 1964. Disc electrophoresis II. Method and application to human serum proteins. *Ann. N. Y. Acad. Sci.*, **121: 404-427.**

Leaback, D.H. 1976. Concentration gradient polyacrylamide gel electrophoresis. In: I Smith (Ed.) *Chromatophoretic and Electrophoretic Techniques II zone Electrophoresis.* William Heinemann Medical Books Ltd., London, **250-271.**

Lowery, O.H., Rosenburangh, N.J., Farr, A.L. and Randall, R.J. 1951. Protein measurement with Foiin-Phenol reagent. *J. Biol. Achem.* **193: 265-275.**

Morton, B.S. 1983. *Feeding and Digestion Bivalvia. The Mollusca* (Ed. A.S.M. Saleuddin and K.M. Wilbur) Vol. 3, Academic Press, New York, pp. **65-147.**

Nagawanshi, Meena 1997. Reproductive physiology of freshwater bivalve molluscs *L. corrianus* from Aurangabad. Maharashtra State. Ph.D. Thesis, Dr. Babasaheb Ambedkar Marathwada University, Aurangabad.

Rodriquez-Romero-F., Garcia-Saez, C., Laquarda-Figueras-A. 1989. Electrophoretic pattern variation in two oyster populations of *Crassostrea corteziensis* from the Mexican coast. Anales del Institute de Ciencias Del Mar. V. Limnologia Univ. National Autonoma de Mexico., **15: 177-184.**

Samuel, P. Philip and Thangavelu, R. 1991. Electrophoretic protein patterns of farm grown and wild edible oyster, *Crassostrea madrasensis* from different locations. *J. Mar. Biol. Ass. India*, **33(1** and **2): 346-349.**

Sarver, S.K. and Boudenslager, E.J. 1991. The genetics of California (USA) populations of the blue mussel: Further evidence for the existence of electrophoretically distinguishable species or subspecies. *Biochemical Systematics and Ecology*, **19(3): 183-188.**

Sreenivasan, P.V. 1984. Protein pattern of the ovary, developing embryo and free veligar in the snail, *Cerithium eoralium* (Kiener). *J. Mar. Biol. Assoc. of India,* **26(1** and **2): 157-159.**

Suzuki, T. 1990. Electrophoretic and chromatographic patterns of haemolymph proteins of the pearl oyster, *Pinctada funcata. Bull. of National Research Inst. of Aquaculture* **18: 7-12.**

❑❑❑

35

Food Alters Inorganic Ions in Moth Larvae

R.K. Gautam* and Shivarti Gautam
Department of Zoology, St. John's College, Agra–282 002
*School of Life Sciences, Dr. B.R. Ambedkar University, Agra

Introduction

The fact that plants and insects have lived in close associations, under conditions that led to many types and degree of interdependence. Because of the survival value of some of these associations, progressive evolution in the direction of greater interdependence has taken place. The larvae forms of moths have emerged as notorious destroyers of their host plants. They exhibit perfect camouflage and form serious pests of young leaves, flowers, buds and other food storing parts of their hosts. Inorganic elements like calcium, sodium, magnesium, phosphorus and iron are very essential for the normal growth and functioning of various organs and systems of the animal body. Among them the calcium and iron are the most important calcium is important in maintaining the all frame work and also work as a co-factor in blood clotting process. Iron is an essential part of haemoglobin. Any change in these elements create physiological disorder in the animal body (Gautam *et al.*, 2001). Hence they have been selected for the present study.

Materials and Methods

Eggs and initial instar stages of the larvae were collected from their host plants and kept for further growth or rearing in wooden cages measuring 30x25x18 cms with wire meshes on five sides, a wooden base, a cardboard door. Freshwater leaves from the host plants were provided everyday. The changes in the stages of the larvae were carefully noted were allowed to pupate and undergo metamorphosis for the confirmation of species and the other larvae were taken out at the mature stage. Killed, dissected out the alimentary canal and histochemical studies were made for the amount of inorganic ions (calcium and iron) in the various regions of alimentary canal, majorly divided into Foregut, Midgut and Hindgut.

The alimentary canal was dissected out fixed in neutral formaline, paraffin sections were prepared and subjected to the following histochemical tests:

[1] Calcium – Alizrin red-S-method (Dahl, 1952).

[2] Iron – Perls methods for ferric iron (Pearse, 1961).

Results and Discussion

Comparative study of *Exelastis catomosa, Margaronia Caesalis – Sonders and Dichoerocis punctiferalis Guen* for localisation of inorganic ions (calcium and iron) in various regions of alimentary canal show following dissimilarities.

Localisation of Calcium

Localisation of calcium in different parts of alimentary canal of all the three moth larvae is almost dissimilar.

Exelastis catomosa

Pharynx and oesophagus showed dull activity. Crop did not at all react. Proventriculus also showed moderate activity and so did the midgut region. Heum and colon of hindgut reacted in a moderate and dull activities respectively. But rectum did not show any response.

Margaronea caesalis Sonders

Pharynx, oesophagus and midgut reacted in a dull activity. Proventriculus showed moderate activity. But crop did not react at all. Heum and colon showed dull activities respectively, rectum exhibited nil activity.

Dichoerocis punctiferalis Guen

Pharynx and oesophagus of foregut showed moderate activity. Crop was dull in activity but proventriculus expressed strong activity. Midgut reacted vigorously to the stain, whereas ileum and colon of hindgut showed moderate activity and rectum exhibited a dull activity. Details are given in Table 1.

Table-1 Localisation of Calcium in the alimentary canal of different moth larvae

Larvae	Host	Stomodaeum (foregut)				Mesenteron	Proctodaeum (Hindgut)		
		Pharynx	Oesophagus	Crop	Proventriculus	Midgut	Heum	Colon	Rectum
Exelastists catomosa	Papilionanceae Spp.	±	±	-	+	+	+	±	-
Margaronea caesalis Sonders	Cucurbitaceae Spp.	±	±	-	+	±	±	±	-
Dichoerocis punctiferalis Guen	Euphorbitaceae Spp.	+	+	±	+ +	+ + +	+	+	+

+++ Vigorous activity; ++ Strong activity; + Moderate activity; ± Dull activity; nil activity.

Table 2: Localisation of Calcium in the alimentary canal of different moth larvae

Larvae	Host	Stomodaeum (foregut)				Mesenteron	Proctodaeum (Hindgut)		
		Pharynx	Oesophagus	Crop	Proventriculus	Midgut	Heum	Colon	Rectum
Exelastist catomosa	Papilionanceae Spp.	-	-	-	-	±	-	-	-
Margaronea caesalis Sonders	Cucurbitaceae Spp.	-	-	-	-	±	-	-	-
Dichoerocis puncitiferalis Gurn	Euphorbitaceae Spp.	-	-	-	-	+	-	-	-

+++ Vigorous activity; ++ Strong activity; + Moderate activity; ± Dull activity; nil activity.

Localisation of Iron

Localisation of iron in different parts of alimentary canal of all the three moth larvae is as follows—In all the three moths larvae. The forgut and hindgut did not responded a positive activity but the midgut showed a mild positive, reaction for iron.

Histochemical observations study of two essential trace lements *viz.*, calcium and iron thus made could be helpful in establishing? Metal ions are involved in a number of processes of biological oxidation. Iron is a vital constituent of both porphrin and nonporphyrin enzymes. In animal with red blood, iron is found in association with a protein in the form of haemoglobin but in tissues. It is rarely present. In the present investigation we observed in the traces in the mesenteron only. On the other hand calcium is involved in physiological control and trigger mechanisms and plays a significant role in maintaining the structure and function of cell wall. The reason why larvae *Dichoerocis punctiferalis* Guen reacted more strongly than the other larave can be attributed to the fact that there is a fair amount of calcium citrate present in the food of the larvae.

References

Dahl, L.K. 1952. Simple and sensitive histochemical methods for calcium. *Proc. Soc. Exp. Biolo* New York. **80: 474.**

Harper, H.A. 1963. *Review of Physiology Chemistry,* 9th ed., Lange Medical Publications, Los Angles, California. p. **437.**

Pearse, A.G.E. 1961. *Histochemical Theoretical and Applied.* J. and Chrchill Ltd., London. p. **942.**

Prosser, C.L. 1973. *Comparative Animal Physiology.* 3rd ed., W. B. Saunders. Philadelphia p. **46.**

Gautam, R.K.; Gautam, Kalpna and Tejeshwari, L. 2001. Pestisides alter inorganic ions in channa punctatus. *J. Batcen.* **13: 221-224.**

❑❑❑

36

Study of Fluctuation of Groundwater Level in Somni Stream Watershed, Patan Block, Durg District, Chhattisgarh

Prashant Shrivastava and Anupama Asthana

Department of Geology, Govt. Arts and Science College, Durg

Department of Chemistry, Govt. Arts and Science College, Durg

Introduction

Somni stream watershed is situated in the northern part of Patan Block, Durg district Chhattisgarh. It cam be easily approached both by rail and road. The national highway No. 6 and Mumbai- Howrah rail route (via Nagpur) runs along northern boundary of the watershed area. The Somni stream watershed falls in Survey of India Toposheet No. 64G/8 and G/12. It is bounded by latitudes 21°5'-21°14'N and longitudes 81°23'-81°34' E.

Materials and Methods

The groundwater structures present in the area and dug wells, dug cum bore wells and tube wells. According to Survey report of PHE Department (1989) there are 274 dug wells and 284 tube wells are present in the study area. Most of the dug wells are private and vary in depth between 5 to 18 meters bgl. The PHED bore wells normally range in depth in between 60 and 100 m bgl. 38 dug wells distributed over the entire watershed area were selected as observation wells. The pre-monsoon and post-monsoon static water levels have been recorded from these wells. The well inventory data is tabulated in Table 1.

Table-1: Well Inventroy Data of Pre-Monsoon and Post-Monsoon Season, Somni Mala Basin

S. No.	Name of Village	Name of Owner	Diameter of well in meter	Height of the parapet in meters	Total depth in meter	Depth of water table (in meter)		R.L. of water table (in meter)		Water table fluctuation (in meter)	Remarks lined/unlined and location
						Post-monsoon	Pre-monsoon	Post-monsoon	Pre-monsoon		
1	Urla	Mr. Marar	3.00	G.L.	15.10	3.70	12.95	284.740	275.490	9.25	Lined/Near Tank
2	Kukda	Govt. well	2.50	1.0	13.90	4.30	11.80	282.690	275.190	7.50	Lined/Near Road
3	Charoda	Public well	1.90	0.50	7.20	3.10	5.65	290.990	287.340	2.55	Lined/Near Jyoti School
4	Devbaloda	Jagdish	2.90	0.50	11.80	4.50	9.65	285.670	279.020	5.15	Lined/Near Temple
5	Sirsa	Kharag Singh	1.80	0.90	15.00	4.85	13.20	-	-	8.35	Lined/Near Road
6	Purena	Public well	2.40	0.30	7.90	3.60	5.85	-	-	2.25	Lined/Near Village
7	Somni	Public well	2.10	-	11.0	4.40	10.45	-	-	6.05	Lined/Near Market
8	Nardhi	Bisahu Shah	2.10	0.40	14.65	4.20	12.75	278.990	270.140	8.55	Lined/Near Bari
9	Aundhi	Public well	2.90	0.60	15.60	4.80	11.85	283.150	276.100	7.05	Lined/Near Village
10	Bhatagaon	Public well	3.20	0.40	13.40	4.10	11.75	286.366	278.610	7.65	Lined/Near Village
11	Pahanda	Govt. well	2.50	0.40	11.50	4.80	10.35	277.920	272.370	5.55	Lined/Near Village
12	Khamaria	Public well	2.80	0.60	14.50	3.20	11.75	283.170	274.620	8.55	Lined/Near Village
13	Motipur	Gram Panchayat	3.60	0.60	9.50	4.15	8.55	286.625	282.225	4.40	Lined/Near Village
14	Batang	Public well	2.50	G.L.	10.50	4.50	10.10	282.270	276.640	5.60	Lined/Near Village
15	Morid	Rajiya Bai	3.00	0.70	9.50	3.40	7.40	303.690	299.690	4.00	Lined/Near Canal
16	Amlidih	Khaman	3.00	1.00	12.50	3.10	11.00	285.090	277.190	7.90	Lined/in Village
17	Mohka khurd	Harlal Dau	2.85	G. L.	13.10	4.90	8.35	298.075	295.590	3.45	Lined/in Village
18	Mohka kalan	Public well	2.50	0.20	9.00	2.60	7.95	296.410	291.060	5.35	Lined/Near Temple
19	Ganiyari	Leela Satnami	3.20	0.65	13.75	2.75	10.60	293.160	283.510	7.85	Lined/Near Temple
20	Pachperi	Hulsa Satnami	3.00	0.90	14.90	2.30	11.75	291.520	282.070	9.45	Lined/in Village

S. No.	Name of Village	Name of Owner	Diameter of well in meter	Height of the parapet in meters	Total depth in meter	Depth of water table (in meter)		R.L. of water table (in meter)		Water table fluctuation (in meter)	Remarks lined/unlined and location
						Post-monsoon	Pre-monsoon	Post-monsoon	Pre-monsoon		
21	Pathandor	Govt. well	3.00	0.90	9.60	2.50	6.75	286.620	282.370	4.25	Lined/in school
22	Aunri (B)	Public well	2.80	1.00	14.00	3.90	12.95	286.940	277.890	9.05	Lined/in Village
23	Kurudih	Ashwani	2.50	0.60	14.50	2.50	12.85	279.135	267.285	10.35	Lined/in Village
24	Parsada	Bughas Ram	2.50	G.L.	12.00	3.15	11.45	274.215	265.915	8.30	Lined/in field
25	Ameri	Public well	3.20	1.00	10.15	2.80	9.00	293.160	286.760	6.20	Lined/in Village
26	Karsa	Public well	1.40	G.L.	12.75	3.95	12.30	284.695	276.345	8.35	Lined/in Village
27	Parewadih	Chamariya	1.30	G.L.	6.35	2.00	5.95	298.835	294.085	3.95	Lined/in Village
28	Jamgaon	Public well	2.50	0.50	13.30	4.40	10.85	294.305	287.855	6.45	Lined/in Village
29	Gabnra	Public well	2.70	0.60	11.80	4.15	9.55	298.525	293.125	5.40	Lined/in Village
30	Ghugwa	Manrakhan	2.10	0.70	17.90	3.10	15.60	292.405	279.895	12.50	Lined/in field
31	Patora	Bhuneswar	3.60	-	13.30	3.10	12.90	310.630	300.830	9.80	Lined/near canal
32	Chunkatta	Chanu	1.05	-	5.19	1.11	5.00	305.290	301.400	3.89	Lined/on Road
33	Murpar	Vishal	0.95	-	7.10	2.00	6.00	301.410	296.310	4.00	Lined/on Road
34	Dhaur	Bishram Singh	3.65	-	10.64	3.41	9.90	290.670	284.780	6.49	Lined/in Field
35	Bendri	Kanhaiya	1.85	-	10.10	1.45	9.00	289.100	280.650	7.55	Lined/in Field
36	Rakhi	Govt. well	2.95	0.90	14.00	2.00	12.12	304.630	254.510	10.12	Lined/in Field
37	Achanakpur	Bhagelan	3.14	0.87	11.13	2.02	9.20	295.040	287.160	7.18	Lined/near Tank
38	Raweli	Public well	3.11	8.82	14.21	2.86	13.20	304.535	295.955	10.34	Lined/at garden

(a) Post-monsoon Water Table

The post-monsoon static water level of the Somni stream watershed area range in between 271.85 and 308.90m above means sea level. The highest post-monsoon SWL is recorded from the village Parsada. The SWL values are plotted on the watershed map and the water table contours are constructed (Figure 1). It is clear from the figure 1 that the general flow of groundwater during post-monsoon is towards NE. The close contour spacing in the areas between Kukda, Parsada, Ghughwa and Karsa villages indicate the groundwater gradient is steep in these protein of the watershed area.

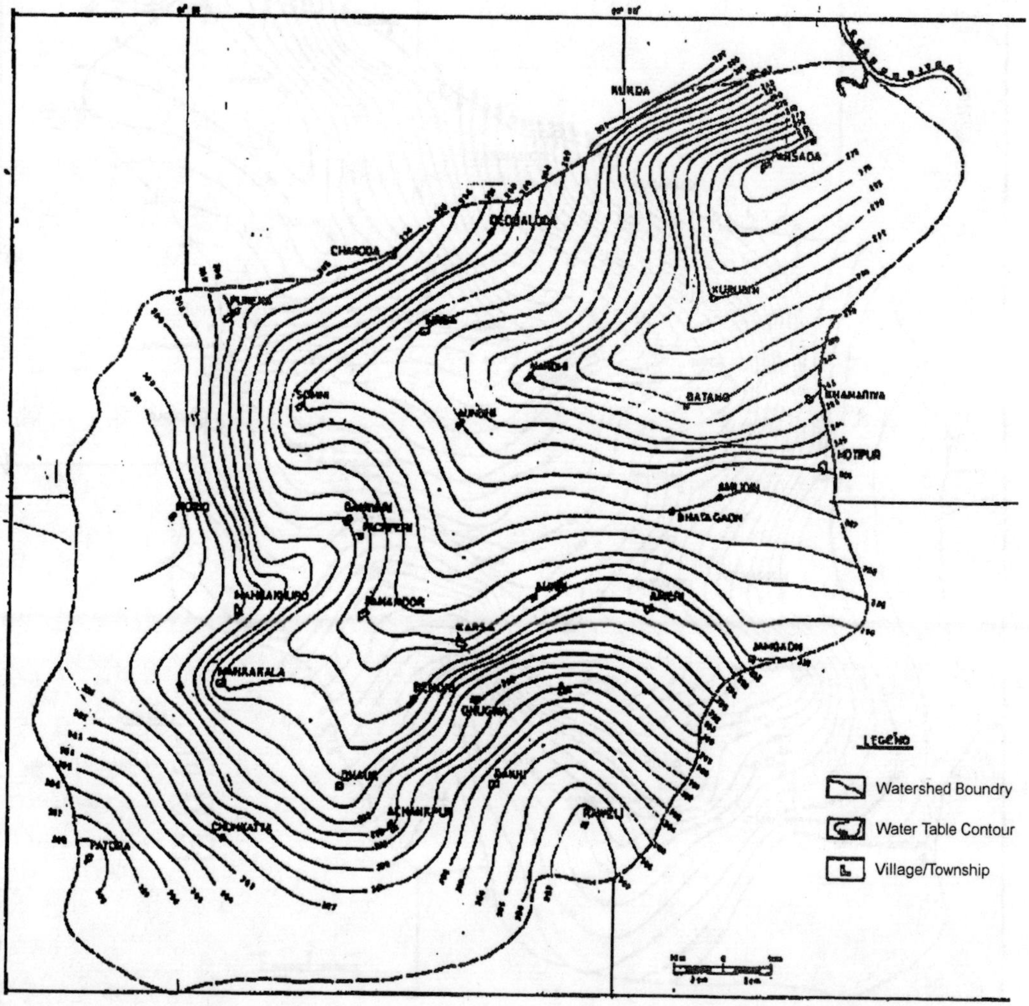

Fig. 1: Post-Monsoon water table contour map of Somni Nala Basin

(b) Pre-monsoon Water Table

The pre-monsoon highest SWL 299.1 m above and lowest 265.5m above msl are recorded from the villages Patora and Parsada respectively. The pre-monsoon water table contour map is presented in Figure 2. The general flow of ground water is in NE direction. The close spacing in water table contours in between villages Purena-Somni-Sirsa-Charoda and near Ghughwa, Aunri indicate the steep gradient of groundwater around these villages.

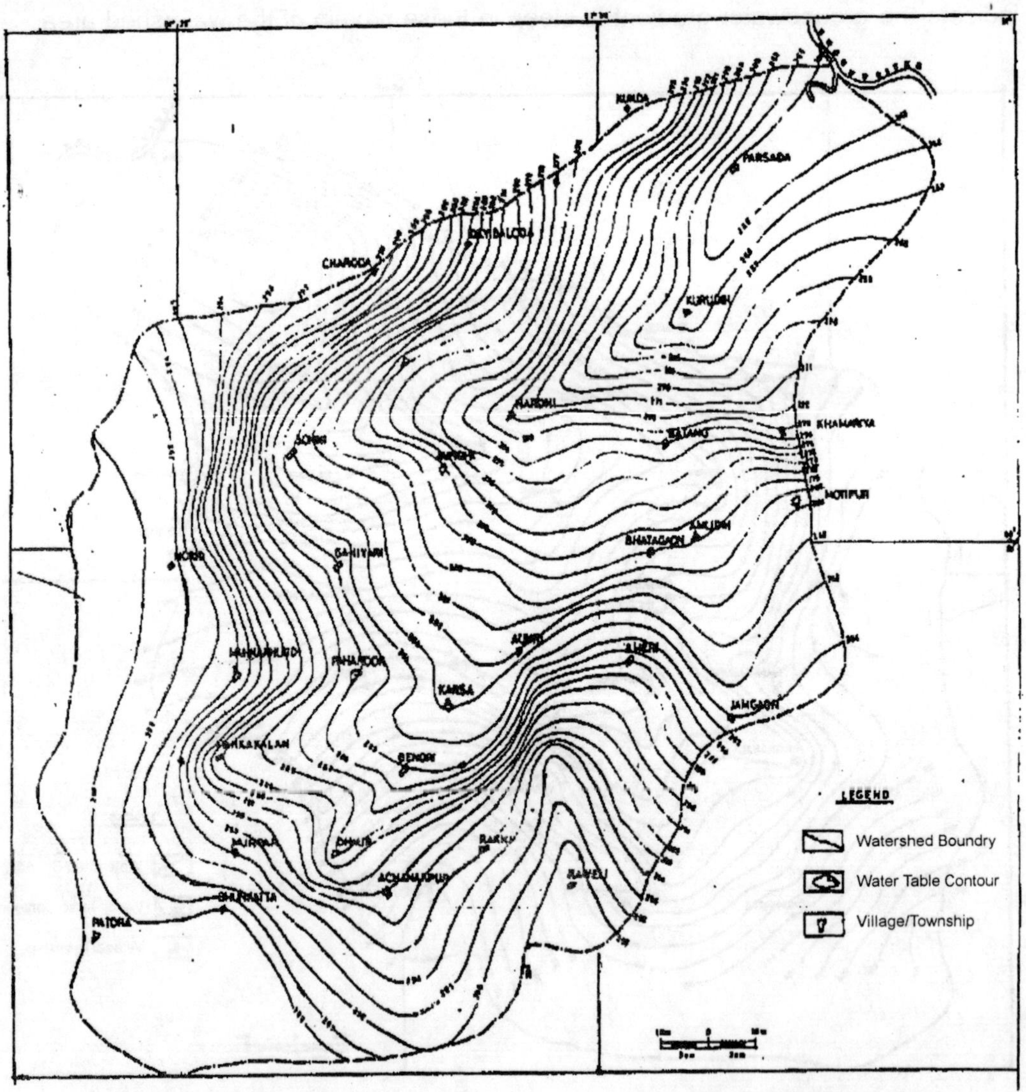

Fig. 2: Pre-Monsoon water table contour map of Somni Nala Basin

Result and Discussion

Groundwater fluctuation

The post-monsoon SWL in the watershed area is generally lowered during pre-monsoon by 4-10 m. The water table contours constructed on the basis of the difference in the pre-monsoon and post-monsoon SWL are presented in Figure 3. It is clear from the Figure 3 that the groundwater level fluctuation is greater in wells of Ghughwa (max. 11 meters), Rakhi, Rawell, Patora as compared to the fluctuations observed in wells of NW portion of the area between Mohkakalan-Morid-Purena-Chardoa and in the wells of SE part of the area near Gabnra-Ameri-Jamgaon and Motipur. The SWL fluctuation in this area is low up to 5 meters.

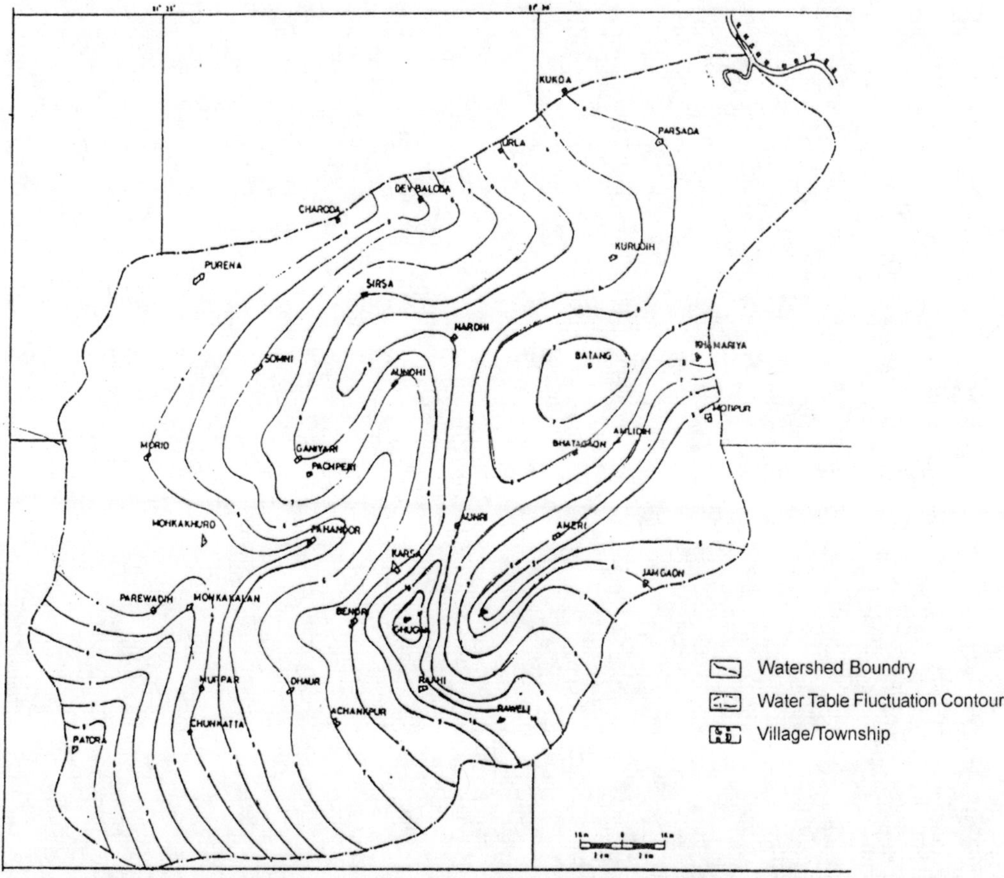

Fig. 3: Water table fluctuation map of Somni Nala Basin

References

Shrivastava, P.K., 1998. Impact of Bhilai Steel Plant Effluent on Quality of Groundwater of Somni Nala Basin, District Durg, Madhya Pradesh, India, pp. **66-98.**

State Groundwater Board of M.P., 1979, Geohydrological Survey Report of the Patan and Gurur Block, Durg District, M.P., pp. **1-81.**

❑❑❑

37

Physico-chemical Studies on Pollution in River Sengar at District Etawah (U.P.)

K.K. Saxena*, Raj Narayan and Yogesh Babu Dixit

P.G. Department of Zoology, Janta Mahavidyalaya, Ajitmal, Auraiya (U.P.)

Introduction

The water quality in terms of physico-chemical characteristic was studied to assess the river water suitability for drinking purpose. Most of the Indian rivers and their tributaries are being used as sites for disposal of demestic sewage and industrial waste leading to the grave problem of water pollution.

The river Sengar originates from and flows through agricultural lands known as Gangetic plains and confluences with the river Yamuna. The river has its stretch of about 300 km. As per geographical preview, Shikohabad and several village discharge their domestic and industrial waste and thus make this river highly polluted. The river reduces to a slow spill channel during summers. Moreover, pollution load is increasing at a fast rate due rapid industrialisation and urbanisation.

This investigations on the river Sengar were conducted fortnightly at five sampling stations (A to E) it showed the effect of pollution on the physico-chemical aspects of the river water at district Etawah (U.P.) in summer, rainy and winter seasons (2003). A few physico-chemical parameters were studied in the field and rest were done in the laboratory.

Materials and Methods

The river Sengar passes through varied topographical and meteorological conditions, hence it provides heterogeneous ecosystem. There are definite interrelationships between plankton and the physico-chemical factors of the river Sengar and an analysis of data reveals the fluctuations thereof and their bearing on the plankton population.

Five sampling stations A to E were selected on the basis of varied ecological condition of the exploration Zone due to domestic and industrial wastes. Station A was setup near the entry point in district Etawah at Dhanua village. Station 'B' was setup near the confluence of its tributary 'Sirsa' river at Chakwa which is about 30 Km. away from the first sampling station, while station 'C' near the village Morhi which is about 20 Km. ahead from sampling station 'B' and the station 'D' is located near village Jua. The distance covered upto this station is about 15 Km. from station 'C'. The sampling site 'E' was set up near Dibiyapur. It is about 15 km. away from the sampling station 'D'.

The water samples were collected fortnightly by immersing a wide mouthed bottle at the sub-surface level. Samples were collected during morning hours in between 8.30 to 10.30 A. M. Brown sample bottle having 250 Ml capacity were used for collecting samples for dissolved oxygen. The sample of the water for physico-chemical characteristics were analysed following Standard Methods (APHA *et al.*, 1995) using HACH model DR-EL/4Kit with a digital pH meter.

Results and Discussion

The growth reproduction and the development of biota is influenced by physico-chemical factors, *i.e.* temperature, transparency, pH, dissolved oxygen, free carbon dioxide.

During the present investigation period, the water temperature of the river stretch explored ranged between 18.5°C to 33.0°C. The minimum was recorded during winter and the maximum during summer season. The minimum and maximum temperature of the river shows a clear correlation with atmospheric temperature. Chacko and Ganapati (1949) reported about the water temperature which followed the atmospheric temperature being hotest during May and June and lowest during December and January. Transparency is an important limiting factor in the population and distribution of fresh water plankton. Transparency in the river sengar was low in rainy season due to heavy rains while it increased in winter and summer at station 'B' and 'D' respectively. Joshi *et al.* (1973 b) found that the transparency of the water of the river Bhagirathi at Garhwal Himalaya did not show wider annual variations. Colour of the Sengar river water was muddy rainy season while green colour was in winter and summer seasons. Green colour indicated fish transparency and correlated with increased plankton production.

The pH plays an important role on the biotic life of a river system as has been suggested by Verma *et al.*, (1977); Rai and Munshi (1979) while Chakraborty *et al.* (1959) reported that pH value of the river Yamuna at Allahabad during 1954-95. The value of Hydrogen ion concentration was low (7.0) in monsoon season and highest value was (8.7) in winter season.

Dissolved oxygen plays a vital role in supporting aquatic life in the running water, which is susceptible to the slight environmental changes in D.O. concentration in the river water may be due to polluted affluent, which lowers the dissolved oxygen. In the present investigation the D.O. concentration varied from minimum of 4.0 mgl to a maximum of 8.5. mg/l. The high value of D.O. was recorded during the winter season where temperature was low, and low value of D.O. was observed during the summer season when the temperature was high and river flow was also low. This finding about D.O. are in conformity with George *et al.*, (1966) and Saxena *et al.* (1966). Carbon dioxide is one of the essential constituents of an aquatic ecosystem and the abundance of it exerts certain specific effects upon aquatic biota. The river Sengar exhibited maximum carbon dioxide as in summer but low concentration were recorded during winter season. In the present investigation carbon dioxide was found to be inversely proportional to the dissolved oxygen. The same correlation between carbon dioxide and dissolved oxygen was reported earlier by Ray *et al.* (1966) and Prakash *et al.* (1978).

The BOD and COD are the significant parameters of pollution (Odum, 1971). The higher load of them produces obnoxious conditions (Verma *et al.*, 1977) and in the open drains they are directly concerning the productivity (Bewtra and Radhacharan, 1968). The value of BOD was minimum in winter and maximum in summer at all the five stations.

The chemical oxygen demand in the river Sengar was higher than that of BOD. The maximum values were recorded in summer and minimum were in rainy season. In river Yamuna COD was also found higher than BOD by Joshi and Bisht (1993) from which it may be concluded that same degree of biodegradable oxygen demanding pollutants are present in the water.

Table-1: Physico-chemical characteristics of water of river Sengar at various sampling stations in different seasons (2003)

Parameters	Seasons	Sampling stations				
		A	B	C	D	E
Temperature (°C)	S	29.8	31.0	30.0	29.5	33.0
	R	22.0	23.0	22.8	22.4	24.0
	W	18.5	19.1	18.8	19.0	20.0
Transparency (cm)	S	37.0	35.5	36.0	38.0	35.5
	R	21.0	20.5	20.5	19.0	17.5
	W	50.0	50.5	49.0	49.5	47.5
Colour	S	Green	Green	Green	Green	Green
	R	Muddy	Muddy	Muddy	Muddy	Muddy
	W	Green	Green	Green	Green	Green
pH	S	7.6	7.3	7.5	7.2	7.6
	R	7.0	7.5	7.8	7.2	7.1
	W	8.0	8.7	8.3	8.1	8.2
Dissolved oxygen (Mg/l)	S	4.9	5.0	4.8	4.2	4.0
	R	7.6	5.0	5.3	6.1	5.3
	W	8.5	6.6	6.0	6.5	7.4
Carbon dioxide (Mg/l)	S	14.0	9.8	13.6	10.9	14.1
	R	9.8	13.2	11.9	10.3	12.6
	W	8.3	13.0	10.8	10.5	10.2
B.O.D. (Mg/l)	S	10.6	10.2	10.8	16.5	14.5
	R	7.7	9.9	7.6	8.5	8.2
	W	5.5	5.2	4.4	4.3	4.3
C.O.D (Mg/l)	S	44.1	41.5	40.8	42.1	39.8
	R	26.0	38.5	35.2	37.5	38.8
	W	32.0	30.0	33.2	33.0	31.0

S = Summer; R = Rainy; W = Winter

References

APHA, AWWA, WPCF: 1995. *Standard Methods for the Examination of Water and Wastewater.* American Public Health Society Pub. Washington, D.C., 19th. Edn., p. **874.**

Bewtra, J.K. and Radhacharan 1968. Further observations on progression of biochemical Demand. *Environ. Hlth.*, Vol. **10: 51-60.**

Chacko, P.I. and Ganapati, S.V. 1949. Some observations on the Adyar river with special reference to its hydrobiological conditions. *Indian Geogr. J.*, **24(3): 25-49.**

Chakrabarty, R.D.; Roy, P. and Singh, S.B. 1559. A quantitative study of the physico-chemical conditions of the river Yamuna at Allahabad in 1954-55. *Indian J. Fish* **6(1): 186-203.**

George, M.G.; Qasim, S.Z. and Siddiqui, A.Q. 1966. A limnological survey of the river Kali with special reference on fish mortality. *Environ. Hlth.* **8(4): 262-269.**

Joshi, B.D.; Pathak, J.K.; Singh Y.N.; Bisht, R.C.S. and Joshi, N. 1993 b. Phytoplankton production in the snow bed river Bhagirathi in the Garhwal Himalaya. *Him. J. Env. Zoo.*, Vol. **7: 60-63.**

Joshi, B.D. and Bisht, R.C.S. 1963. Some aspects of physico-chemical characteristic of western Ganga canal near Jawalpur at Haradwar. *Him. J. Env. Zool*, **7: 76-82.**

Odum, E.P. 1971. *Fundamentals of Ecology*. W.B. Saunders Co.; Philadelphia. 3rd Ed., p. **574.**

Prakash, C.; Rawat, D.C. and Grover, P.P. 1978. Ecological study of river Yamuna (Poiya Ghat to Jawahar Bridge) *IAWPC. TECH. Ann.* **5: 32-45.**

Rai, D.N. and Munshi, J.S.D. 1979. Observation on diurnal changes of some physico-chemical factors of three tropical swamps of Darbhanga (North Bihar). *India, Comp, Physiol. Ecol.* Vol. **4(2): 52-55.**

Ray, P.; Singh, S.B. and Sehgal, K.L. 1966. A study of some aspects of ecology of the rivers Ganga and Yamuna at Allahabad (U.P.) in 1958-59. *Proc. Nat. Acad. Soc. India (B)*, **36(3): 235-272.**

Saxena, K.L.; Chakrabarty, R.N.; Khan, A.Q.; Chattopadhyaya, S.N. and Harish Chandra 1966. Pollution studies of the river Ganga, near Kanpur. *Environmental Health*, **8: 270-285.**

Verma, S.R.; Tayagi, A.K. and Dalela, R.C. 1977. Studies on the characteristics and disposal problem of industrial effluents with reference to ISI Standards (Part II). *Ind. J. Environ. Hlth*, **19(3): 165-175.**

❑❑❑

38

Trimmatostroma—A New Pathogen on *agave*

Radhika Devi, Dr. P.R. Sushama, Dr. P. Surekha Rani
Mycology and Plant Pathology Lab, Department of Botany,
Osmania University College for Women, Koti, Hyderabad–500 195

Introduction

Agave angustifolia Haw. is in ornamental plant which is commonly grown in the college campus at Osmania University College for Women, Koti., Hyderabad (Andhra Pradesh), everywhere for its beautiful ever green large foliage which is perennial and has aesthetic value.

A perusal of literature confirms that it is found on dead leaves in Maharashtra (Rao *et al*., 1987), on dead woods in Madhya Pradesh (Sharma and Sushama, 1990) and from soil (Manoharachary *et al*., 1977), but there is no previous record of this pathogen on living pants in India.

Therefore, it is a new record of fungi on living plants from Hyderabad. (A.P.)

Materials and Methods

Fresh diseased/infected leaves of *agave angustifolia* Haw was collected from the college campus and symptoms were identified surface sterilisation of the diseased leaf material was done by using sterilised distilled water. The sterilised leaf bits of infected tissues were placed at equidistance on agar plate under aseptic condition to isolate the fungal pathogen on Potato Dextrose Agar medium (Riker and Riker, 1936). *Pure cultures* were obtained by single spore isolation technique and sub-culture was maintained on PSA slants. Morphological character of fungal spores and culture was studied and identification was done with the help of standard manual (Ellis, 1971).

Infection started from lower leaves in the form of small irregular/regular, reddish brown spots surrounded by an indistinct pale olive green or yellow halo. individual spots varied in size from minute pinpoint to about one centimeter in diameter, scattered over entire leaf blade. When the infection was severe, coalescing of spots produced irregular blotches on the leaf blade and finally killed the leaves. Killing of leaves usually started from tip or margin of the leaf lamina. The necrotic spots were light brown in the centre portion with dark reddish brown margin on the periphery.

Trimmatostroma

Sporodochia pulvinate some time confluent, dark blackish brown to black. Mycelium emmerced, stroma present, erumpent, often large brown, pseudoparenchymatus. Conidiophores macronematous or semi-macronematous meristematic, growth in length taking place by a lying down of cross walls and elongation of cells just behind the apex, unbranched or occasionally branched, usually short, straight of flexuous, closely packed together forming sporodochia, pale to mid brown, smooth rugulose or verruclose, conidiogenous cells integrated, terminal, roughly cylindrical, fragmenting. Conidia formed in branched basipetal chains, schizogenous, dry, simple, cylindrical, rounded at the apex, ellipsoidal, clavate, pyriform, sub-spherical etc.; pale to dark brown, smooth or verruculose with transverse and often longitudinal or oblique septa.

T. salicis Corda

Conidia curved or bent, cylindrical, rounded at the ends, or clavate, often forked or variously branched, with upto 13 transverse septa and sometimes one or a few longitudinal or oblique septa, pale to mid brown, walls and septa darker, smooth and spore measurement ranges between 11μ – 40μ x 5μ – 10μ.

Pathogenicity test of the isolated fungi was conducted by artificial inoculation, using blotter techniques (moist blotter method). Fungal spore mass from 7 to 10 days old culture was inoculated on sterile healthy leaves of *Agave angustifolia* plants and was incubated at room temperature for a week. Similarly control, without fungal spore mass was maintained.

Results and Discussion

The results for Pathogenicity test conducted *in vitro* showed the characteristic leaf spots identical to the natural symptoms in the field. Resolution yielded the original fungal pathogen, thus establishing its pathogenicity. The leaves which were served as control did not show any symptoms.

Thus the entire pathological survey reveals that the diseased caused by *T. salicis* on *Agave angustifolia* is not recorded earlier from the living plants. Therefore, the survey found a new pathogen record from Hyderabad A. P.

References

Ellis, M.B. 1971. *Dematiaceous Hyphomycetes*. Commonwealth Mycological institute, Kew, England, p. **41.**

Rao, V.G. and Varghese, K.I.M.; Hasiza, Rajak and Singh. *Perspective in Mycological Research*. International bio science series. Vol. 12 N. Delhi: Today and Tomorrow Printers and Publishers, **1-5**, 1987.

Riker, A.J. and Riker, R.S. 1936. Introduction to Research on Plants Diseases. John, S. Switco. Inc. St. Louis, Chicago, New York, Indian polis.

Sharma, N.D. and Neema Sushma 1990. New host record of fungi from India. *Indian Phytopathology*. **43(3): 477-481.**

❑❑❑

39

Impact of Mercuric Chloride on the Biochemical Composition of Blood of *Clarius batracus* (Linn.)

A. Parithabhanu and M.A. Subramanian
P.G. and Research Department of Zoology, C.N. College, Erode–368 004

Introduction

At present, there is an increasing concern over heavy metal pollution as it endangers aquatic ecosystem (Ramakrishna Rao, 1992). Rapidly expanding industrial potential in India has led to various research works in the field of heavy metal pollution in Indian rivers (Bhuwaneswaran *et al.*, 1999). Of all the heavy metals, mercury is the most toxic one and many mercuric compounds are included in the blacklist by all International capitals (Tayler *et al.*, 1972). Many freshwater bodies in India are shown to be polluted by mercury (Agarwal and Kumar, 1978). Metelev *et al.* (1971) have opined that even traces of mercuric chloride in water bodies may be highly toxic to the animals. The heavy metals are found to affect the physiology and biochemistry of fishes (Jha, 1991; Janarthana Reddy *et al.*, 1992; Dhana Packiam *et al.*, 1998). Histological changes caused by heavy metals have also been studied in fishes by Richardson *et al.* (1974), Bergeri and Patel (1982) and Pavlos and Mahagen (1987).

The impact of pollutants can be well ascertained by analysing the blood, as the blood is the reflector of the whole body pathophysiology. Therefore, the haematological parameters are of great value in understanding functional status of the animals exposed to toxicants. Investigators like Pandey *et al.* (1984), Sambasiva Rao and Ramana Rao (1985), Sakthivel and Sampath (1987) and Seker and Christy (1996) have made studies on hematological parameters with reference to pesticides in fishes. The adverse effects of industrial effluents on various blood parameters are let known by Haniffa *et al.* (1986), Subramanian *et al.* (1988) and Varadaraj *et al.* (1993). Much importance has also been given to show the adverse effects of heavy metals on the blood of many fresh water fishes (Balabanova, 1999; Mitra *et al.*, 2001; Ramalingam *et al.*, 2001; Joshi *et al.*, 2002). However, studies pertaining to the toxicity of mercuric chloride on the biochemical composition of blood in fishes are scanty. Therefore the present work has been carried out to evaluate the toxicity of mercury on various biochemical parameters in the blood of common fresh water teleost, *Clarius batracus.*

Materials and Methods

The fishes measuring 15-18 cm in length and weighing about 10-15 gms procured from a local fish farm and inorganic mercuric chloride supplied by Sarabai Chemical Ltd., Mumbai were used. A standard mercuric chloride solution was prepared by dissolving 10 mg of salt in 10 litre of water. From this standard solution, different concentrations (0.02, 0.04, 0.06, 0.08, and 0.10 ppm) were prepared. After determining the LC_{50} 96hr value of the heavy metal to the fishes, a group of 10 laboratory-acclimatised fishes were introduced into each concentration for 3 weeks with appropriate controls. The blood from the fishes were collected and subjected to biochemical estimations, such as total free amino acids (Yemm and Cocking, 1957), total proteins (Gornall *et al.*, 1949), total free sugars (Roe, 1955), glycogen (Kemp and Kits, 1975) and total lipids (Barnes and Black Stock, 1973).

Results and Discussion

As per Table 1, the glycogen content, total proteins and total lipids decrease in the blood of *C. batracus* under the toxicity of mercuric chloride whereas the levels of total free sugars and total free amino acids increase. In animals, any stress could inflict excessive energy demand, which is immediately fulfilled by blood glucose. As a result, the blood sugar level could increase as observed in the present experimental fishes. As a consequence, the glycogen reserve in the blood could be subjected to glycogenolysis with the resultant depletion. This is in conformity with the findings of Mukhopadhyay and Dehadrai (1980) and Natarajan (1989) in pesticide-treated fishes. The reduction in the glycogen content in the blood of present experimental fishes could

Table–1: Biochemical consituents in the blood of control and treated *C. batracus*. (Each value is mean ± SD of 5 observations)

Concentrations of $Hgcl_2$ (ppm)	Biochemical consituents (mg/100 ml)				
	Total free amino acids	Total Proteins	Glycogen	Total free sugars	Total lipids
Control	450.00 ± 16.73	69.00 ± 6.63	16.80 ± 1.49	149.00 ± 6.63	392.68 ± 15.43
0.02	472.00 ± 13.29 NS (+4.88; r = 0.72)	59.80 ± 1.32* (13.33; r = 1.00)	13.60 ± 1.35* (–19.04; r = 0.58)	164.00 ± 10.48 NS (+10.06; r = 0.35)	355.01 ± 25.11 NS (–9.59; r = 0.88)
0.04	512.00 ± 11.66* (+ 13.77; r = 0.61)	51.60 ± 1.35* (–25.21; r = 0.51)	11.00 ± 0.89* (–34.52; r = 0.69)	196.00 ± 10.74* (+31.54; r = 0.08)	328.34 ± 22.24* (–16.38; r = 0.75)
0.06	572.00 ± 13.26* (+27.11; r = 0.90)	43.00 ± 1.67* (–37.68; r = 0.54)	9.20 ± 1.16* (–45.23; r = 0.98)	236.00 ± 10.62* (+58.38; r = 0.38)	276.47 ± 19.64* (–29.59; r = 0.70)
0.08	68.00 ± 7.48* (+35.11; r = 0.63)	31.80 ± 1.16* (–53.91; r = 0.10)	8.00 ± 1.26* (–52.38; r = 0.89)	246. 00 ± 11.12* (–65.10; r = 0.97)	194.19 ± 20.04* (–50.54; r = 0.99)
0.10	628.00 ± 13.26* (+85.55; r = 0.90)	29.00 ± 2.36* (–57.97; r = 0.95)	6.60 ± 0.01* (–60.71; r = 0.93)	266.00 ± 10.19* (–78.52; r = 0.08)	179.86 ± 16.87* (–54.19; r = 0.80)

* Significant at P>0.05; NS = Not Significant

\+ Indicates increase over control; – Indicates decrease from control

also be sue to the inhibition of glycogenesis, as also observed by Kabeer *et al.* (1978) in bivalves treated with pesticide.

In the present investigation, mercuric chloride causes reduction in total proteins in the blood of *C. batracus*. This is suggestive of degradation of proteins with the resultant increase of total free amino acids. This corraborates the findings of Kabeer *et al.* (1978) in pesticide-treated molluscs and Joyce Shoba Rani and Janaiah (1991) in *C. batracus* under pesticide toxicity. Nagabhushanam *et al.* (1983) has reported that the free amino acids serve as supplementary energy source under the condition of emergency during chronic stress.

The decrement of total lipids in the blood of the species under study indicates the utilisation of lipids to meet the energy demand during the stress caused by mercuric chloride. According to Srinivas *et al.* (1991), the endogenous fat is the only source for energy requirements in animals during prolonged stress. This falls in agreement with the findings of Jebakumar (1993) and Govindan (1994) who have reported the decrement of lipids in the tissues of fishes during stress caused by insecticides.

Thus the metal intoxication has disturbed the normal functioning of cells with the resultant alterations in the fundamental biochemical mechanisms in fishes. This would in turn result in the mortality of fishes on chronic exposure to the heavy metal.

References

Agarwal, M. and Kumar, H.D. 1978. Physio-chemical and physiological assessment of two mercury polluted effleunts. *Ind. Environ. Health*, **20: 141-155.**

Balabanova, L.V. 1999. Toxic effects of heavy metals on the immuno competent cells of fish. *Report of World Fish Convention*. 1999.

Barnes, and Black Stock, J. 1973. Estimation of lipids in marine animals and fish. Detailed investigation of the Sulphophosphovanillin method for total lipids. *J. Exp. Mar. Biol. Ecol.*, **12: 103-118.**

Bergeri, K.V. and Patil, H.S. 1982. Behaviour responses of freshwater fishes Barbus aurilus and Lepidocephalichithy ountes to heavy metals. Proind. Sym. on Physiological Response of Animal to Pollution.

Bhuvaneswaran, N., Santhalakashmi, G. and Rajeswari, S. 1999. Water quality of river Adyar in Chennai city. The river a boon or a bane, *Indian. J. Env. Protection.*, **19(6): 412-415.**

Dhanapackiam, P., Manohar, L. and Sampoorani, V. Median lethal effects of mercury and cadmium on oxygen consumption and gill histopathology of common crab, Cyprinus carpio. *J. Ecobiol.*, **10(3): 193-198.**

Gornall, A.G., Bardawill, C.J. and David, M.M. 1949. Determination of total serum proteins by means of biuret reaction: *J. Biol. Chem.*, **177: 551-556.**

Govindan, V.S., Jacob, L. and Devika, R. 1994. Toxicity and metaboilic changes in *Gambusia affinis* exposed to Phosphamidon. *J. Ecotoxical. Env. Monit.*, **4(1): 1-6.**

Haniffa, M.A., Murugasan, A.G. and Prochelvi, M. 1986. Haematological effects of distillery and paper mill effluent on *Heteropneustes fossilis* (Blood) and *Sarotherodon mossambicus* (Peters). *Proc. Indian Acad. Sci. (Anim. Sci.)* **95: 155-161.**

Janarthana Reddy, S., Kalarani, V., Therakanandha, B., Reddy, D.C. and Ramamurthi, R. 1992. Changes in energy metabolism of the fish *Labeo rohita* in relation to prolonged lead exposure and recovery *J. Ecotoxicol. Environ. Monit.* **8: 43-45.**

Jebakumar, S.R.D., Kumaraguru, A.K. and Jayaraman, J. 1993. Accumulation and disspation of phosphomidon in the tissues of freshwater fish *Oreochromis mossambicus. Comp. Physiol. Ecol.*, **18(1): 12-17.**

Jha, B.S. 1991. Alteration in the protein and lipid contents of intestine, liver and gonads in the lead exposed freshwater murrel, *Channa puntatus* (Bloch). *J. Eco. Biol.*, **3(1): 29-34.**

Joshi, P., Deepa Harish and Manjushri Bose. 2002. Effect of lindane and malathian exposure to certain blood parameters in a freshwater teleost fish *Clarias batrachus. Polu. Res.* **21(1): 55-57.**

Joyce Shoba Rani, V. and Janaiah, C. 1991. Ammonia metabolism in freshwater teleost *Clarias batrachus* (Linn.) on exposure to trichlorofon. *Bull. Environ. Contam. Toxicol.* **46: 731-737.**

Kabeer, A.I., Begum, M.D., Sivaiah, S. and Ramana Rao, K.V. 1984. Sub-lethal toxicity of malathian on the proteases and free amino acid composition in the liver of teleost *Tilapa mossambica* (Peters). Tocicol. Letter. **20: 59-62.**

Kemp, A. and Kits, V.A.J.M. 1975. A colorimetric micromethod for the determination of glycogen in tissues. *J. Biochem*, **56: 646-648.**

Metelev, V.V., Kanrew, A.I. and Dzasokhova, N.G., 1971. *Water Toxicology*, Vodnaya Tokshikologiyhs Kolos Publishers, pp. **67-101.**

Mitra, J., Samir Banerjee and Nirmal Kumar Sarkar. 2001. A study of food-chain transport and haematotoxicity of Lead. *Poll. Res.*, **20(2): 163-166.**

Mukhopadhyay, P.K. and Dehadrai. 1980. Biochemical changes in the air breathing eat fish *Clarias batrachus* (Linn.) exposed to malathion. *Environ. Pollut. Ser.* A. **22: 149-158.**

Nagabhushanam, R., Kedarkar, M.S. and Sarojini, R. 1983. *Textbook of Animal Physiology*, Oxford and IBH Publishing Co., New Delhi.

Natarajan, G.M. 1989. Changes in the carbohydrate metabolism during acute and chronic exposure of air breathing fish *Channa straistus* to xydemeton methyl (Metasystax), *Comp. Physiol. Ecol.*, **14(14): 181-184.**

Pandey, B.N., Prasad, S.S., Chanchal, A.K. and Singh, S.B. 1984. Effect of some pesticides on the oxidative metabolism and haematology of Clarias batrachus (Bloch). *Proc. Sem. Eff. Pest. Agi. Fau.*, **13-18.**

Paulose, P.V. and Mahajan, C.L. 1987. Effect of temperature on toxicity and accumulation of mercury compounds in a fish *Gambusia affinis. Com. Physiol. Ecol.*, **12(1): 52-54.**

Ramalingam, V., Vimaladevi, V., Narmadaraji, R. and Prabakaran, P. 2000. Effect of lead on haematological and biochemical parameters in freshwater fish *Cirrhina mrigala. Pollu. Res.*, **19(1): 81-84.**

Ramakrishna Rao, T. 1992. Missing the words for the trees. An overview of aquatic Ecotoxicology in India. Abstract No. 81. Loth European conference on Ecotoxicology Amsterdams. The Netherlands.

Richardson, M.E., Fox, M.R.S., Jr. Fry, B.E. 1947. Pathological changes produced in Japanese guail by ingestion of cadmium, *J. Nutr.*, **104: 323-338.**

Roe, J.H. 1955. The determination of sugar in blood and spinal fluid with anthroma reagent, *J. Biol. Chem.,* **212: 335-343.**

Sakthivel, M. and Sampath, K. 1987. Effects of dye-stuff effleunts on different levels of protein and blood parameters of *Cyprinus carpio. Proc. Nat. Symp. of Environ. Biol.*, Department of Zoology, St. Xavier's College, Palaymakotti, Tamil Nadu, India.

Sambasiva Rao, K.R. and Ramana Rao, K.V. 1986. Independent and combined toxicity of cabaryl and phenthoate on some haematological parameters of murrel *C. Punctatus* (Bloch). *Ind. Zool.*, **10: 29-31.**

Sekar, P. and Christy, I. 1996. Haematological changes in the freshwater cat fish *Mystus vittatus* exposed to sublethal concentration of phosphomidon. *J. Ecobiol.*, **8(1): 25-28.**

Srinivas, Md. Rafi, G., Prasad, T.A.V. and Reddy, D.C. 1991. Effect of thermal acclimation on the lipid metabolism in a freshwater field crab, *Ozoitelphusa senex* (Fabricius). *J. Ecolbiol.* **3(2): 134-147.**

Subramanian, M.A., Shahul Hameed, and Varatharaj, G. 1988. Haematological changes in the fish *Macrones keletius* (Gunther) after exposure to the tannery effleunt. *The Indian Zoologist*, **12(162): 71-74.**

Tayler, 1972. A review of the lethal and sublethal effects of mercury on aquatic life. *Residue Rev.*, **72: 33-70.**

Varatharaj, G., Subramanian, M.A. and Nagarajan, B. 1993. The effect of sublethal concentration of paper and pulp mill effleunts on the haematological parameters of *O. mossambicus* (Peters) *J. Environ. Biol.* **14(4): 321-326.**

Yemm, E.W. and Cocking, E.C. 1957. The determination of amino acid in the ninhydrin, *Analyst*, **80: 209-213.**

❑❑❑

40

Flavobacterium ferrugineus as an Ecofriendly Biopesticide against Larvae of *Pericallia ricini* (Arctidae: Lepidoptera)

S. Dinakaran and S. Anbalagan

P.G. Department of Zoology, Madura College, Madurai–625 011

Introduction

Total eradication of agricultural pests from agro-ecosystems in impossible and scientifically unnecessary. More preciously regulation of their population density below the economic injury level is possible technically and meaningful. This solution is partially achieved by synthetic chemical insecticides. However, they are equally responsible for several effects such as environmental contamination, biological magnification in food chains, development of resistance in pests and harmful effects on non-target organisms. Hence the need for alternative control methods that are cost-effective and ecofriendly is a greater at present. Biological control using variety of natural enemies offer an excellent alternative to chemical pesticides and it forms a major component of current Integrated Pest Management programmes (Jeyaraj, 1985; Ananthakrishnan, 1992; Narayanan, 1995). The utility of insect pathogenic microbes especially bacteria appears to be gaining acceptance and popularity in applied entomological researches (Falcon, 1971; Morris, 1982; Dinakaran and Krishnan, 2002). *Pericallia ricini* commonly called, as wooly bear is a serious polyphagous pest during its larval stages. It infests a wide range of edible crops such as potato, castor, banana, pumpkin, cotton, sesamum, moringa and brinjal (Nayar *et al*., 1976; Chockalingam, 1979). In the present research, we isolated and taxonomically characterised a pathogenic bacterium, *F. ferrugineus* from naturally diseased larvae of *P. ricini*. Pathogenicity of pure culture was evaluated against the larval stages of same pest under laboratory conditions and the results are presented in this communication.

Materials and Methods

Naturally infected third and fifth instar larvae of *P. ricini* were collected from the castor plantations in previously autoclaved glass vials, sealed immediately and transported to laboratory in cold box. They were surface sterilised in 0.01 per cent mercuric chloride for few minutes, washed several changes in sterile water. The whole body homogenate of larvae was prepared in sterile water and streaked onto nutrient agar plates under aseptic conditions. Most consistently growing colonies were sub-cultured and maintained in agar at 37°C. Among them, a yellow colour colony was found to be more consistent in occurrence and prominent in growth. This isolate was subjected to twelve standard biochemical tests to identify at species level by following Buchannon and Gibbons (1974). Colony-morphological characters were also considered for identification. Freshly moulted third, fourth and fifth instar larvae of *P. ricini* from stock culture were starved for 5 hours and used as test larvae for pathogenecity test. As a preliminary study, bacterial suspension from pure culture incubated at 37°C for 24 hours is directly used as pesticide formulation without any quantification. The suspension was sprayed onto healthy castor leaves using an automiser and the starved larvae were then fed on the sprayed leaves. Freshly treated leaves were given daily. A control was maintained in leaves treated with only nutrient broth. The experiments were replicated thrice each comprising ten larvae. Observation on percentage mortality over control were made upto 96 hours and converted into probit mortality values.

Results and Discussion

The morphological and biochemical features of the pathogenic isolate suggest that it would be *F. ferrugineus*. Mortality data of third, fourth and fifth instar larvae of *P. ricini* fed on pure culture of this pathogen are given in Table 1. Third instar is more susceptible to *F. ferrugineus* when compared to fourth and fifth instar larvae. It shows 100 per cent mortality after 48 hours treatment when compared to 86-90 per cent in other two stages. Isolation and characterisation of pathogenic bacteria from naturally diseased insects and then evaluation of pathogenecity of their pure culture as biopesticide is a most widely accepted protocol in microbial pesticide research (Jayaraj, 1985; Mohamed *et al.*, 1992; Dinakaran and Krishnan, 2002) as well as in understanding the aetiology of microbial diseases in productive insects (Nagendran *et al.*, 1999). For instance, Mohamed Ali *et al.* (1992) have isolated three species of bacteria (*Bacillus thuringiensis, Serratia marcescens* and *Pseudomonas aeruginosa)* from naturally infected larvae of *Hyblaea purea* and tested their pathogenecity against the same pest. Mortality of pests after treatment with bacterial pesticides varies with species and age of the larvae. First two instars of *Heliothis armigera* were more susceptible to biobit, a commercial formulation of *B.thuringeiensis* than the late instars (Douressamy and Regupahty, 1993). Similarly *B. thuringiensis*, *S.marcescens* and *P. aeruginosa* have caused 100, 70 and 33.3 per cent mortality respectively in larvae of *H.puera* (Mohamed Ali *et al.*, 1922).

Table-1: Mean, standard deviation and percentage for mortality in larval stages of *P. ricini* exposed to *Flavobacterium ferrugineus*

Instar	Exposure Duration (hours)					
	24		48		72	
	Mean ± S.D.	% Mortality	Mean ± S.D.	% Mortality	Mean ± S.D.	% Mortality
III	6.67±0.58	66.6	2.67±0.58	93.2	0.67±0.58	99.8
IV	7.67±0.58	76.7	1.67±0.58	93.2	0.67±0.58	99.9
V	7.67±0.58	76.7	1.33±0.58	89.9	1.0±0	99.9

As in the present study, oral feeding of pathogenic bacteria through leaves was found to be superior than spraying the larvae topically (Kalla *et al.*, 1997; Roychoudhary *et al.,* 1995). Therefore it could be expected that *F.ferrugineus* could be sprayed over the plant to suppress *P. ricini.* The consistent occurrence of *F. ferrugineus* in field-collected larvae of *P. ricini* suggests that this pathogen may play an important role in containing the pest in nature. So far many species of *Bacillus* such as *B. papillae*, *B. cereus, B. sphaericus* and *B. firmus* were reported as entomopathogenic against lepidopteran (Burges, 1981; Varma and Mohamed, 1986). *Streptococcus* sp. was reported to involve in flacheric disease of silkworm (Krishnaswami *et al.*, 1978). Inspection of dead larvae in the present study reveals milky disease in which haemolymph turns to white fluid. So far several species of *Bacillus* were reported to cause milky disease in many insect pest including larval Lepidoptera (Dutky, 1963; Steinhaus, 1964). Field trial as well as evaluation of pathogenecity to other pests are two major avenues needed research in future with reference to *F. ferrugineus* biopesticides.

References

Ananthakrishana, T.N. 1992. *Emerging Trends in Biological Control of Phytophagous Insects*. Oxford and IBH, New Delhi, p. **239.**

Buchannon, R.E. and Gibbons, N.E. 1974. *Bergy's Manual of Determinative Bacteriology*. 8th Ed., Williams and Wilkin, Baltimore.

Burges, H.D. 1981. *Microbial Control of Pests and Plant Diseases: 1970-1980*. Academic Press, New York, p. **939.**

Chockalingam, S. 1979. A study on food consumption and utilisation in *Pericallia ricini* (Lepidoptera: Arctidae). *Comp. Physiol. Ecol.*, **4: 78-80.**

Dinakaran, S. and Krishnan, N. 2002. Bioefficacy of a bacterial pesticide (*Streptococcus rehi*) against larvae of Pericallia ricini (Arctidae: Lepidoptera).

Douresswamy, S. and Regupathy, A. 1992. Efficacy of *Bacillus thuringiensis* and endosulfan against larvae of *Heliothis armigera* Hubn. *J. Appl. Zool. Res.*, **4: 45-47.**

Dutky, S.R. 1963. Microbial pesticides. In: *Insect Pathology: An Advanced Treatise* (Ed. E.A. Steinhaus), Academic Press, New York, pp. **75-116.**

Falcon, L.A. 1971. Use of bacteria for microbial control. In: *Microbial Control of Insects and Mites* (Eds. H.D. Burger and N.W. Hussey), Academic Press, London, pp. **67-96.**

Jeyaraj, S. 1985. History and development of microbial control. In: *Microbial Control of Pest Management*. TNAU. Coimbatore, pp. **1-18.**

Kalla, S., Joshi, K.C. and Pant, N.C. 1997. Effectiveness of three toxins of *Bacillus thuringiensis* against shisham defollator *Plecoptera reflexa*. *J. App. Zool. Res.*, **8: 115-116.**

Krishnaswami, S., Narasimhanna, M.N., Suryanarayanan, S.K. and Kamaraj, S. 1978. *Silkwork rearing-Sericulture Manual II.* FAO and Oxford and IBH, New Delhi, pp. **108-109.**

Morris, O.N. 1982. Bacteria as Pesticide: forest application. In: *Microbial Viral Pesticides* (Ed. E. Kurstar), Marcel Dekkar, New York, pp. **239-288.**

Mohamed Ali, M.L., Verma, R.V. and Sudheendrakumar, V.V. 1992. Microbial pathogen associated with forest insect pests of Kerala. In: *Emerging Trends in Biological Control of Phytophagous Insects* (Ed. T.N. Ananthakrishnan), Oxford and IBH, New Delhi, pp. **119-128.**

Nagendran, N., Thiruvalluvan, M. and Krishnan, N. 1999. *Pseudomonas* sp. associated with Flacherie affected silk work. *J. Appl. Zool. Res.*, **10: 181.**

Narayanan, K. 1995. Microbial control of inset pest: A corner stone in sustainable agriculture. In: *Plant-Microbe-interaction in Sustainable Agriculture* (Eds. R.K. Bhel, A.L. Khurana and R.C. Dogra). CCS Publisher, New Delhi, pp. **173-182.**

Nayar, K.K., Ananthakrishnan, T.N. and David, B.V. 1979. *General and Applied Entomology*. Tata McGraw Hill, New Delhi, pp. **265-266.**

Roychoudhary, N., Joshy, K.C. and Pandey, D.K. 1994. Effect of three variety of toxins of *Bacillus thuringiensis* against teak leaf foliator *Ectectona machaeractis* Walker (Pyralidae). *J. Appl. Zool. Res. Res.*, **30: 57-60.**

❑❑❑

Effect of Isopod Parasite, *Cymothoa indica* on Pearl Spot, *Etroplus suratensis* (Bloch) from Parangipettai Coastal Waters (South-East Coast of India)

M. Rajkumar, P. Perumal, P. Santhanam and N. Veerappan
Centre of Advanced Study in Marine Biology, Annamalai University, Parangipettai–608 502

Introduction

Several facultative and obligatory parasitic members of the order Isopoda are deleterious to fishes, capable of exhibiting tremendous destructive activity (Trilles, 1979; Maxwell, 1982). Parasites interference in the biology of fishes as they affect their behaviour, health and distribution (Rohde, 1993). Isopods associated with fish constitute a significant population of crustacean ectoparasites, yet little is known about their biology (Bunkley-Williams and Williams, 1998).

To date, about 450 isopod species have been identified as parasites in freshwater and marine teleosts, as well as elasmobranchs (Kabata, 1985; Moller and Andress, 1986). Most are cymothoids flabelliferans, a group of isopods with a short free-living planktonic phase. Cymothoids are exclusively parasitic and the presence of a few adults can cause damage to hosts. They are protandric hermaphrodites, living on the skin, in the gill chambers, or in the mouth of the fish the position is thus highly specific (Bear, 1951 and Trilles, 1969). They feed on host blood, plasma in wounds and fish tissue, sometimes resulting in massive tissue damage (Bunkley-Williams and Williams, 1998).

Diseases are a most serious limiting factor in the aquaculture production and knowledge of the fish diseases and parasites is highly essential for a successful aquaculture, particularly in a country like India, which has long and highly productive coastal waters. But, a glance at the relevant literature shows that the tropical coastal and brackish-waters, which likely to contain many unknown parasites and pathogens, are poorly documented.

Isopoda absorb their nourishment directly from the host's body and thus depend upon their hosts for feeding (Pillai, 1958). Segal (1987) observed the death of host fish, *Meniida beryllina* died due to the attack of isopod parasite, *Nerocila acuminate* which lived the physical damage produced by the isopod parasite, upon the host has been studied by a few workers. Mini Nair (1987) however, found that the feeding intensity of fishes was unaffected due to the infestation by isopods.

In spite of all these reports, works pertaining to pathological effects of isopod parasites on the physiology of host fishes are scanty particularly from Parangipettai environments. Hence the present attempt was made to study the effect of isopod infestation on (host) fish, with reference to reduction of the respiratory surface area and loss of weight in the pearl spot, *Etroplus suratensis*.

Materials and Methods

The samples of fish, *E. suratensis* were collected from Vellar estuary (Lat. 11° 29'N, long. 79° 96'E) and examined thoroughly for he presence of isopod parasites. *E. suratensis* is an Asian cichlid (Family: Cichlidea) endemic to Southern India and the Island of Sri Lanka and they are popularly called as Pearl spot (Green chromide) (Keshava *et al.*, 1988). It is an oval-shaped fish, laterally compressed fish with a pointed head. The body colour is creamy olive green to greenish brown. The body is marked with six to eight slighty diagonal transverse bars which may at times, be indistinct. Each scale has a golden spot and the fins are the same colour as the body. The anal fin may have some blue iridescence. The back of the dorsal fin is a reddish-orange colour. Their face lacks the spangling of the body and is instead a metallic apple green, which is very pronounced on the forehead and radiating into their eye. The dorsal fin is long and the base extends to the caudal peduncle. The soft rays are more elongated than the hard front rays of the dorsal fin. The total length of the fish was ranged from 400-460 mm. It is a common species occurring throughout the year (Mills and Lambert,

2004). The site of attachment, orientation of parasites (on the host) and the number of parasites in each location were recorded.

To study the effect of infestation on the pearl spot fish (*E.suratensis*), the length and weight data were analysed. The fishes were categorised as infested and uninfested. The average length and weight of infested and uninfested fishes were determined. The loss of weight and percentage loss of weight of male and female fishes were calculated. Respiratory surface area of gill arch was measured to find out the influence of parasites on the gill surface area. The arches of uninfested and infested fishes were carefully dissected out and blotted to remove the moisture. Then the total surface area of gill arch of both infested and uninfested fish was compared, and the differences were noted and considered as the reduction of respiratory area due to infestation.

Results

Effect on weight of host fish

The effect of *C.indica* infestation on the weight of *E.suratensis* is presented in (Tables 1 and 2) (Figs. 1 and 2). It is evident from the results that the infestation affected the weight was 6.7 g when compared to that of parasitised fish. Further the weight loss was more pronounced in female (49.57%) than in males (33.23%).

Table-1: Effect of *Cymothoa indica* in relation to the body weight (g) of *Etroplus suratensis*

Nature of Infestation	Mean body weight	Reduction of weight	Percentage reduction of weight
Uninfected	87.97	18.24	6.72
Infected	66.64	10.40	

Table-2: Percentage weight (g) variation in male and female of *Etroplus suratensis*

Particulars	Sex of host	Uninfected	Infected
Mean weight	Male	54.72	30.48
	Female	41.60	35.56
Loss of weight	Male	-	10.13
	Female	-	17.63
Percentage loss of Weight	Male	-	33.23
	Female	-	49.57

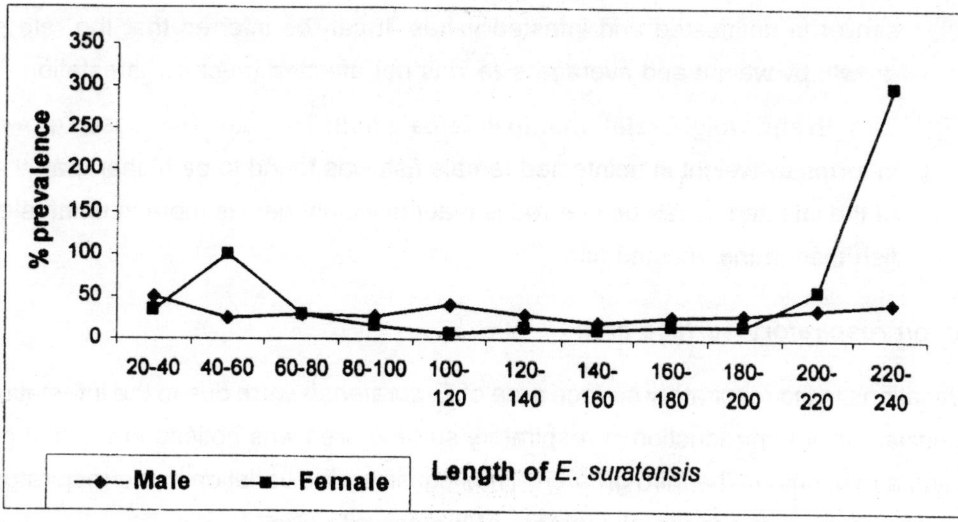

Fig. 1: The percentage prevalence in relation to size and sex of *Etroplus suratensis*

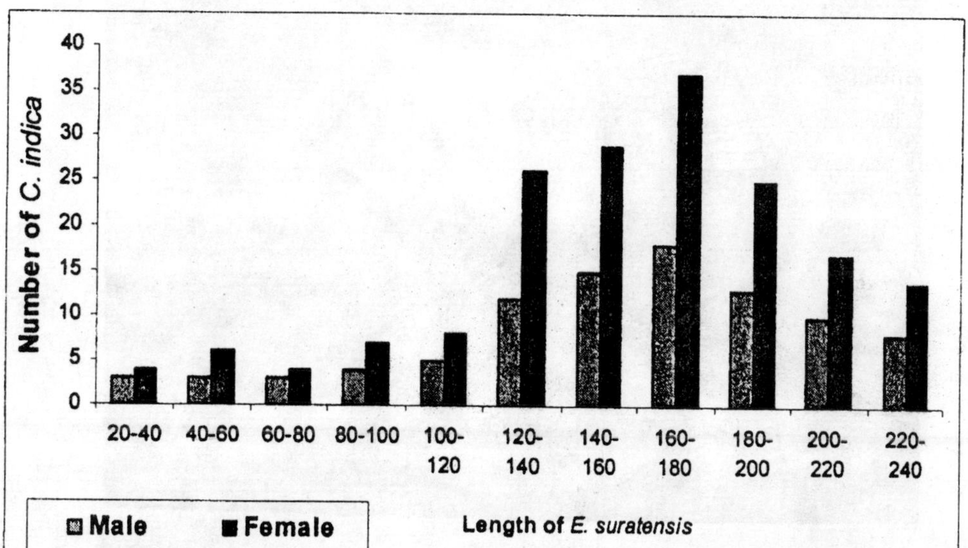

Fig. 2: The number of male and female *C. indica* in relation to length of *Etroplus suratensis*

Effect on length and weight relationship

The effects on length ad weight relationship of *E. suratensis* by the infestation of *C.indica* in both male and female fishes were compared and the results are as follows:

(i) **Length and weight relationship in male fish:** The rate of growth is almost similar in uninfested and infested fishes. It can be inferred that the rate of growth by weight and average size was not affected much by infestation.

(ii) **Length and weight relationship in female fish:** The rate of increased growth in terms by weight in uninfested female fish was found to be higher than that of the infested. It can be inferred is that the weight gain is more in uninfested fish than in the infested fish.

Effect on respiratory surface area

Variations in the respiratory surface area of *E. suratensis* were due to the infestation of *C.indica*. Maximum reduction in respiratory surface area was noticed in the first gill arch and a minimum in the third gill arch. Thus considerable variation in the respiratory area was observed due to the attachment of the parasite (Fig. 3).

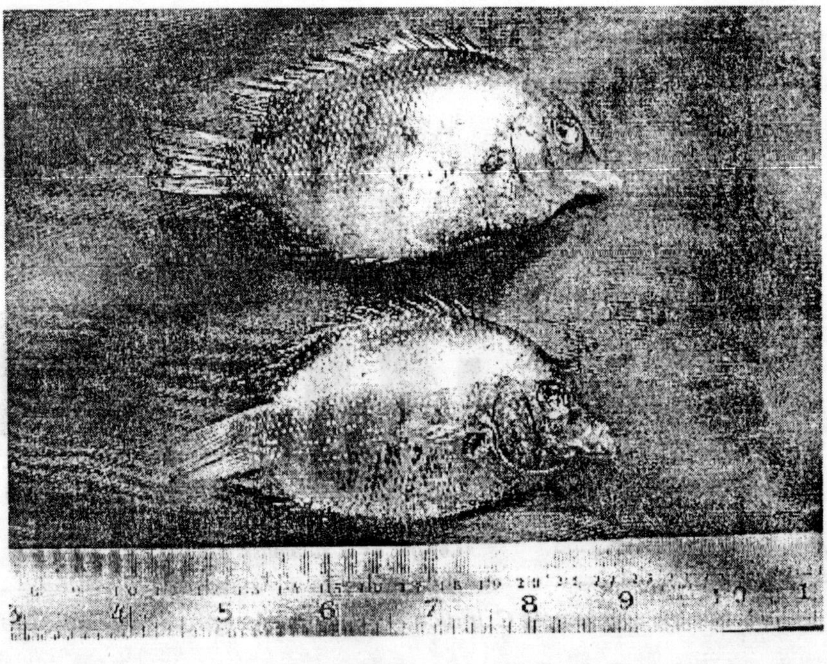

Fig. 3: *Cymothoa indica* parasite attach on mouth of the *Etroplus suratensis*

Gross lesions

The gross lesions were observed in the bronchial chamber, buccal cavity and body surface is reported.

(a) **Branchial chamber:** Infested fishes has extremely pale gills indicating severe anemia. Gill rakers were seriously lost, apical edges damaged and out of gill lamellae heavily destroyed. Some secondary gill lamellae were fused or thickened. Between the gill lamellae calus-like thickening were observed. Further, the gill lamellae of the first and second gill arches were found eroded towards posterior position due to *C. indica*. Due to the lodging of parasites at the gill clefts, the gill arches showed lesions with a wide depression. The parasites further stimulated over secretion of the mucus production.

(b) **Buccal cavity:** Gross lesions observed in the buccal cavity of infested fishes showed small pin-holes in the cartilage support of the gill arches (developed), through which dactyls of peropods of isopod were penetrated claws are dig into the host tissues. The result in usually a localised destruction of the epidermis and inflammatory response around the attachment area was noticed.

(c) **Body surface:** Isopods make frequent shifts in position on the host causing a serious wound. They often move about as they feed. The activities also stimulate mucus secretion, epidermal proliferation and dilation of dermal capillaries. Mucus cells increase in number in the epidermis peripheral to the wound. Due to increased mucus secretion, there are signs of inflammation in the dermis beneath the wound. The digestive secretion apparently causes significant damage o the host's tissues.

Discussion

E.suratensis is commonly distributed and occurring throughout the year in the Parangipettai coastal waters. They are commercially important and contribute to sizeable market landings from the brackish waters of the southeast coast of India. *E.suratensis* being the largest have many desirable features which make them ideal fishes for aquaculture. Its wide salinity tolerance, ability to breed in confined waters, fast rate of growth, good body weight, tasty flesh, highly adaptable feeding habits, robust and sturdy body and good market price are some of the favourable characteristics for selection of this fish as a candidate species for brackish water aquaculture. In some parts of South India, *E. suratensis* is considered a delicacy and priced as much as Rs. 30 per kg which is comparable to the market value of other high quality table fishes (Keshava *et al.*, 1988).

Infestation causes serious problems to host animals either directly or indirectly affecting the physiological status of host. Loss of weight has been probably the most common effects on crustacean infestation. The preset study also revealed that there was significant weight loss in the infested pearl spot fish, *E.suratensis* thereby proving the fat that infestation causes weight loss fishes.

Kabata (1984) expressed that the general effects as loss of weight can be attributed to more than one cause and the most obvious one could be the loss of food reserve drawn from the various depots and other tissues to help in coping with the ranges if infestation. Low weight of infested fish might results from failure to grow normally. However, the present study on the length and weight relationship proved that the infestation alter the growth of the fish. It shows that the increase in length of fish is not in accordance with the increase in weight of fish due to infestation. Trilles (1979) found that several species of European cymothoids were capable of slowing down the growth in their hosts, although they did not affect weight-size ratios of the fish. However, Maxwell (1982) reported that the significant difference in the length and weight relationship of infested and uninfested fishes of Jack Mackeral, *Trachurus decliuis* when infested by *Cerotophea impricatus*.

The present study showed significant reduction of respiratory surface area because of the infestation of cymothoids in the branchial chamber was noticed. The dorsal surface of the parasite was always in close contact with the first gill arch, causing more atrophy of the first gill filament. The pressure exerted towards second to fourth gill arches were comparatively lesser than the first gill arch and the damage was also less.

Kroger and Guthrie (1972) stated that male, *Olencira praegustor* damaged the gill of juvenile Atlantic Menhaden, *Brevoorita tyrannus*. Stephensen (1976) stated that the gill damage as owing to buccal parasites. Bowmen (1960) found that the gill filaments of Hawallan Moray eel, *Gymnothorax euroslus* were missing in the anterior or posterior most regions of gill arches due to the attachment of *Lirneca puhi*. Overstreet (1978) reported that the similar males and juveniles of isopod. *Olencira praegustator* parasitic on Gulf menhaden cause change on the gill filaments through feeding.

In the present observation the parasites (*C.indica*) mostly attach with buccal region and the branchial chamber of the (host-fish) first gill arch. *C.indica* first described from Bangkok, Thailand has been recorded by Chilton (1924) from Chilka Lake and is stated to infest the mouth of *Gobius giuris*, causing deformation of the host's tongue.

Subsequently, Panikkar and Aiyar (1973) recorded *C.indica* on *Etroplus maculatus, E.suratensis* and *Glossogobius giuris* from the Adyar estuary, Chennai (India). Evangeline (1963) reported the occurrence of this parasite, in the Adyar fish farm, from a variety of hosts, *viz*., *E. suratensis, Tilapia mossambica, Macrones gulio, Gobius giuris, Polynemus tetradactylus, Pomadasys hasta* and *Sphyraena abtusata*. It is thus evident that *C.indica* exhibits a wide geographic and host distribution. Jayadev Babu and Sanjeeva Raj (1948) noticed the infestation in several regions like the chin, nape and pectoral fin base and the buccal cavity. But, Ravichandran *et al*. (1999) found that the dorsal source of the parasite facing the first gill arch where the parasites were attached in *Joryma brachysoma*. The damage of gill filaments thus was not only for feeding but also the pressure exerted by the dorsal side of the parasite.

The gross size and shape of parasites can act as physical irritants, which may be responsible for the presently observed damages of the branchial tissues. The reduction in the surface area was thus due to several factors such as mode of attachment, movement, size and duration of stay of the parasites.

Pale gills damaged in gill rakers and erosion of gill lamellae are the severe gross lesions were observed as a consequence of isopod infestation. Pale gills of infested fishes indicate anemia, which may be due to the loss of blood and the construction of branchial circulation by the attachment of parasite and of the haemophageous nature of the branchial cymothoids (Romestand and Trilles, 1977; Romestand, 1979). The calus like thickening observed in the gill arch and gill filaments may be due to the constant irritation caused by the body and appendages of the parasite. Erosion and thickening are the two unique morphological changes noticed owing to the infestation. These changes are mainly due to the heavy pressure exerted by the parasite and also by their feeding nature. Kabata (1985) observed destruction of host tissues as a result of the pressure exerted by the parasite body, when present in the gill cavity. Longer stay of parasite within the gill camber may also prevent and obstruct the normal growth of the gill arches. This may be the reason for the erosion of gill arch and fusion of gill lamellae. More mucus secreted in the infested gills may be due to the reaction to irritation created by the parasite. Romestand and Trilles (1977) observed that the secretion of mucus on the ligament surface as a reactive response of the host against infestation. As a consequence of the attachment of peropods host tissues were compressed and eroded at the attachment sites, which were surrounded by an inflammed peripheral welt of peropods.

Lesions associated with reproduction of parasite to the host are related to the direct activity of the parasites. A marked increase in the size of the parasite may be seen with the development of marsupium filled with juvenile isopods. This can significantly increase the pressure atrophy caused by the presence of the parasites. These reproduction related activity in the host increases the change that the parasite will serve as a vector for microbial parasites such as hematozoans (Smith, 1975). Uninfested fishes were more so active than the infested, and were quite still or swam feebly than the infested fishes, resulting in low rate of activity.

Assessment of the general effects of parasites on the condition of their hosts is best with numerous difficulties. The main difficulty in garging the extent of damage inflicted on the fish lies the normal condition of the fish. Certain reports suggested that the effects of parasitisation might become apparent after a latent period of fairly long duration (Reichembach-Klinke *et al.* 1968). The parasitic infections also result in the abnormal behaviour of the host. The parasitic infections change the blood picture and also the parasitic condition of many species leads to the secondary infections by bacteria and fungi. Some reports (Venkataraman and Sreenivasan, 1952 and 1954) indicated that the environmental factors may influence the bacterial floras of the skin and gills.

In general, parasitic infection of fishes mainly depends upon a host of factors such as age, size, sex, maturity, stage, behaviour, feeding and breeding, lifecycle, physico-chemical and particularly environment factors. The negative impact of parasites on host's growth survival has been demonstrated for several parasite-host systems, both in aquaculture and in natural populations (Sindermann, 1987). However, host-parasite relationships are generally very complex and difficult to clarify. With the exception of case of mass mortalities caused by outbreaks of parasites, assessment of the effects of parasite infection in natural fish populations is particularly difficult because of the presence of predators or scavengers which rapidly remove the moribund or dead fish.

It appears that *C. indica* occurs in a limited range of teleost fishes and that it takes shelter in the host mainly for the purpose of breeding (Misra and Nandi, 1986). In the present study, the pearl spot fish, *E. suratensis* parasitised by the isopod, *C. indica* showed loss of weight, loss of fat content, changes in the water content of various tissues and reduction in respiratory surface area of the host fish.

References

Baer, J.G.: Ecology of the family Anchuridae (Crustacea: Isopoda) with remarks on certain morphological peculiarities. *J. Linn. Soc (Zool)*, **36: 109-160** (1951).

Bowmen, T.E.: Description and notes on the biology of *Lironeca puhi*, sp. (Isopoda: Cymothoidae). Parasite of the Hwaaiian murrey eel, *Gymnothorax eurostus* (Abbott.). *Crustacenana* L. **82-91** (1960).

Bunkley-Williams, L. and E.H. Jr. Williams: Isopods associated with fishes: A synopsis and corrections. *J. Parasitol.*, **84(5): 893-896** (1998).

Chilton, C.: Fauna of the Chilka Lake II. Tanaidacea and Isopoda. *Mem. Indian Mus.*, **5: 877-878** (1924).

Evangeline, G.: Occurrence of the isopods fish parasite, Cymothoa indica in the Adyar Fish Farm, Chennai. *Madras J. Fisheries*, **1**, (1963).

Jayadev Babu, S. and S. Sanjeevaraj: Isopod parasites of fish of Pulicat Lake. *Proc. Symp. Coastal Aquaculture* (Finfish) **3: 818-823** (1984).

Kabata, Z.: *Diseases caused by metazoans: Crustaceans in diseases of marine animals*. Vol. IV, Part I. Introduction Pisces Biologische Anstalt, Helgoland by O. Kinne, **321-399** (1984).

Kabata, Z.: *Parasites and disease of fish cultured in tropics. Isopoda*, Taylor and Francis; London pp. **265-271** (1985).

Keshava, P. Santh Joseph and M. Mohan Joseph: Reproduction of Pearl-spot, *Etroplus suratensis* (Bloch) in the Nethravati-Gurpur estuary, Mangalore. In: M. Mohan Joseph (Ed.), *The First Indian Fisheries Forum*, Proceeding Asian fisheries Society, Indian branch, Mangalore, pp. **237-241** (1988).

Kroger, R.L. and J.F. Guthrie: Incidence of parasitic Isopods, *Olencira praegustator* in juvenile Atlantic Menhaden. *Copeia,* **2: 374-379** (1972).

Maxwell, J.G.H.: Insfestation of jackmackeral, *Trachurus dacliuis* (Jengus), with cymothoid isopod, *Ceratothoa imbricatus* (Labricus) in Southeastern Australian waters. *J. Fish. Biol.*, **29(3): 341-349** (1982).

Mills Dick and Lambert Derek: *The Aquarium Fish Handbook*. The completer reference from anemone fish to Zamora Woodcats. 1st Edition for the United Kingdom (2004).

Mini Nair: Studies on the systemics and ecology of isopods of the southwest coast of India. Ph.D., Thesis, University of Kerala (1987).

Misra, A. and N.C. Nadi: A new host record of *Cymothoa indica* Schioedte and Meinert *(Crustacea: Isopoda)* from Sundarbans, West Bengal. *Indian J. Fish.*, **33: 229-231** (1986).

Moller, H. and K. Anderss: *Diseases and parasites of marine fishes*. Moller, Kiel (1986).

Overstreet, R.M.: Marine maladies worms germs and other symbionts from the Northern gulf of Mexico Mississippi, Alabama Sea Grant Consortium, p. **140** (1978).

Panikkar, N.K. and R.G. Aiyar: On a cymothoan parasite on some brackishwater fishes from Madras. *Curr. Sci.*, **5(8): 429-430** (1937).

Pillai, N.K. Tanaidocea and isopds of Travancore. Ph.D. Thesis, University of Kerala, p. **152** (1958).

Ravichandran, S., A.J.A. Ranjit Singh, N. Veerappan and T. Kannupandi: Effect of isopod, Parasite *Joryma brachysoma* on *Ilisha melastoma* from Parangipettai Coastal Waters (South East Coast of India). *Ecol. Env. and Cons.*, **5(2): 95-101** (1999).

Reichenbach-Klinke, W., F. Braun, H. Held and S. Riedmuller: Vorlaufige Ergebnisse Vergelichender-physiologischer undersuchungen an coregonen vershiedener obserbayerischer seen (Fettgehatt, Blutbild, Fermentspiegel, Parasitisierung). *Arch Fischereiwiss* **19: 114-130** (1968).

Rohde, K.: *Ecology of Marine Parasites*, 2nd Edn. CAB International, Wallingford, pp. **1-298** (1993).

Romestand, B.: Etude ecophysiologique des parasitoses a Cymothadiens. *Ann. Parasitol. Hum. Comp.* **54(4): 423-1979.**

Romestand, B. and J.P. Trilles: Influence decymothoiddicus (crustacea, Isopoda, Flabellifera) succrtaines constand hematologiques despoissons hotes. *Zeitschrift fur. Parasitenkunde*, **52: 91-95** (1977).

Segal, E.: Behaviours of juvenile *Neroclia acuminata* (Isopoda: Cymothoidae) during attack and attachment and feeding on fish prey. *Bull. Mar. sci.*, **41(2): 351-360** (1987).

Sindermann, C.J.: Effects of parasites of fish populations: practical considerations. *Int. J. Parasitol.*, **17: 371-382** (1987).

Smith, F.G.: *The Pathology of Fishes*. Chap. 6, Crustacean parasites of marine fishes. **189-203** (1975).

Trilles, J.P. (1969) Recherches sur les isopodes 'Cymothoidae' des cotes francaises. Apercu general et comparatif sur la bionomie et la sexualite de ces Crustaces. *Bull Soc Zool Fr* **94(3): 433-445.**

Trilles, J.P.: Les cymothoidae (Isopoda, Flabellifera, parasites de poissons) du Rijksmuseum van Naturlijke Historie de leidern II. Afrique, Amerique et regions indo-ouest-pacifiques. *Zool meded, Leide.*, **54: 245-275** (1979).

Venkataraman, R. and A. Sreensivasan: A preliminary investigation of the bacterial flora of the mackerals of the West Coast. *Ind. J. Med. Res.*, **40: 524-533** (1952).

Venkataraman, R. and A. Sreenivasan: Bacteriology of inshore and of mackerals off Telicherry (Malabar). *Proc. Nat. Inst. Sci.*, **20: 651-655** (1954).

□□□

Index